8.15.96

Solid State Chemistry

Solid State Chemistry

Compounds

Edited by

A. K. Cheetham
Professor of Materials
University of California, Santa Barbara

and

Peter Day
Director
The Royal Institution of Great Britain, London

CLARENDON PRESS · OXFORD
1992

Oxford University Press, Walton Street, OX2 6DP

Oxford New York
Athens Auckland Bangkok Bombay
Calcutta Cape Town Dar es Salaam Delhi
Florence Hong Kong Istanbul Karachi
Kuala Lumpur Madras Madrid Melbourne
Mexico City Nairobi Paris Singapore
Taipei Tokyo Toronto
and associated companies in
Berlin Ibadan

Oxford is a trade mark of Oxford University Press

Published in the United States by
Oxford University Press Inc., New York

© Oxford University Press, 1992

First published 1992
Reprinted 1993, 1996

A catalogue record for this book is available from the British Library

Library of Congress Cataloging in Publication Data
Solid state chemistry: compounds / edited by A. K. Cheetham and Peter Day.
Includes bibliographical references and index.
1. Solid state chemistry.   I. Cheetham, A. K.   II. Day, P.
QD478.S6334   1992   541'.0421—dc20   91–12173
ISBN 0–19–855166–5

Printed and bound in Great Britain by
Biddles Ltd, Guildford and King's Lynn

Preface

The continuing worldwide search for new and useful materials has ensured that the solid state is one of the major growth areas of chemistry and there is a widely perceived need for good, up-to-date textbooks in the area. This book, like the previous volume which dealt with Techniques, is aimed at final-year honours and postgraduate students who may be planning a career in the field. As with Volume 1, we chose a multiauthor approach in order that our account should be more authoritative, and we are delighted and encouraged by the very positive response from colleagues who were invited to contribute. The book deals first with bonding in solids, and then focuses on several classes of important inorganic materials. Whilst we have been able to cover many key areas, including superconductors and zeolite catalysts, our coverage is not as comprehensive as this wide-ranging subject deserves. Significant omissions that we hope to fill in a subsequent edition include optoelectronic and magnetic materials and solid electrolytes. Nevertheless, we hope that readers will find this a useful and interesting book, and that it will be perceived as a valuable complement to Volume 1.

Oxford   A.K.C.
December 1991   P.D.

Contents

Contributors

P. A. Cox
Inorganic Chemistry Laboratory, University of Oxford,
South Parks Road, Oxford OX1 3QR, UK

P. Day
The Royal Institution of Great Britain,
21 Albemarle Street, London W1X 4BS, UK

J. Etourneau
Laboratoire de Chimie du Solide du CNRS, Université de Bordeaux I,
351, Cours de la Libération – 33405 Talence, Cedex, France

A. J. Jacobson
Department of Chemistry,
University of Houston, Houston, TX 77204, USA

J. M. Newsam
Biosym Technologies Inc.,
10065 Barnes Canyon Road, San Diego, CA 92121, USA

C. N. R. Rao
Solid State and Structural Chemistry Unit, Indian Institute of Science,
Bangalore – 560012, India

K. J. Rao
Solid State and Structural Chemistry Unit, Indian Institute of Science,
Bangalore – 560012, India

A. Simon
Max-Planck-Institut für Festkoerperforschung,
Heisenbergstrasse 1, D-7000 Stuttgart 80, Germany

A. W. Sleight
Department of Chemistry,
Oregon State University, Corvallis, Oregon 97331-4003, USA

1 Electronic structure of solids

P. A. Cox

1.1 INTRODUCTION

Modern inorganic chemistry is based firmly on the ideas of electronic structure which are provided by the quantum theory of atoms and molecules. The approach used by most chemists is necessarily very approximate, since although quite accurate quantum-mechanical calculations can now be performed on small molecules, the results of these are expressed in terms which are too sophisticated to extend easily to larger systems. Thus, inorganic and organic chemists tend to use the language of approximate molecular orbital theory, and to think in terms of simple models of the filled and empty orbitals of a molecule in order to interpret its structure and properties. This brief account of the electronic structure of solids follows the chemists' approach, and is rather different from the treatment normally given in textbooks on solids. Such books tend to be written by physicists for students of physics, and depend heavily on the properties of free electrons moving in a crystal. As a result, these accounts are somewhat uncongenial to the average chemist, both because they start from a point of view which is too mathematical, and because they stop short of describing the types of complex solid which are of current chemical interest. In the present treatment, solids are discussed using the same simple models, derived from linear combinations of atomic orbitals, as are used in molecular chemistry.

The most obvious feature of the electronic levels in solids is the presence of *energy bands*, and Fig. 1.1 suggests how these can be regarded as an extension of orbitals in molecules. The widely spaced orbital levels of free atoms combine in a molecule to form bonding and antibonding combinations. As the number of atoms in a molecule is increased, the molecular orbitals become closer together in energy, until in a solid they merge together to form a continuous range of levels. Just as with a molecule it is filled and empty molecular orbitals which are of interest, so in a solid it is the nature of the energy bands—their energies, atomic character, and the number of electrons which occupy them—which controls the chemical bonding and many of the interesting properties. For some purposes, a slightly more elaborate picture is useful, which is provided by the *density of states function*, $n(E)$. As the molecular levels converge with an increasing number of atoms, at some energies there will be a closer spacing of orbitals than at others. The density of states gives a measure of the number of levels available for

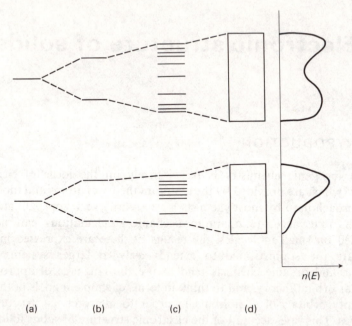

Fig. 1.1. Origin of electronic energy bands in solids. Orbitals of (a) an atom, (b) a small molecule, (c) a large molecule, becoming (d) the density of states in solid.

electrons at different energies per unit volume of the solid. Most of the following account will not concentrate in detail on the density of states, however, but will be concerned with coarser features of electronic structure, such as the widths of the bands and of the energy gaps between them, where $n(E)$ is zero.

1.1.1   Energy levels and chemical bonding

The appropriate atomic orbitals used to construct the energy bands in solids will clearly depend on the atoms involved and the nature of the chemical bonding between them. As a general rule, however, the width of a band—like the separation between molecular orbital energies—depends on the degree of overlap between the orbitals which make it up. Tightly bound core orbitals (or the 4f orbitals of the lanthanide elements) give rise to very narrow bands, of less than 0.1 eV in width. Conversely, the very strong overlap between the principal valance orbitals in a metal gives a broad band which may be more than 10 eV wide.

From a simple chemical point of view solids are divided into four classes:

1. In *molecular* solids such as nitrogen (N_2) and benzene (C_6H_6) molecules retain their identity with little change from the gas phase. The weak van

der Waals forces which hold them together hardly influence the electronic structure. Thus, the energy bands correspond closely to the filled and empty molecular orbitals of the individual molecules. Spectroscopic measurements do show some interesting effects which result from the intermolecular forces, but these are generally small.

2. In *ionic* solids such as sodium chloride or nickel oxide, chemical bonding is provided by the transfer of charge from one atom to another. The energy bands are best regarded as being made up from the atomic orbitals of anions and cations. For example, in NaCl the 3p levels of the chloride ion make the top filled band, while the sodium 3s orbitals give rise to the bottom empty band.

3. *Covalent* solids such as diamond, on the other hand, are bound by the overlap and sharing of electrons between orbitals on adjacent atoms. A carbon–carbon bond in diamond is very similar to one in a saturated hydrocarbon. The most appropriate view of the electronic structure is to think of the bands as constructed from bonding molecular orbitals (making up filled bands) and antibonding orbitals (forming empty bands).

4. In *metallic* solids, bonding is also provided by overlap between orbitals, but in this case to provide a delocalized cloud of electrons, rather than localized electron-pair bonds as in covalent solids. In simple metals such as sodium, the overlap of atomic orbitals on adjacent atoms is so strong that it gives rise to bands which are much broader than the original energy separation of different atomic orbitals. In metallic sodium, the 3s, 3p, 3d, and even higher orbitals merge into a single wide band. The atomic orbitals have lost their individuality, and it is in many ways better to think of the electrons in a metal as moving freely, with very little attraction to the atomic cores.

This classification scheme is of course an over-simplification, and most of the solids which are of interest to chemists may have properties characteristic of more than one group. For example, ionic solids may contain complex ions which are themselves covalently bonded, and which retain much of their own electronic structure, as in a molecular solid. Many transition metal compounds of a predominantly ionic nature are also metallic, and may show some features of metallic bonding. Examples of some of these will be considered in later sections. It is important first, however, to understand the principles governing the electronic structure of simple cases, as then it is quite easy to extend the ideas to cover more complex examples.

1.1.2 Metals, insulators, and semiconductors

One of the most fundamental divisions of solids is into metals and non-metals. In a metal there is a partially filled band of electrons, whereas

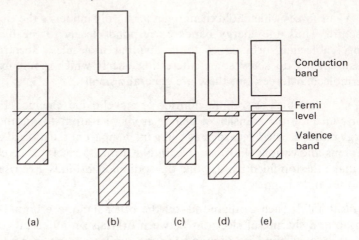

Fig. 1.2. Energy bands of (a) a metal, (b) a wide-gap insulator, (c) an intrinsic semiconductor, (d) an n-type, and (e) a p-type semiconductor.

an insulator or a semiconductor has an energy gap between the top filled level, called the *valence band*, and the bottom empty level, the *conduction band*. This is illustrated in Fig. 1.2. At absolute zero, electrons will occupy the lowest available states in accordance with the exclusion principle. In a metal, there will then be a sharp boundary between the top filled and bottom empty level, called the *Fermi level*. At higher temperatures, some electrons will be thermally excited. Due to the indistinguishable nature of electrons, the Boltzmann distribution used for molecular energy levels is not applicable, and the thermal distribution of electrons is governed instead by the *Fermi–Dirac function*:

$$f(E) = [1 + \exp\{(E - E_F)/kT\}]^{-1}.$$

For energies E well below the Fermi energy E_F, the value of $f(E)$ is close to one, showing that the available levels are nearly fully occupied. Conversely, for energies much higher than E_F, $f(E)$ is very small, and levels here are nearly empty. The transition region between these extremes covers an energy span of about $2kT$ either side of the Fermi energy. (At room temperature, kT is about 0.024 eV.)

Solids with an energy gap E_g many times larger than kT are insulators, since only very few electrons are excited across the gap, either to carry current themselves, or to leave mobile holes in the filled band. In a pure solid, the number of such carriers is proportional to $\exp(-E_g/2kT)$, so that electrical conductivity shows an activation energy equal to half the band gap. For appreciable conductivity, the band gap must be smaller than 1 eV or so. Such solids are *intrinsic semiconductors* (Fig. 1.2). Even with a semiconductor such as germanium ($E_g = 0.7$ eV), however, the conductivity is usually dominated by impurities. As explained later, the effect of these is to introduce additional electronic levels within the band gap, thus considerably lowering the activation energy for the formation of free carriers. In an *n-type semiconductor*, a narrow band of occupied (donor) levels is present just below the conduction band. In the *p-type* case, the impurities introduce empty levels just above the filled valence band; it is now easy to excite electrons thermally from the valence band, leaving holes which can migrate and conduct electrical current.

Evidence for the energy gap in insulators and semiconductors comes not only from electrical conductivity measurements, but also from absorption spectra, at wavelengths ranging from the infrared to the ultraviolet. Apart from IR absorptions due to vibrational excitation, photons with energies less than the band gap E_g cannot be absorbed by the solid. A strong absorption edge is observed when the photon energy is equal to E_g. Thus solids such as NaCl ($E_g = 9$ eV) and diamond ($E_g = 5$ eV) appear transparent as single crystals, or white in powdered form, due to multiple reflection of light at microcrystal surfaces. Solids with a band gap less than 3 eV begin to absorb visible light, and become progressively yellow, red, and black as the gap decreases through the visible spectrum. Semiconductors with smaller band gaps do not transmit in the visible region, but do so at lower photon energies in the infrared. For example, germanium becomes transparent at a wavelength greater than 1.8 µm.

In fact the lowest excited states of solids may be more complicated than the simple band-to-band excitation picture just described. It is often possible to observe excitation to energy states just below the band gap. In these states, the electron does not completely escape into the conduction band, but remains trapped in the electrostatic attraction of the hole left behind in the valence band. The onset of absorption may therefore give a slightly incorrect estimate of the band gap. A more accurate value can be obtained by observing *photoconduction* due to the electron and hole produced by the excitation. Such photoconduction can only occur at energies sufficient to allow the electron and hole to escape from one another, thus corresponding to a true excitation across the band gap.

In contrast to non-metallic solids, metals show no gap in their absorption spectra, and remain opaque down to very low photon energies. The optical properties of metals, in fact, cannot be understood in terms of excitation of single electrons, but depend on the strong electrodynamic interaction between electrons in the solid. This is the origin of the high reflectivity of metals. The same is true for non-metals, at photon energies well above the band gap. Thus, semiconductors such as silicon are reflective in the visible region and have a metallic appearance, although they are transparent in the infrared.

In a non-metallic solid, the Fermi level no longer has the significance of a sharp boundary between filled and unfilled levels. However, it is still meaningful as it defines the thermodynamic chemical potential for electrons. When solids are in electrical contact, a condition for equilibrium of electrons between different phases is that the Fermi levels should coincide. The Fermi level must always fall between the top filled level and the lowest empty one, and in a pure solid is approximately mid-way in the band gap. However, impurity doping will alter the Fermi level. This is shown in Fig. 1.2, where the Fermi levels of the different types of solid have been placed at the same level.

1.2   SIMPLE NON-METALLIC SOLIDS

Apart from molecular solids, which are not considered further in this chapter, the simplest solids to understand in chemical terms are the ionic and covalent classes, and the intermediate situation of heteropolar bonding with both ionic and covalent contributions.

1.2.1   The ionic model of electronic structure

The simple chemical view of an ionic compound such as sodium chloride, NaCl, suggests that the valence band, corresponding to the top filled levels, will be formed from the chlorine 3p atomic orbitals to which an electron is added in making the Cl^- ion, and that the conduction band, being the lowest empty level, will be made up of the sodium 3s orbitals which are occupied in the neutral atoms but are emptied in making Na^+. The excitation of an electron across the gap ($E_g = 9$ eV in NaCl) thus corresponds essentially to a charge transfer from chlorine to sodium. We can understand why such compounds are insulators, since with the elements in their normal oxidation states, the anion orbitals are completely filled and must make a filled band, whereas the cation orbitals are completely empty.

In spite of the simplicity of the picture just described, detailed calculation of the energies of the valence and conduction bands, and of the gap, involves some quite complex factors which are illustrated in Fig. 1.3. The atomic orbital energies of the free atoms (as estimated, for example, from atomic ionization energies) fall in the correct order, with the anion energies lower (more negative, because more tightly bound) than those on the cation. However, formation of the ions reverses this order, as shown in Fig. 1.3(b). This is because the extra electron repulsion in the anion allows the electrons to be more easily removed. The latter effect is particularly marked with anions that have multiple charge: for example the O^{2-} anion is unstable by 7 eV in the gas phase, with respect to O^- and an electron. Even with single charges the separated ions are unstable with respect to neutral atoms. For example, in the case of NaCl, the ionization potential of Na is about 1 eV higher than the electron affinity of Cl. It is only when the ions are placed

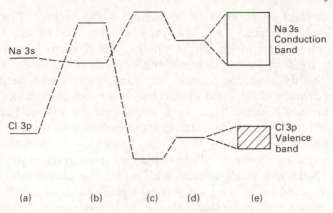

Na 3s Conduction band

Cl 3p Valence band

Na 3s

Cl 3p

(a)   (b)   (c)   (d)   (e)

Fig. 1.3. Energy levels for sodium chloride showing (a) Na 3s and Cl 3p atomic levels, (b) change in gas-phase ions, (c) influence of Madelung potential, (d) polarization energy, and (e) overlap to form the valence and conduction bands.

together in a lattice that the ionic charge distribution is stabilized, and the correct order of orbital energies is obtained. Each ion now experiences an electrostatic potential—the *Madelung potential*— due to the other ions in the lattice. For anions, with near-neighbour cations, the potential is positive, and stabilizes the filled levels. (The value of the potential is 9 eV in NaCl, but may be much more in compounds with multiple-charged ions.) Each cation is surrounded by anions, and the resulting negative potential destabilizes its orbitals by a corresponding amount.

In sodium chloride, the combination of atomic and Madelung terms would lead to a predicted band gap of 17 eV—too large by nearly a factor of two! As Fig. 1.3 shows, however, there are two more factors to be considered. The first is that whenever an electron is placed in an empty band or removed from an occupied one, the extra charge in the lattice produces a significant polarization of the surrounding ions. An idea of the energy involved can be found by considering the change which results when a charge Q, spread out over a sphere of radius R, is placed in a medium with a dielectric constant ε. The polarization of the surrounding medium causes an energy lowering given by:

$$E_{pol} = (Q^2/2R)(1 - 1/\varepsilon).$$

The difficulty in applying this to ions in solids is in knowing what the value of R is appropriate: in fact a value equal to the ionic radius gives a reasonable estimate. More accurate calculations treat explicitly the polarization of the ions surrounding the charge, with the longer-range effects being treated by a uniform dielectric constant. Polarization energies may be 2 eV or more. Since as a result it becomes easier both to add and to remove an electron, the effect of polarization is represented in Fig. 1.3 by lowering the energy of the empty level and raising that of the filled one.

The final effect which must be considered is the overlap of orbitals on adjacent ions in the lattice. Of course, the overlap of filled shells is the origin of the repulsive term in the lattice energy, which keeps ions at a finite separation. Its effect on the electronic energy levels is to broaden them into bands. In highly ionic compounds, the band width comes largely from overlap of orbitals of the same kind. Thus in a series of fluorides of alkali and alkaline earth metals, it has been found that the valence band width, derived from the F 2p orbitals, decreases smoothly with the F–F distance in the lattice. In most halides, the valence bands are quite narrow (about 2 eV in NaCl), but the more diffuse oxide ions overlap more strongly with each other, and give bands around 5 eV wide. The cation orbitals which make up the conduction band are generally much more extended, and give bands which are much wider than valence bands in simple ionic compounds.

Although the detailed calculation of band gaps is quite complicated, it is interesting that the final values correlate quite closely to the differences of orbital energies in the neutral atoms. The difference between free atoms and ions is largely compensated for by the Madelung potential in the lattice, and so the final energies are much less sensitive to the real charge distribution than one might think. As is the case with lattice energy calculations, the ionic model of electronic structure has some self-compensating features, which make it a useful working approach even in compounds that must deviate markedly from fully ionic character. It is a useful rule that either an increase of ionization energy on the cation, or a decrease on the anion, leads to a reduction in the band gap. It is interesting to compare NaCl ($E_g = 9$ eV) with AgCl ($E_g = 4$ eV), and the series AgCl, AgBr, and AgI, where the well-known change of colour from white to yellow indicates a reduction of the gap into the visible region of the spectrum.

1.2.2 Covalent solids

The simplest covalent solids are formed by the group 14 elements with the diamond structure. A chemical picture would suggest that each atom has sp^3 hybridization, forming bonding and antibonding orbitals by interaction with its neighbours. The valence band is composed of sp^3 bonding combinations, fully occupied in the ground state, and the conduction band of the unoccupied antibonding combinations. As in the molecular orbital treatment of a compound such as methane, however, it is necessary to consider in more detail the s and p atomic orbitals which make up these levels.

Figure 1.4 shows an energy-level picture appropriate to a solid having tetrahedral bonding between atoms with s and p valence orbitals. On the extreme left are the orbital energies of the isolated atoms. As the atoms come together, these orbitals overlap to form bands. The lower band of s orbitals can accommodate two electrons per atom, and the p band a total of six. At a critical value of the interaction strength, the bands cross, and this point

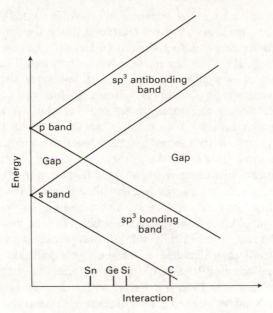

Fig. 1.4. Energy levels of a tetrahedrally bonded solid, as a function of the strength of interatomic overlap. The approximate positions of C (diamond), Si, Ge, and Sn are shown.

can be thought of as the bonding strength necessary to compensate for the atomic excitation energy involved in forming hybrids. To the right of this point, there is a lower band of bonding orbitals and an upper antibonding band, each holding four electrons per atom. It will be seen that the band gap depends crucially on the bonding strength, relative to the s–p energy separation in the atoms. The approximate positions of the tetrahedral solids carbon, silicon, germanium, and tin are shown in Fig. 1.4. It appears that the reduction in E_g (C–5 eV, Si–1.1 eV, Ge–0.7 eV, Sn around zero) results from the weakening of the interaction down the group. The same effect is also shown in the decline in the bond energies. It is interesting to note that for an element to the left of the cross-over point, only *two* electrons per atom—those in the p band—contribute to the bonding. The full s band contributes no net bonding as states with bonding character, in the lower half of the band, are cancelled by antibonding levels in the top half. Such an element is effectively divalent in the solid. In this condition however, the tetrahedral structure would not be stable, and it would be preferable to go to a more close-packed metallic structure where more bonding could be obtained. Tin, which is just at the cross-over point, does indeed have another metallic allotrope. However, it appears that the way in which the Group 14 elements move towards this critical point (and presumably pass it in the case of lead, which is unknown in the diamond structure) is related to the chemical trend which gives increasing divalent character down the group.

The non-metallic elements to the right of Group 14 show a progressive decrease of coordination number in their solid structures. Thus phosphorus forms P_4 molecules as well as the three-coordinate layer structures shown by the lower elements in Group 15; Group 16 has two-coordination in S_8-ings and the helical structures of Se and Te; and the halogens form diatomic molecular lattices. This change exactly parallels that shown in molecular compounds of the elements in their low oxidation states, and must happen for the same reason. As the number of valence electrons increases, it becomes necessary to accommodate some of them in non-bonding orbitals. Just as in molecules, therefore, the highest filled levels of the valence bands in these solids must be made up largely of non-bonding orbitals of lone-pair type. As in Group 14, the band gaps are smaller for the elements lower in each group. The reason is basically similar, that weakening of the electron-pair bonding leads to a decrease of the energy separation between non-bonding and antibonding levels. At the same time, the later groups show more structural flexibility than is possible with the diamond lattice. As the near-neighbour bonding weakens, it becomes more favourable to strengthen the formally non-bonding interactions. This itself broadens the bands and helps to reduce the gap.

1.2.3   Intermediate bonding

The previous two sections have covered the extremes of ionic and covalent bonding in solids. Very many simple binary compounds lie between these limits, with a mixture of ionic and covalent characteristics. Bonding is provided partly by a covalent sharing of electrons between the atoms, but with a higher proportion of the density in the bonding orbitals being located on the more electronegative atom. It is convenient to think of such cases by using a simple molecular orbital (MO) model appropriate to a diatomic molecule, where each atom has just one valence atomic orbital (AO) (Fig. 1.5). If E_A and E_B are the unperturbed AO energies, and V_{AB} the interaction which comes from their overlap in the molecule, then the energies of the

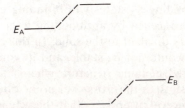

Fig. 1.5. Molecular orbitals for a model diatomic molecule.

bonding and antibonding molecular orbitals which result can be obtained approximately from the secular equation:

$$\begin{vmatrix} E_A - E & V_{AB} \\ V_{AB} & E_B - E \end{vmatrix} = 0.$$

The roots of this are easily found to be:

$$E = (E_A + E_B)/2 \pm \{V_{AB}^2 + (E_A - E_B)^2/4\}^{1/2}.$$

The energy gap between the lower bonding and the upper antibonding level can be written as

$$\Delta E = \{E_c^2 + E_i^2\}^{1/2}$$

where E_c is the gap $(2V_{AB})$ which could be present in the homonuclear, purely covalent situation, and E_i is the ionic part $(E_A - E_B)$ which comes from the energy difference of the original AOs.

In a solid, the sharp MO levels will be broadened into bands, and so unfortunately the simple formula can no longer be used for the energy gap between the top of the valence band and the bottom of the conduction band. The formula should be useful, however, to estimate the *average* excitation energy involved. Phillips and Van Vechten showed how such an average could be found for a wide variety of simple solids, by identifying a suitable peak in the absorption spectrum. In their empirical scheme, the covalent energy gap E_C was found from the Group 14 solids C, Si, Ge, and Sn. The values can then be used directly in isoelectronic series, such as Ge, GaAs, ZnSe, and CuBr. For other cases the value of E_C, as it depends on interatomic overlap, can be fitted to a decreasing function of the atomic separation. Then the ionic part E_i can be found for binary solids and used to obtain the atomic orbital energies E_A and E_B. These energies can be regarded as *electronegativities*, and indeed correlate quite satisfactorily with other electronegativity scales, such as those of Pauling or Mulliken.

The simple MO model not only gives the energy splitting between bonding and antibonding orbitals, but also predicts the distribution of charge between the atoms, when only the bonding orbital is occupied. The *fractional ionicity* of the bond can be written as

$$f_i = E_i/\Delta E = E_i/\{E_i^2 + E_c^2\}^{1/2}.$$

This varies between 1, when the gap is entirely determined by the ionic part, and the bonding charge is entirely located on one atom, and zero, in the homonuclear limit. For solids normally regarded as ionic, the values of f_i are satisfyingly high (for example, 0.94 for NaCl).

One of the most remarkable features of the Phillips–Van Vechten model is the correlation between the ionicities of simple AB compounds and their crystal structure. Although the ionic arguments based on radius ratios are qualitatively appealing, it is well known that they are unable to predict in

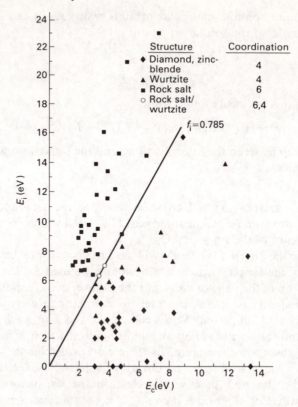

Fig. 1.6. Phillips–Van Vechten plot for binary AB solids with six- and four-coordination. The line corresponds to the critical ionicity $f_i = 0.785$. (From J. C. Phillips, *Rev. Mod. Phys.* **42**, 317 (1970).)

detail the change-over from the six-coordinate NaCl structure to the four-coordinate (zinc blende and wurtzite) ones. Figure 1.6 shows a plot of a large number of AB solids. The horizontal and vertical axes are the E_c and E_i parameters, determined spectroscopically as explained above. A straight line through the origin on this plot corresponds to compounds which have a constant ionicity, f_i. The line drawn is that for $f_i = 0.785$, and it separates almost perfectly the compounds adopting the NaCl structure (which have ionicities higher than the critical value) from those with the blende structure (which have lower values). It appears that the wurtzite lattice, which has a Madelung constant slightly higher than the zinc blende structure, is favoured by compounds just on the borderline.

Although the Phillips–Van Vechten approach is more successful than the conventional ionic one at rationalizing crystal structures, it must be pointed out that more sophisticated models of ionic size have been shown to work much better in this respect than normal ionic radii. Thus the question of whether crystal structure is really controlled by size or by bonding type, is not one that can be given a simple answer.

1.3 METALS

1.3.1 Simple metals

In simple pre-transition metals, the overlap of AOs is so strong that the band widths are considerably larger than the separation of AO energies. Bands from the different valence AOs overlap and merge into a single broad band, in which the electrons behave more as if they can move freely through the solid. The result is rather like a gas of electrons, but the properties are very different from those of a normal molecular gas. In a conventional gas at ordinary temperatures and pressures, the quantization of the translational energy levels can be ignored, since with molecules at thermal energies, there are many available levels for each molecule. There are two reasons why the electron gas in a metal is different. The number of electrons per unit volume is some four orders of magnitude greater than in a normal gas, and also the smaller mass of the electron leads to a much greater energy separation of the quantized translational levels. As a result the properties of the electron gas are dominated by the exclusion principle, which demands that each orbital level can only accommodate two electrons with opposite spin.

As shown in solid-state physics textbooks, the particle-in-a-box energy levels may be used to derive the following formula for the density of states per unit volume, $n(E)$, for free electrons:

$$n(E) = 4\pi(2m/h^2)^{3/2}E^{1/2}.$$

Photoelectron and other forms of spectroscopy show that this square-root form is quite closely followed by many simple metals. By integrating and inverting this equation, it is possible to find the energy of the top filled level, above the bottom of the band, as a function of the electron density N. The result is

$$E_{max} = h^2/2m(3N/8\pi)^{2/3}.$$

Some values calculated from this equation are shown below, and compared with experimental estimates in brackets:

Na: 3.2 (2.8), Mg: 7.2 (7.6), Al: 12.8 (13.8).

The agreement is very good. All these values are much larger than the thermal energies at normal temperatures, showing that only electrons very close to the Fermi energy can be thermally excited into higher levels. This is illustrated in Fig. 1.7, and is completely different from a gas even of light molecules such as hydrogen: at room temperature and one atmosphere pressure the value of E_{max} would be seven orders of magnitude smaller than

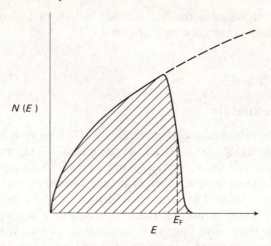

Fig. 1.7. Density of states curve for a free-electron metal, showing the thermal excitation of electrons predicted by the Fermi–Dirac distribution.

for an electron gas and all molecules would be in thermally excited translational states.

It appears that most of the electrons in a metal are effectively trapped by the exclusion principle in levels far below the Fermi level, and so cannot contribute to the free-electron properties. Such properties are determined, therefore, not so much by the total electron density as by the density of states at the Fermi level, $n(E_F)$. Some important properties of a metal are:

1. The *specific heat*, which varies linearly with temperature:

$$C_v = \pi^2/3k^2n(E_F)T.$$

This is in contrast to the vibrational contribution, which at low temperatures is proportional to T^3. Thus the free-electron contribution comes to dominate the specific heat in a metal at low temperatures, and is one of the best ways of measuring $n(E_F)$.

2. The *magnetic susceptibility*, which is temperature independent, and is dominated by the paramagnetic Pauli contribution:

$$2\mu_{B^2}n(E_F).$$

This is quite different from the $1/T$ Curie behaviour from localized unpaired electrons (see the chapter by Hatfield in Volume 1).

3. The *electrical conductivity* is more complicated, as it is a dynamic effect. The simple free-electron formula is

$$N e^2\tau/m$$

where τ is the relaxation time for scattering by impurities and lattice vibrations. Because of the latter effect, the conductivity of a metal falls with increasing temperature, as the vibrational amplitudes of atoms in the lattice increase and scatter electrons more strongly.

The success of the simple free-electron model in simple metals depends on the very large overlaps between the diffuse valence AOs in these solids. However, the appealing simplicity of the theory makes it convenient to use even in cases where such a condition is not strictly fulfilled. This can be done in a mathematical way, by making the electron mass a variable parameter. In solids with narrow bands, coming from weak overlap of orbitals, electrons are less mobile and behave very much as if they have a larger *effective mass* than free electrons. Conversely, broad bands lead to a small effective mass. Many of the free-electron formulae can now be carried over, with the genuine electronic mass (m) replaced by an effective mass (m^*).

1.3.2 Transition metals

Transition metals differ from simple metals, by virtue of the occupied $(n-1)$d valence orbitals. Although the $(n-1)$d orbitals have energies similar to the ns and np orbitals, they are rather smaller and overlap relatively weakly with surrounding atoms. This is particularly true of the 3d orbitals in the first transition series.

Figure 1.8(a) shows the formation of the energy bands of a metal in the first transition series. The 3d orbitals give quite a narrow band, which is overlapped by the much broader free-electron-like band produced by the diffuse 4s and 4p orbitals. The figure shows schematically the position of the Fermi level for an element early in the series (e.g. titanium), one near the middle (e.g. manganese) and one near the end (e.g. nickel). By the time copper is reached, the 3d band is full, and the Fermi level lies in the free-electron region, as in a pre-transition metal. This simple picture is rather approximate, since the 3d orbitals contract with the increasing effective nuclear charge across the series. In the later elements the d band becomes narrower, and moves to lower energy relative to the s–p band.

The narrow d band gives the transition metals a higher density of states at the Fermi level than in simple metals. This is not generally reflected in the conductivity, however, as the effective mass is quite high, and the electron mobilities are lowered by other effects. One of the most significant properties of the later transition elements iron, cobalt, and nickel, however, is their *ferromagnetism*. A band picture of a ferromagnetic metal is shown in Fig. 1.8(b), where a different density of states curve is plotted for spin-up and spin-down electrons. In a normal paramagnetic metal the number of spin-up and spin-down electrons is the same, and they have identical energy distributions. In a ferromagnetic metal, however, some electrons have

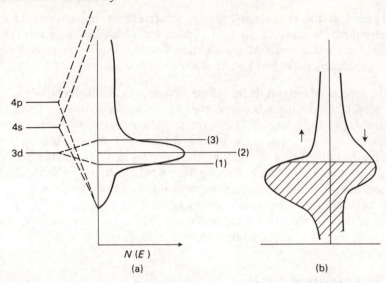

Fig. 1.8. (a) Bands in a transition metal. Levels marked 1, 2 and 3 show respectively the position of the Fermi level for an element early, middle and late in the transition series. (b) Splitting of spin-up and spin-down bands in a ferromagnetic metal.

transferred from one spin to the other. There is an energy penalty to pay for this, as the majority-spin electrons must occupy higher energy states in the band. The compensating feature is that each atom has a greater number of parallel-spin electron pairs than in the normal situation. The favourable exchange contribution thus leads to a smaller repulsion between electrons. There are two reasons why the balance between these factors tends to favour ferromagnetism when the electrons occupy a narrow band.

1. The high density of states means that there are plenty of levels available for the majority-spin electrons, without having to occupy states too high in energy.

2. Electron repulsion is especially important in the small AOs which make narrow bands.

The result of the two latter factors is that the elements iron to nickel, right at the end of the first transition series where the bands are at their narrowest, are the ones that show ferromagnetism. In fact, chromium and manganese also have magnetic moments, but they are aligned in a more complicated, antiferromagnetic, ordering. The band picture cannot explain them in a simple way.

1.3.3 Solids under high pressures

Many solids which are non-metallic under normal conditions become metals subjected to high pressures of some hundreds of kilobars. One of the

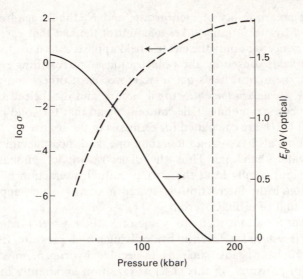

Fig. 1.9. Energy gap and electronic conductivity of solid iodine under pressure. (From H. G. Drickamer and C. W. Frank, *Electronic transitions and the high-pressure chemistry and physics of solids*, Chapman and Hall, London (1973).)

best-known examples is that of the molecular solid iodine, which becomes metallic at about 170 kbar. Figure 1.9 shows how both the conductivity and the band gap determined from absorption spectroscopy vary with pressure. The gap appears to decrease continuously with pressure, becoming zero at the transition. A simple explanation for such behaviour is that increasing pressure in a solid increases the overlap between adjacent atoms or molecules. Thus the bands, which in iodine are made up of the filled and empty MOs of the I_2 molecules, broaden. The gap between the bottom of the conduction band and the top of the valence band diminishes. In some other cases the transition to a metal is accompanied by a structural change to a more close-packed lattice, which gives broader bands. This happens, for instance, with the semiconductors germanium and silicon, which both show rather abrupt transitions in conductivity, at about 100 kbar and 160 kbar, respectively.

There is another approach to the transition from insulating to metallic behaviour, which starts by calculating the dielectric constant ε of a solid formed from polarizable entities such as atoms or molecules. The Clausius–Mosotti formula for ε is

$$(\varepsilon - 1)/(\varepsilon + 2) = R/V.$$

V is the molar volume, and R is the molar refractivity, given by:

$$R = 4\pi\alpha N_A/3$$

where α is the polarizability of each group, and N_A the Avogadro number. The Clausius–Mosotti formula takes account of the fact that each atom or molecule experiences not only the electric field applied externally to the solid, but also local fields caused by the polarization of surrounding regions. As R/V increases, these local fields act to reinforce each other, and when this ratio reaches 1, a *polarization catastrophe* occurs, and the dielectric constant is predicted to become infinite. This can only mean that the solid is metallic. When values for R/V are calculated for elements in the solid state, it is found that most non-metals have values less than one, and most metallic elements have R/V greater than one. Thus the dielectric model emphasizes that metallic behaviour results from the strong mutual interaction of atoms in the solid, although the interaction is treated in a way which appears very different from that of orbital overlap.

It is clear that for atoms with any value of R, the ratio R/V can be made equal to 1 if V is reduced sufficiently by applying a high pressure. Even solids with very large band gaps, such as molecular hydrogen, must become metallic at high pressures: in this case, a transition apparently takes place at around 2 Mbar.

1.4   TRANSITION METAL COMPOUNDS

Solid compounds of the transition metals cover an extremely diverse range, from metallic alloys to the molecular solids formed by organometallic compounds. This section discusses the electronic structure of compounds with fairly electronegative elements such as halides and chalcogenides, where the ionic model can be used as a first approximation. For compounds with a d^0 electron configuration, exactly the same considerations apply as with the simple ionic solids formed earlier, except that the lower part of the conduction band is formed from d, rather than s orbitals. The band gaps are usually somewhat less than in similar compounds of pre-transition metals. Thus TiO_2 and WO_3 have gaps around 3 eV. However, the really interesting properties of transitional metal compounds arise when the d levels are partially occupied. The solids may then be metallic or semiconducting, and may have interesting optical and magnetic properties.

1.4.1   Metallic compounds

The easiest compounds to understand are those where d electrons partially fill the conduction band, giving metallic properties. It is interesting to compare WO_3 (non-metallic), with ReO_3 and $NaWO_3$ (d^1, metallic), or TiO_2 and $SrTiO_3$ (both non-metallic) with VO_2 and $LaTiO_3$ (both d^1 and metallic at high temperatures, although they show interesting transitions at low temperatures). In the ternary oxides, the orbitals of the pre-transition element

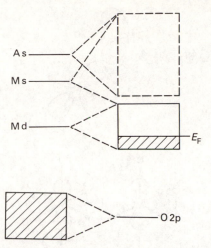

Fig. 1.10. Energy levels for a metallic ternary oxide such as AMO_3 where M is a transition metal, and A a pre-transition element.

are generally at considerably higher energy, as shown in Fig. 1.10. As with the binary compounds, any electrons in excess of those required to fill the oxygen 2p valence band will occupy the transition metal d levels.

Metallic transition metal oxides have a considerable d band width, which must arise from overlap of the d orbitals in the solid. Some of the possible sources of d-electron band width are considered in Fig. 1.11. In compounds with a high proportion of metal atoms in the structure, the d orbitals can overlap directly, to give metal–metal bonding. This occurs with some low-oxidation-state oxides, such as TiO, VO, and NbO. They have structures based on the NaCl lattice, but with very high proportions of vacant sites, either arranged in an ordered way (as in NbO), or disordered (in TiO and VO). It appears that the vacancies allow the lattice to contract, thus leading to stronger overlap between the d orbitals and broadening the band. Presumably the metal–metal bonding compensates the considerable loss of ionic binding energy which must result from the vacancies. The most obvious examples of metal–metal bonding, however, arise in the so-called metal-rich compounds (described in detail in Chapter 4), such as Sc_2Cl_3 and ZrCl, where the transition element has an oxidation state much lower than normal. The sructures of such compounds are dominated by extended metallic bonding, in which the excess d electrons are used. There are commonly chains or sheets of linked metal octahedral, quite reminiscent of the metal–metal bonded clusters such as $Mo_6Cl_8^{4+}$.

Metal–metal bonding may also occur in the dioxides such as VO_2 and MoO_2 which have the rutile structure. Although the metal atoms are on average further apart now, the structure has edge-shared metal–oxygen octahedra which allow closer approach and overlap along the *c* axis.

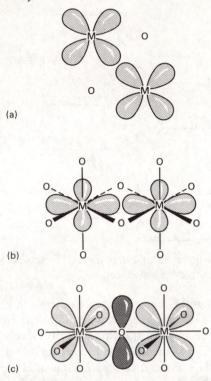

Fig. 1.11. Origin of d-electron band width. (a) Direct overlap between t_{2g} d orbitals in a monoxide such as TiO; (b) Interaction along chain of edge-sharing octahedra in rutile dioxides; (c) Indirect overlap by π bonding of t_{2g} orbitals with an intervening oxygen atom in perovskites.

However, as the proportion of metal atoms decreases, the possibilities of direct overlap also diminish, and in compounds such as ReO_3 and the related perovskites, it is almost certain that the transition metal atoms are too far apart for appreciable metal–metal bonding to occur. The d band widths, which lead to metallic properties in many of these solids, must come from an indirect overlap of d orbitals, via an intervening metal atom in the structure. This is also illustrated in Fig. 1.11. In ReO_3 and most metallic perovskites, the occupied d levels form the lower, t_{2g} part of the d band, which has only π-type interaction with the oxygen p orbitals. However, in compounds such as the d^7 metal $LaNiO_3$, σ bonding with the e_g d levels must also occur. The important point in all these compounds is that metallic character is *not* a manifestation of direct metal–metal bonding, but is rather an effect of metal–oxygen covalency. This covalency has been observed directly in measurements by photoelectron spectroscopy on compounds such as ReO_3; the formally oxygen 2p valence band is actually strongly mixed with the atomic orbitals of the metal. Similarly, the conduction-band states must be considered as an antibonding combination of metal and oxygen levels. The same is true, of course, in discrete transition metal molecules and complex ions.

1.4.2 Electron repulsion: the Mott–Hubbard gap

The simple band approach suggests that any compound with partially filled d orbitals should form a metallic solid. Yet any chemist knows that this is not the case; a great majority of halides, as well as many oxides and compounds with less electronegative ligands, form non-metallic solids, which often have the magnetic and spectroscopic properties associated with incompletely filled levels. A typical example is nickel oxide. Pure NiO is green, and shows d–d transitions associated with the octahedral ligand-field splitting of the 3d orbitals, just like an isolated ion such as $Ni (H_2O)_6^{2+}$. Similarly, the magnetic properties show the presence of two unpaired electrons on each Ni^{2+}, although there is quite a strong interaction between neighbouring ions, leading to antiferromagnetic ordering of the spins at lower temperatures. In compounds such as this, the d orbitals appear to be *localized*, rather than forming a conduction band as in metallic compounds. It is generally accepted that the reason for the non-metallic behaviour lies in the strong electron repulsion between electrons in d orbitals. In the band model, electron repulsion is not completely neglected, but it is assumed that it can be averaged out, to give an effective potential in which all electrons move. In many transition metal compounds, this is no longer a valid approximation.

Figure 1.12 shows how the ionic model would be used to predict the electronic levels of an oxide such as NiO. The oxygen 2p and metal 4s levels are just as they would be in a pre-transition metal oxide like CaO. For the partially filled 3d levels however, there are two important electronic energies to be considered, which are quite different. One is the energy involved in *removing* an electron:

$$M^{2+} \rightarrow M^{3+}: \quad 3d^n \rightarrow 3d^{n-1}$$

and the other is the energy involved in *adding* an electron to the ion:

$$M^{2+} \rightarrow M^+: \quad 3d^n \rightarrow 3d^{n+1}.$$

For free ions the energy difference between these processes is the difference between the second and third ionization potentials, and is around 15 eV for ions in the first transition series. From a slightly different point of view, it is the energy required to remove an electron from one 2+ ion and place it on another one:

$$2M^{2+} \rightarrow M^{3+} + M^+.$$

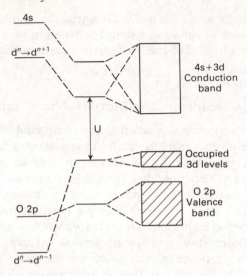

Fig. 1.12. Energy levels of a non-metallic oxide such as NiO, showing different energies for $d^n \to d^{n-1}$ ionization and $d^n \to d^{n+1}$ attachment processes.

The origin of this energy is the extra electron repulsion which results from placing an extra electron on one ion. It is clear that a collection of ions in the gas phase would not form an electronic conductor, in spite of the partially filled d levels, because of the large repulsion energy which would have to be provided before electrons could be moved.

In the solid state, two effects familiar from the discussion of ionic solids reduce the large energy gap. The first is the polarization energy, which always makes it is easier to move electronic charges in a lattice. After polarization is included, the repulsion energy (represented as the difference beween the 3d ionization and attachment levels in Fig. 1.12), is often called the *Hubbard U*, and the effect of electron repulsion in causing a band gap the *Mott–Hubbard* splitting. The final band gap is smaller than U, by an amount which depends on the width of the d bands. As shown in Fig. 1.12, the overlap of the d levels may be quite small, and the occupied d band quite narrow. However, in many cases the upper d level is probably overlapped by the 4s band, which is much broader because the 4s orbitals are more diffuse and overlap strongly.

It can be seen from Fig. 1.12 that the band gap, which is about 3.8 eV in NiO, arises because the width of the 3d bands is insufficient to overcome the Hubbard repulsion energy U. In a metallic oxide, however, the bands must be broader, and then the occupied and empty levels overlap in energy and give a single partially filled band. The properties of the solid therefore depend crucially on the relative magnitude of U and the band width. For the 4f orbitals in lanthanides, direct spectroscopic measurements have been performed, which give values for U of 5–7 eV. These are much lower than the free-ion values of nearly 20 eV, demonstrating the importance of the polarization terms in the solid state. The 4f orbitals are highly contracted and have very small band widths (probably around 0.01 eV, except for cerium, where they may be considerably larger). As a result, the 4f levels are highly localized in solid compounds of the lanthanides, and are quite well approximated for many purposes as unperturbed atomic orbitals. Similar measurements of U for transition metal compounds are harder to make but a variety of indirect evidence suggests values of 3–5 eV for the 3d orbitals.

The complexity of the properties of transition metal compounds arises partly because the d band widths are often of similar magnitude to the Hubbard U. Thus a series of compounds may often show a bewildering variety, changing from metallic to non-metallic and back again. Some useful generalizations are the following:

1. Metallic compounds are commoner among the early transition metals, because the d orbitals are more diffuse and overlap more strongly. This is exemplified by the series TiO, VO (metals), MnO, FeO, CoO, and NiO (non-metallic).

2. For the same reason, metallic compounds are commoner for elements of the second and third transition series.

3. Most halides are non-metallic since the d bands are too narrow, but in combination with less electronegative elements the d bands broaden for two reasons: indirect effects from metal–ligand covalency become more important, and also structures (such as NiAs and related ones) tend to be adopted in which direct metal–metal overlap is easier.

4. Metallic compounds are commoner where the transition metal either has an unusually low oxidation state (when metal–metal bonding is favoured), or an unusually high one (when metal–ligand covalency is stronger).

5. The presence of a high-spin electron configuration, with parallel electron spins, tends to stabilize a localized, non-metallic state, since the favourable exchange energy would be lost in a metal. Thus d^5 high-spin ions such as Mn^{2+} have notably less tendency to form metallic compounds.

1.4.3 Band gaps due to ligand-field splittings and crystal distortions

Not all non-metallic compounds of transition metals show the paramagnetic and other properties associated with unpaired electrons. There are a number of effects which can produce gaps of a more conventional kind, between a completely filled band and an empty one. The most obvious is the ligand-field splitting. In an octahedral site in a crystal, the d orbitals split into a lower t_{2g} and an upper e_g level just as in an octahedral complex. If the widths of the t_{2g} and e_g bands are not too large, there may be a gap between them. In this case, a low-spin d^6 compound, in which the t_{2g} band is just full, can be expected to be non-metallic. This happens with $LaCoO_3$ and $LaRhO_3$, although in the cobalt compound the low-spin state is only just stable, and a transition to metallic character takes place at high temperatures.

Ligand-field splitting effects are also important in solids with other coordination geometries. A well-known example is the d^8 electron configuration in square-planar coordination where, if the ligand-field splitting is sufficiently large, there is a gap between the upper filled d level and the empty $d_{x^2-y^2}$ orbital. This is very common in the solid-state chemistry, as in the coordination chemistry, of platinum. Another interesting example is found in the series ZrS_2, NbS_2 and MoS_2. The d^0 Zr compound is non-metallic and the d^1 Nb sulphide metallic, as expected. However, MoS_2, with a d^2 configuration, is non-metallic. In the layer structure adopted by MoS_2, the metal has a trigonal prismatic, rather than the more usual octahedral, coordination. With such coordination, it appears that a single d_{z^2} level is lowest in energy, and that the band so formed does not overlap with ones from the higher d orbitals (see Fig. 1.13). Thus the two d electrons in MoS_2 just fill the lower band.

There are also more subtle ways in which the crystal structure can give rise to a gap in the middle of a partially occupied band. An example is found in VO_2, which at higher temperatures has the regular rutile structure, and is metallic, with the $3d^1$ configuration. Below a transition temperature of 340 K, a distortion occurs, which leads to an alternation of the $V-V$ distances along the c axis. The d band splits into a lower part, which is strongly

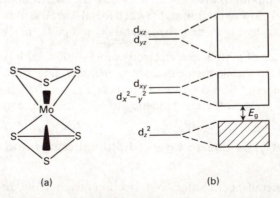

(a) (b)

Fig. 1.13. (a) Molybdenum disulphide structure, showing the trigonal prismatic coordination of Mo; (b) ligand field levels and resulting energy bands, showing the gap above the d_{z^2} band.

bonding between the close vanadium atoms, and an upper part which is antibonding. The lower band can hold only two electrons per *pair* of vanadium atoms, so that VO_2 becomes non-metallic in its distorted form. Although this picture is probably oversimplified, and neglects important electron repulsion effects, it is an example of a more general phenomenon, the *Peierls distortion*. It can be shown that a metal which results from overlap of orbitals in only one dimension in the lattice is unstable. Below a certain temperature, a lattice distortion will inevitably occur, which tends, as in VO_2, to trap the electrons in regions which have more bonding character, and to produce a band gap. The Peierls theorem, which is important in understanding the properties of low-dimensional compounds, such as those with chains of platinum atoms, is discussed in Chapter 2.

1.5 DEFECTS AND IMPURITIES

All real solids contain defects and impurities. Since they often give rise to extra levels within the band gap, the presence of defects in non-metallic solids may often dominate the electrical and optical properties. Indeed, the applications of semiconductors in electronic devices depend on the orbitals associated with impurities which are deliberately introduced. In metals, on the other hand, defects have a more subtle effect. Generally they do not give rise to major alterations in electronic behaviour, but act rather to lower the mobility of electrons, thus reducing the conductivity.

1.5.1 Doped semiconductors

When a dopant such as phosphorus is introduced into a covalent semiconductor, it substitutes at a regular tetrahedral lattice site. The impurity thus uses four valance electrons to form bonds to its neighbours in the lattice. Since phosphorus has one extra electron, however, this must be accommodated in some orbital other than the full valence-band states. To a first approximation, it will go into the antibonding conduction band. However, phosphorus also has an extra nuclear charge, and as the electron moves away from the impurity atom, this extra positive charge in the lattice will act as an attractive potential. The situation is rather like that of a hydrogen atom, but with two important differences.

1. The electron is travelling in the conduction band, not in free space. The way in which the kinetic energy is modified is taken into account by replacing the free electron mass by an effective mass m^*, as explained previously.

2. The potential energy of attraction to the positive centre is reduced by the dielectric constant ε of the intervening solid.

With these modifications, the formula for the binding energy of the hydrogenic levels is

$$E_n = -e^4 \, m^*/2(\varepsilon \hbar n)^2$$

and the radius of the electronic orbital is

$$r_n = \varepsilon(n\hbar)^2/e^2 \, m^*.$$

The two modifications noted above are very important in semiconductors such as silicon and germanium, because carriers in the conduction bands have a very low effective mass, and the dielectric constants are very high. Both of these effects lower the binding energy and increase the orbital radius. In fact, the reason why this *effective mass approximation* works at all is that even in the ground state the orbital has a radius much larger than the lattice spacing. Thus the electron does not feel the detailed microscopic lattice structure of the semiconductor.

For the electronic properties, the important energy difference is that between the lowest state with $n = 1$, and $n = \infty$, which corresponds to the electron being fully ionized into the conduction band. This gives the energy required for thermal excitation of the electron out of its bound *donor level*, into a conducting state. For donors such as phosphorus in silicon, typical ionization energies are around 0.045 eV, and since this is so much smaller than the band gap of 1.1 eV, such donor electrons will normally dominate the conductivity of doped solids.

In a p-type semiconductor, elements such as boron, to the left of Group 14, are present at some lattice sites. Since there are now fewer electrons, this replacement must leave some unfilled levels or *holes* in the valence band. The smaller nuclear charge of the impurity attracts these holes, just as in the case of the donor levels, and a similar hydrogenic model can be used. Thus p-type doping gives rise to acceptor levels just above the conduction band. Ionization energies of acceptors are similar to those of donors, and now represent the energy of excitation of an electron from the valence band into the acceptor level.

As well as greatly increasing the electrical conductivity of a semiconductor, impurity levels are important as they fix the Fermi level away from its mid-gap position. In an n-type material, the Fermi level is held close to the conduction band edge by the presence of occupied donor levels, and with p-type doping, it is near the valence band edge. The different Fermi levels have important consequences when materials of different types are placed in contact and lead, for example, to the rectifing properties of a p–n junction.

As the concentration of impurities increases, the associated levels will start to overlap and form a narrow band. This will not immediately give rise to metallic properties, however, because the electrons remain localized by repulsion effects, similar to those in transition-metal compounds. By considering the balance of electron repulsion with the predicted band widths, Mott showed that a doped semiconductor should become metallic when the donor concentration n satisfies the equation

$$n^{1/3}r = 0.25$$

where r is the hydrogenic radius of the donor orbital. A similar equation can be derived from the polarization catastrophe model described in Section 1.3.3. For donors in silicon, the transition takes place when the concentration is around 10^{18} atoms per cm^3, which is much higher than the doping levels normally used in solid-state devices. The Mott formula has been shown to work for a surprisingly wide range of solids. Most ionic and molecular solids, with larger band gaps than silicon, have considerably lower dielectric constants. The donor radius is then much smaller, and correspondingly higher dopant concentrations are required to reach the semiconductor-to-metal transition. For example, metal atoms can be doped into a solid argon matrix, and form n-type impurity levels. In this case, concentrations of around 20–30 per cent are required for metallic behaviour. Other applications include solutions of sodium in liquid ammonia, which show metallic character at higher concentrations.

1.5.2 Defects in ionic solids

The highly extended nature of impurity levels in semiconductors, and the corresponding success of the hydrogenic model, depends on the small effective masses and the high dielectric constants in these materials. The same conditions are not generally fulfilled in more ionic solids. Defect levels are now more localized, and are situated at energies further removed from the band edges. The classical example of a defect in an ionic solid is the *F centre* in alkali halide crystals. This consists of an electron trapped at a vacant halogen site. F centres have been studied by ESR, and some hyperfine structure is observable from surrounding ions, indicating that the electron spreads out a little from the defect. The major part of the electron density, however, is localized in the vacancy, and is illustrated in Fig. 1.14. The cations

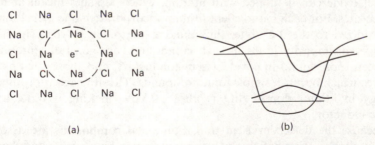

(a)

(b)

Fig. 1.14. F centre in an ionic solid. (a) Approximate distribution of ground-state electron in anion vacancy. (b) Potential well, with ground- and excited-state wavefunctions.

surrounding the vacancy provide a potential well, in which the electron has a ground state resembling a 1s atomic orbital, and excited states similar for example to atomic 2p levels. The colour of the F centres comes from the excitation of the electron to these higher states, giving an optical absorption band which in NaCl, for example, peaks at 2.7 eV. At slightly higher photon energies, the electron can be excited into the conduction band, and photoconductivity is observed.

Apart from the more localized nature of the defect levels, there is another important difference between ionic and covalent solids. An electron in an ionic solid causes a significant distortion of the surrounding lattice, as it attracts cations and repels anions. Thus an electron moving in the conduction band of a solid such as NaCl carries this lattice distortion around with it. The electron with its associated polarization field is known as a *polaron*. The formation of a polaron may appreciably lower the mobility of the electron, and cause it to have a larger effective mass than that of a bare electron in the conduction band.

Transition metal compounds often have quite high concentrations of defects. In many cases, they are associated with the possibility of variable oxidation state. Thus the d^0 oxides TiO_2, $SrTiO_3$, and WO_3 are easily reduced, and become n-type semiconductors. In the ground state the electrons probably reside on the d orbitals of transition metal ions close to the defects, from where they can be excited into the conduction band. p-type behaviour is possible in cases where the transition element can be oxidized. For example NiO readily takes up excess oxygen, forming nickel vacancies in the lattice. Some Ni^{2+} must be oxidized to Ni^{3+}, which is equivalent to making holes in the occupied 3d levels in NiO. In the ground state, the holes are bound electrostatically to the Ni vacancies.

Similar effects may be produced by doping. The d^0 oxides may be doped with alkali metals, as in Na_xWO_3. For sufficiently large x, the electrons denoted by the sodium delocalize into the W 5d band, giving a metal as discussed previously. However, for x less than about 0.3, the electrons appear to be bound to impurity sites, and the compound is an n-type semiconductor. Nickel oxide can be doped with lithium, which replaces nickel at normal lattice sites. For each monovalent lithium incorporated, one nickel ion must be oxidized to the $3+$ state, thus making the oxide p-type. The binding energies of impurity levels in such compounds are generally intermediate between those found in covalent semiconductors, and those such as the F centres which occur in simple ionic compounds. Thus the acceptor ionization energy in lithium-doped NiO is about 0.3 eV, making it quite a good semiconductor.

Because the defect levels in transition-metal compounds are associated with a change of oxidation state, there may also be a significant change of ionic radius of the ion which has the extra electron or hole. This in turn may produce a lattice distortion which helps to trap the carrier at a particular site. This phenomenon is really an extreme form of the lattice distortions which form polarons in ionic crystals, and is called a *small polaron*. For example, Na_xWO_3 with a low value of x probably contains distinguishable W^{5+}, with longer distances to the surrounding oxygens than for the majority W^{6+} species. In a similar way, the Ni^{3+} ions in doped NiO will have a smaller radius than Ni^{2+}. Although it is generally agreed that such small polaron formation may accompany the trapping of electrons or holes at impurity sites, there is still considerable dispute about their role in the electronic conduction in many semiconducting oxides. In some cases it appears that the lattice distortion round the hole may persist when it is thermally excited away from the impurity site. Conduction would then occur, not by simple motion of holes in a band, but by hopping of a hole accompanied by a strong lattice distortion. In principle, it should prove possible to distinguish the two mechanisms by transport measurements, and p-type MnO is a compound where it appears that conduction occurs by small polaron hopping. However, in other cases, such as NiO, a definitive answer is not yet available.

1.5.3 Highly disordered solids

Solids with a high degree of lattice disorder are of considerable technological interest. Examples include amorphous semiconductors. In these cases it is no longer adequate to consider isolated defects in an otherwise crystalline environment, but it is necessary to start by abandoning the idea of long range order in the solid. Similar considerations apply to the liquid state. In the treatment of electronic structure presented so far, very little reference has in fact been made to long-range order. It is apparent that the most important factor controlling the electronic energy levels is the *local* environment of atoms in the solid. Thus in ionic solids, the local Madelung potential dominates in controlling the position of the valence and conduction bands. Similarly, with covalent and metallic materials, it is the overlap of atomic orbitals between near neighbours in the solid which is important. So long as the local structure is unperturbed, the existence of long-range order is irrelevant to the gross features of the electronic levels. It appears that in amorphous silicon and germanium, most atoms remain tetrahedrally bonded to their neighbours, and that the disorder arises from the random fashion in which the tetrahedra link up. Thus it would be anticipated that in amorphous semiconductors the gap between the bonding valence band and the antibonding conduction band would be preserved. This is very nearly correct, although the electronic properties of these materials indicate that they are some electronic levels present in the gap. Such states probably arise at sites where the local tetrahedral bonding is not maintained. A three-coordinate silicon atom will have one electron left in a non-bonding orbital, a so-called *dangling bond*. Similar electronic states are observed at surfaces.

The number of dangling-bond states in amorphous semiconductors may be reduced considerably by combining them chemically—for example by hydrogenation, which restores tetrahedral coordination to the unsaturated valencies.

Although disorder may not always alter the gross features of electronic energy levels, it does appear to have some more subtle effects, and may be apparent in properties, such as conductivity, which depend on the mobility of electrons. The disorder will give rise to a random variation of the potential in which conduction electrons move. The lowest energy states in the band may become localized in regions where the potential is lowest, and this phenomenon of *Anderson localization* is thought to be important in some examples of semiconductor-to-metal transitions which occur in disordered solids. The most clear-cut examples, are found with doped semiconductors, but Anderson localization, coming from the disordered potential, has been suggested as a factor in the transitions which occur in some transition-metal oxides, such as Na_xWO_3 and $La_{1-x}Sr_xVO_3$. Both of the latter compounds show transitions from semiconducting to metallic behaviour as x increases.

1.6 BIBLIOGRAPHY

Adams, D. M., *Inorganic solids*, John Wiley, London (1974).

Coles, D. R. and Caplin, A. D., *The electronic structures of solids*, Edward Arnold, London (1976).

Cox, P. A., *Valence-shell photoelectron spectroscopy of solids*, in (ed. Day, P.), *Emission and scattering techniques*, NATO ASI C73, D. Reidel, Dordrecht (1981).

Cox, P. A., *Electronic Structure of Inorganic Solids*, Oxford University Press (1987).

Dalven, R., *Introduction to applied solid-state physics*. Plenum, New York (1980).

Drickamer, H. G. and Frank, C. W., *Electronic transitions and the high-pressure chemistry and physics of solids*, Chapman and Hall, London (1973).

Edwards, P. M. and Sienko, M. J., *Chem. Brit.*, **19**, 39 (1983).

Goodenough, J. B., *Prog. Sol. State Chem.*, **5**, 39 (1971).

Harrison, W. A., *Electronic structure and the properties of solids*, W. A. Freeman, San Francisco (1980).

Kittel, C., *Introduction to solid state physics*, 5th edn, John Wiley, New York (1976).

Mott, N. F., *Metal–insulator transitions*, Taylor and Francis, London (1974).

Mott, N. F. and Gurney, R. W., *Electronic processes in ionic crystals*, 2nd edn, Oxford University Press (1948).

Phillips, J. C., *Rev. Mod. Phys.*, **42**, 317 (1970).

2 Chain compounds and one-dimensional physical behaviour

P. Day

2.1 INTRODUCTION

There are three reasons for devoting a chapter to solids containing chains of closely spaced atoms, ions or molecules. Firstly, there are numerous physical phenomena quite specific to lattices having interactions between repeating units in one dimension alone. Secondly, because the topology of a chain is so simple, chemical correlations between structure and properties are more easily arrived at. Finally, the structural principles themselves are more easily discernible when coordinated polyhedra are only joined along a single axis.

Although compounds such as the chain silicates have mechanical properties which arise from their highly directional bonding (for example the fibrous structure of asbestos), it is their unusual optical, magnetic and electrical conduction properties which give chain compounds their greatest interest. The best known example is probably KCP ($K_2Pt(CN)_4Br_{0.30}3H_2O$), whose structure consists of planar $Pt(CN)_4$ molecular units, stacked one on top of the other to give chains of Pt atoms separated by a distance (2.89 Å) not very much greater than in Pt metal (2.78 Å). The crystals are copper-coloured, with a metallic-like reflectivity. Because of the CN^- ligands, and the Br^-, K^+, and H_2O separating the Pt atoms, the distance between neighbouring chains is about four times greater than the Pt–Pt separation within a chain. It is not surprising then, that the conductivity parallel to the chains is 10^4–10^5 times greater than that perpendicular to them. The strong reflection too is only seen when the electric vector of the light is parallel to the chain: when it is perpendicular the crystals are pale yellow and transparent. Thus we have the apparently paradoxical situation of a compound which is metallic along one axis and an ionic insulator in the two directions at right angles!

Clearly, not all chain compounds are metallic, and may have much simpler structural formulae that KCP. What, then, are the criteria for metallic behaviour in such a chain, and what kinds of lattice lead to the appearance of chains? These are questions we shall address in this chapter.

As described in Chapter 1, the extent of overlap between orbitals centred on neighbouring metal atoms is a crucial factor determining the electronic behaviour, especially whether the chain will behave as a metallic conductor, or as an insulator with localized (and probably unpaired) electrons. Distance between adjacent atoms is clearly one criterion of orbital overlap, which in turn depends on whether the atoms contributing the highest energy filled levels to the valence shell are in direct contact with each other. In many structures, for instance, transition metal cations with partly filled d-shells are separated by anions or neutral bridging ligands, which provide the pathway for cation–cation interaction. In looking at the structures of chain compounds we shall therefore concentrate on this aspect.

However, before describing representative chain structures we begin by summarizing some of the special features in the physical properties of one-dimensional (1-D) lattices.

2.2 SPECIAL FEATURES OF CHAIN COMPOUNDS

In Chapter 1, Cox introduced the two extreme models used to describe the electronic states of solids, namely the band model and the Mott–Hubbard model. In its simplest form the first of these has much in common with the one-electron molecular orbital model while the second is close to the valence bond or Heitler–London model. The latter is the best starting point for describing the outermost valence shell electrons in an insulator, where the orbitals are almost entirely localized on individual atoms, with only a small overlap between them, as would be the case in say CoF_2 or a molecular lattice containing coordination complexes. Conversely the band model is the best way to describe the electronic states of a conductor. Both present unusual features when applied to 1-D arrays of atoms, principally because of the crucial role played by fluctuations in either the charge or the spin distribution, which serve to destabilize an ordered ground state. Such fluctuations, which may be static or dynamic, lead to the appearance of phenomena such as the Peierls instability, described below, which have no simple analogue in lattices with higher dimensionality.

One of the simplest theorems about a 1-D lattice states that such an array of unpaired spins cannot exist in a long-range-ordered state at any temperature above absolute zero. The reason for this can be seen by considering a chain of atoms each carrying an unpaired spin as shown in Fig. 2.1. If the spins are constrained to a given direction (say up or down) it is called an Ising system. Now denote the state of the ith spin by the integer n_i, which can take values $+1$ or -1 corresponding to an up or down spin. If there is an interaction between neighbouring spins of the form $-Jn_in_{i+1}$ (where

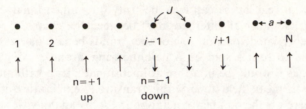

Fig. 2.1. Chain of spins with ferromagnetic exchange J between neighbours.

$J > 0$ is the energy of interaction, the total energy is

$$E = -J \sum_i n_i n_{i+1} = -NJ \qquad (2.1)$$

for a chain containing $N + 1$ sites. At absolute zero the fully ordered state with all the spins pointing up or all down, has the lowest energy, but at a finite temperature T it is the free energy $F = E - TS$ which must take its minimum value, where S is the entropy. Relative to the fully order state, the energy ΔE required to make a single reversal of orientation in the chain is $2J$, but the reversal can occur at any of the N sites with an equal probability. Thus the entropy change is $k \ln N$ and

$$\Delta F = 2J - kT \ln N \qquad (2.2)$$

which, for sufficiently large N, is negative unless $T = 0$. A chain containing spin reversals is therefore thermodynamically favoured over the fully ordered chain or, in other words, there cannot be a phase transition from the ordered to disordered state at any finite temperature. In Section 2.5.1 we shall see how accurately this prediction is obeyed in real examples. No such prediction of comparable simplicity exists in lattices with higher dimensionality because each site at the perimeter of a domain of oppositely oriented spins has more than one neighbour, and the shape of the domain also determines the number of perimeter sites. In fact Onsager showed that a two-dimensional square lattice of Ising spins does have a phase transition from ordered to disordered state at

$$kT_c = 2J/\ln(1 + \sqrt{2}). \qquad (2.3)$$

At finite temperatures the equilibrium state of our chain of spins consists of domains of length $\xi(T)$, called the correlation length, within which the spins are ordered, while the orientations of spins separated by longer distances have no fixed correlation. Clearly the correlation length increases as the temperature falls. A rough estimate of $\xi(T)$ can be found by putting $\Delta F = 0$ in eqn (2.2) when we find

$$\xi(T) = a\, e^{2J/kT}. \qquad (2.4)$$

Of course, no real lattice can ever be truly one-dimensional, since there will always be some small interaction J' between the chains. If we assume that the ξ/a correlated spins on one of the strands behave as rigid units, the energy needed to turn a spin on a neighbouring strand is roughly $(\xi/a)J'$ instead of J', as it would be in a three-dimensional lattice with an exchange constant J'. In mean field theory the transition temperature is given by $kT_c = zJ$, where z is the number of neighbours, so by analogy we would have

$$kT_c = (\xi(T)/a)J' = 2J/\ln(2J/J') \tag{2.5}$$

by substituting from eqn (2.4).

Turning to the other extreme model for the physical behaviour of a solid let us now re-examine the chain of atoms with one electron per atom in terms of the band model. Overlap between the valence orbitals of neighbouring atoms produces a set of wave functions whose energies $E(k)$, as a function of wavevector $\mathbf{k}$, are described by the full line dispersion curve in Fig. 2.2(b) ($\mathbf{k} = 2\pi/\lambda$, where λ is the repeat length of the linear combination wave function along the chain). The density of energy states $n(E)$ as a function of energy (see Chapter 1) is shown as the full line in Fig. 2.2(c), and defines an energy band from 0 to $2E_0$. With one orbital and one electron per atom we have an exactly half-filled band so the Fermi surface is the horizontal line in Fig. 2.2(c), below which all the hatched levels are occupied. For a uniformly spaced chain like the one in Fig. 2.2(a) the wavevector of the

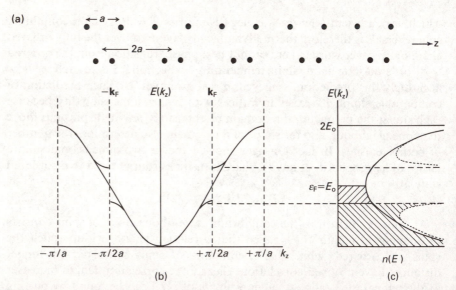

Fig. 2.2. (a) A uniform and a dimerized chain, (b) energy dispersion $E(k)$ versus wavevector $\mathbf{k}$, (c) density of states $n(E)$ versus energy E for uniform and dimerized chain.

highest level in the band, corresponding to $E = 2E_0$ is $\mathbf{k} = \pm\pi/a$. This is called the first Brillouin zone. Suppose, however, that we distort the chain a little, for instance as in Fig. 2.2(a). The repeat distance along the chain is now $2a$ and the new 'boundaries' of the first Brillouin zone are at $\pm\pi/2a$, indicated by the vertical dashed lines in Fig. 2.2(b). Non-vanishing matrix elements of the distortion potential between states of similar $\mathbf{k}$ result in the appearance of two new dispersion curves near the boundaries of the new Brillouin zone, drawn as dotted lines in Fig. 2.2(b). The result is a gap in the density of states at the wavevector k_F where the Fermi surface was in the dispersion curve of the undistorted chain, and a new density-of-state curve also shown by dotted lines in Fig. 2.2(c). Evidently, by comparison with the curve for the undistorted structure, the new one has the Fermi surface at lower energy. Distortion therefore lowers the energy of the occupied state while raising that of the unoccupied ones. Consequently the distorted lattice is inherently more stable than the undistorted one and, as Peierls wrote in 1955, 'it is therefore unlikely that a one-dimensional model could ever have metallic properties'.

Although we have described the Peierls instability in terms of a doubling of the periodicity in a chain with half-filled orbitals, there is no reason why the same effect should not occur whatever the degree of filling of the band. The distorted lattice simply has a periodicity

$$1/2\mathbf{k}_F = a/\pi n_e \tag{2.6}$$

with a band gap Δ which depends on the Fermi energy ε_F and the strength of coupling between the electron gas and lattice vibrations (phonons). If the latter is parameterized by a dimensionless constant λ

$$\Delta \propto \varepsilon_F \exp(1/2\lambda). \tag{2.7}$$

If the average number of valence electrons per atoms n_e is not an integer or a rational fraction then the new unit cell is not an integral multiple of the former one. Such lattices are called *incommensurate*. The Pt salt KCP is a good example of this, as we shall see in Section 2.5.4. How much the atoms in the chain are displaced by the Peierls instability depends on a balance between the electron kinetic energy gained and the elastic energy expended. Since the density of states of the distorted chain now has a gap separating the filled and empty states the chain is no longer a metallic conductor but a semiconductor. The simplified argument we have just given would suggest that metallic behaviour should never be found in one-dimensional compounds. In essense it is the same argument as the one we gave for the instability of the one-dimensional magnet. However, we have omitted to take account of the coupling between the chains, which will always be present in any real system. As in the case of the 1-D magnet, the 1-D electron gas is split up into domains at finite temperature, each one having a correlation length ξ. With increasing ξ and decreasing temperature even a tiny coupling between the

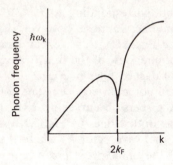

Fig. 2.3. Kohn anomaly in the acoustic phonon dispersion of a chain of atoms.

chains becomes significant, and a transition from a metallic to a three-dimensional ordered insulating state occurs.

We have described the Peierls distortion as if it were static, but the instability of the electron gas at a wavevector $\mathbf{k}_F$ with respect to distortion will also be manifested in the dynamics of the chain; longitudinal vibrations of wavevector $2\mathbf{k}_F$ have a frequency of zero, but away from $2\mathbf{k}_F$ they retain a finite frequency. Thus, the acoustic phonon dispersion curve contains a dip, called a Kohn anomaly, as shown in Fig. 2.3. This has been observed experimentally (Section 2.5.4).

2.3 STRUCTURES OF CHAIN COMPOUNDS

There are several conventional ways of representing the structural elements of inorganic crystal lattices, for example by close packing of spheres, or 'ball and stick' diagrams. Bearing in mind the important role played by anions or molecules bridging between neighbouring cations, the most convenient way of representing structures for our purpose is in terms of polyhedra with the ligand atoms at the vertices and the metal atom in the middle. The relation between the various representations of a chain of edge-sharing octahedra, like that found in the rutile (TiO_2) structure, is shown in Fig. 2.4.

Using the polyhedral representation there are two main variables which control the way that chains are built up: the number of vertices in each polyhedron and the number of vertices shared by neighbouring polyhedra. The former is simply the coordination number (CN) of the central cation, or least electro-negative element in the structure, the latter, which we will call B, may extend up to 3, including zero, as in KCP where the atoms supplying the most loosely bound valence electrons are in direct contact. For convenience we will describe structures with the same number of shared polyhedral vertices together.

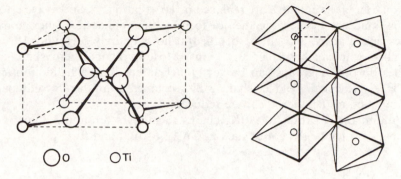

◯ O   ◯ Ti

Fig. 2.4. Representations of the structure of rutile.

2.3.1   *B = 3*

Sharing of three bridging groups between cations only takes place when the coordination polyhedra are octahedra (CN = 6). Then we have linear chains of octahedra sharing opposite faces, a structure occurring in the so-called hexagonal perovskite lattice with stoichiometry ABX_3 (Fig. 2.5). The latter is best thought of as made up from hexagonal close packed layers in which

Fig. 2.5. The unit cell of a hexagonal perovskite ABX_3. Small open circles are B, large open circles are X and dotted circles are A.

one-quarter of the anions X are replaced by large cations A. Successive layers of this kind are stacked together to form a three-dimensional hexagonal-close-packed array, and all the octahedral holes entirely surrounded by X are filled with much smaller B cations. Most hexagonal perovskites are halides but the A groups can be very much larger than a halide ion, e.g. $N(CH_3)_4^+$. The chains of face-sharing BX_6 octahedra are then forced apart, and interactions between them are reduced.

Without the A cations the BX_3 chains themselves are found alone in a number of trihalide structures such as $TiCl_3$, $MoBr_3$ and $RuCl_3$.

2.3.2 *B* = 2

With two bridging groups, the simplest way to joint octahedra together into a chain is by sharing opposite edges, as in the rutile structure of Fig. 2.4. A single chain of AB_6 octahedra joined in this way has the stoichiometry AB_4, and examples occur in nature such as $TcCl_4$ and NbI_4. In the latter though, there is a distortion which probably arises from the Jahn–Teller effect, since the electron configuration is $4d^1$. The rutile structure is not an ideal example of this type, however, because the apical atoms in each octahedron are also shared with a neighbouring chain. Consequently the shortest distance between cations in neighbouring chains is only 30 per cent bigger than the spacing between nearest neighbours within each chain. A much better approximation to one-dimensional behaviour is obtained by replacing the apical atoms of the octahedra with more bulky neutral ligands, X. Crystals of the type AB_2X_2 are formed by many dihalides of the 3d elements as addition compounds with H_2O, NH_3, pyridine, and thiourea; a typical structure is shown in Fig. 2.6. The AB_4 chain can also be an infinite anion,

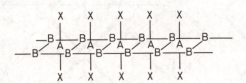

Fig. 2.6. AB_2X_2 structure, e.g. $CoCl_2 2H_2O$ (A = Co, B = Cl, X = H_2O).

in which case the chains are separated by counter cations, usually belonging to Group IA or IIA. Examples are Na_2MnCl_4, Na_2IrO_4, and Ca_2PbO_4.

As well as octahedra, tetrahedra can also be linked by opposite edges into infinite chains (Fig. 2.7), the resulting stoichiometry being AB_2. Such a structure is found in $BeCl_2$ and also, more extensively, in ternary compounds where the chains have an overall negative charge and are separated by more electropositive countercations. Sulphide coordination is frequently tetrahedral in compounds of 3d elements. An example is $KFeS_2$ or, with alternating

Fig. 2.7. Tetrahedra linked through edges.

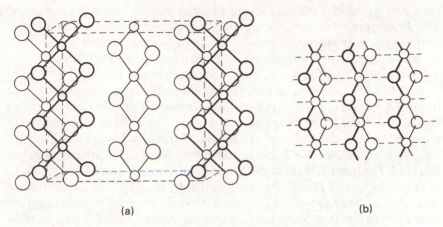

(a)   (b)

Fig. 2.8. Comparison between the structures of (a) $PdCl_2$ and (b) $CrCl_2$.

transition metal cations, $NH_4Cu^IMo^{VI}S_4$. Some non-metal sulphides, like SiS_2, also have this structure.

Low spin d^8 metal ions are frequently 4-coordinate, but with square-coplanar geometry, and so chains formed by sharing the opposite-edges of squares are a feature of the structural chemistry of Pd(II) and Pt(II). The halide $PdCl_2$ has such a structure, but in the oxides PdO and PtO alternate chains of the same type are joined at right angles so the crystals have no obviously one-dimensional physical properties. Tetragonally elongated octa-hedral, approximating to square planar if the axial ligands are sufficiently far away, is a coordination geometry often found with metal ions having $3d^4$ and $3d^9$ configurations, for which regular octahedral coordination would result in a Jahn–Teller unstable ^5E_g or ^2E_g ground state. Thus $CrCl_2$ and $CuCl_2$ have chain structures closely similar to $PdCl_2$, but with the chains packed together in such a way that chloride ions from neighbouring chains lie above and below the centre of each square planar unit. The structures of $PdCl_2$ and $CuCl_2$ are compared in Fig. 2.8.

2.3.3 $B = 1$

The larger the number of equivalent bridging ligands, the higher must be the coordination number of the polyhedra making up the repeating unit in the chain. If only one anion or molecule forms the bridge between neighbouring cations the coordination number of the cation can be as low as two. Two-coordination (linear B–A–B) among metal ions is confined to the nd^{10} configuration in Groups IB and IIB, so compounds like AgNCS and AuI contain chains which can be described by the simple alternation –A–B–A–B–.

Where the coordination number of the central atom is 3 one could envisage planar or pyramidal triangles joined through corners into a zig-zag chain. The BO_3 groups in borates are planar but pyramidal coordination is confined to the B-subgroup ions with ns^2 configurations, which are distorted by the 'inert-pair' effect. Examples of planar triangular units joined in this way are $LiBO_2$ and $Ca(BO_2)_2$, and of pyramidal ones $NaAsO_2$.

With four-coordination, tetrahedral or square planar groups can likewise be joined into chains by corner sharing. Corner-sharing SiO_4 units are, of course, the basic building blocks of dozens, if not hundreds, of silicate minerals. A single chain of corner-linked AB_4 tetrahedra has the stoichiometry AB_3, so metasilicates like Li_2SiO_3 and the whole class of pyroxenes are of this kind. The latter include $MgSiO_3$ and the mineral diopside $CaMg(SiO_3)_2$, in which the silicate chains are held together by the Mg^{2+} in six-coordinate sites and Ca^{2+} in eight-coordinate ones. Asbestos minerals (amphiboles) are actually made of double chains in which the repeat unit is a pair of SiO_4 tetrahedra with the stoichiometry Si_4O_{11}. The framework structures made from SiO_4 units in this way are described in more detail in Chapter 7. Apart from Si, other tetrahedrally coordinated elements forming single chains include P (as in $NaPO_3$), some Group IB examples (K_2CuCl_3 and Cs_2AgI_3) and 3d elements when coordinated to sulphur (e.g. Ba_2MnS_3).

A chain of AB_6 octahedra sharing *trans*-vertices has an overall stoichiometry AB_5 (Fig. 2.9) and this arrangement is found in a few pentafluorides,

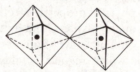

Fig. 2.9. Octahedra linked through vertices.

such as α-UF_5 and BiF_5. As in all the other cases we have mentioned so far, the chain of vertex-sharing octahedra can also exist as an infinite anion, in ternary compounds like K_2FeF_5 and $SrPbF_6$, which is better written as $Sr(PbF_5)F$. Many more examples of the same topology exist with six-coordinate units containing more than one kind of ligand, for instance H_2O as well as halide ions. A good example is $FeF_3 . 3H_2O$, while a well studied case of a one-dimensional antiferromagnet in this category is $CsMnCl_3 . 2H_2O$ where, however, the bridges are not straight but bent. The bridging ligand is different from the equatorial ones in the chain compound $WOCl_4$, which is also interesting in that the oxygen atoms are not midway between the W atoms, i.e. the chain is dimerized: W–O–––W–O–––W–O. Furthermore, the group bridging between adjacent metal ions need not be a single atom, but could be a molecular anion or bifunctional neutral molecule. Equally, the equatorial groups completing the octahedron around the metal ion may come from multidentate ligands like ethylenediamine, $H_2N(CH_2)_2NH_2$. In such a way coordination chemists have built up some very elaborate structures to study how magnetic exchange interactions are affected by varying metal–metal distance and ligand substitution. A typical structure of this kind is $Cu(pyrazine)_2(NO_3)_2$ (Fig. 2.10).

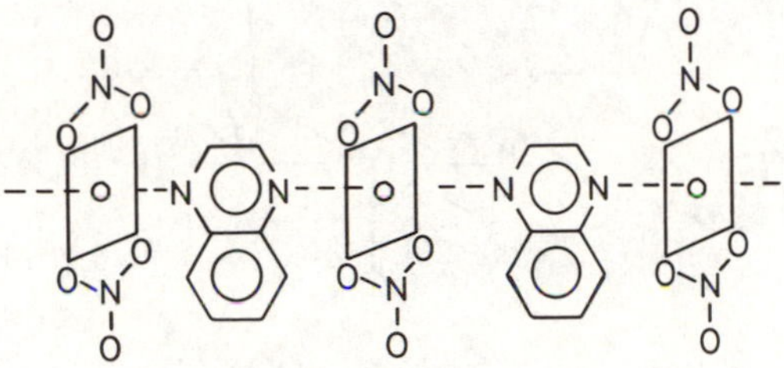

Fig. 2.10. Bis(quinoxaline)Cu(II) nitrate.

With single bridges we come to mixed valency examples for the first time in this section. Mixed valency compounds are ones which contain an element in two different oxidation states in the same lattice. We will describe the consequences of mixed valency for the physical properties of chain compounds in Section 2.4, but for the purpose of surveying structures we note that a frequent result of mixed valency is that the element in question occurs in two crystallographically distinguishable sites. A simple example among chains having single bridges between neighbouring metal atoms is Wolfram's Red Salt (WRS) $[Pt^{II}(C_2H_5NH_2)_4][Pt^{IV}(C_2H_5NH_2)_4Cl_2]Cl_4 \cdot 4H_2O$ (Fig. 2.11). This compound is the prototype for a large series of Pt(II,IV) and Pd(II,IV) salts in which the basic structure of a chain of octahedra sharing *trans*-vertices is dimerized. Instead of the interatomic distances along the chain axis being equal (–Cl–Pt–Cl–Pt–Cl–) they alternate (––Cl–Pt–Cl–––Pt–––Cl–Pt–Cl–––). Thus alternate Pt atoms have two closely spaced *trans*-chlorines and the others have two more distant ones. One could say that half the Pt has octahedral coordination, PtN_4Cl_2, while the other half approaches square planar, PtN_4. The former is characteristic of Pt^{IV} and the latter of Pt^{II}.

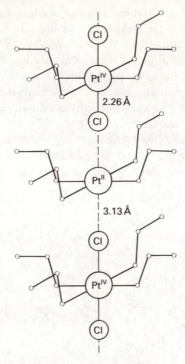

Fig. 2.11. Wolfram's Red Salt. Equatorial ligands are $C_2H_5NH_2$.

2.3.4 $B = 0$

In a crystal lattice composed of molecular units such as coordination complexes, the most favourable circumstances for direct interaction between metal ions, without any bridging ligands between them at all, occur when the complex ions are coordinatively unsaturated. By the last phrase we mean that in the isolated molecule, not all the coordination sites around the central metal ion are occupied by ligands. As will be familiar from standard inorganic chemistry courses the most common examples of this phenomenon are transition metal complexes with nd^8 configurations. In the ligand field exerted by six identical groups forming a regular octahedron, the five d-orbitals split into a lower triply-degenerate set (t_{2g}) and an upper set with two-fold degeneracy (e_g). Labelling the axes x, y, z along the directions pointing towards the ligand the d-orbitals of the t_{2g} set are xy, xz, yz and of the e_g $x^2 - y^2$ and z^2. If two ligands in *trans*-positions (say along z) are pulled away, or replaced by others exerting a weaker ligand field (for example the $Pt(C_2H_5NH_2)_4Cl_2$ in the last section) then the t_{2g} and e_g sets are split in the manner shown in Fig. 2.12. When there are eight electrons to be distributed amongst the orbitals we see that there are two possible arrangements. When the two e_g orbitals are degenerate, or nearly so, we will find

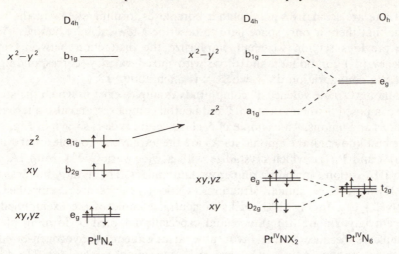

Fig. 2.12. Ligand-field splitting and site symmetry of Pt(II) and Pt(IV) in Wolfram's Red Salt. The arrow is the intervalence electron transfer transition.

one electron in each, and the ground state is a spin triplet. When the gain in orbital energy is sufficient to overcome the energy lost by putting two electrons in the same orbital then two electrons will be found instead in the z^2 orbital, which becomes steadily lower in energy as the ligand field along the z-axis gets weaker. Consequently many d^8 metal complexes are diamagnetic and have a square planar arrangement with only four ligands. The z^2 orbital has its greatest electron density along the z-axis so qualitatively one could say that there is 'half a lone pair' on either side of the plane containing the ligands. Transition metals and oxidation states which gave rise to square planar complexes are listed in Table 2.1. Clearly, if square planar complexes are stacked on top of one another, the central metal atoms can come into very close contact. Another example of coordinative unsaturation is the nd^{10} configuration which frequently gives linear complexes, i.e. with two *trans*-ligands only.

Table 2.1 *Elements and oxidation states forming square planar complexes. Dotted lines surround those ions known to form metal chain compounds*

Co(I)	Ni(II)	Cu(III)
Rh(I)	Pd(II)	Ag(III)
Ir(I)	Pt(II)	Au(III)

There are so many square planar complexes forming stacks in the solid state that there is only space here to describe a few salient examples. As in the previous section we shall emphasize the distinction between single valency (d^8) compounds and those with mixed valency, corresponding to electron configuration d^{8-x}, where x is often about 0.3.

Amongst single valence d^8 compounds examples exist in which the stacks are composed of anionic, cationic, and neutral complexes, and also alternating cations and anions. An example of each structure type is shown in Fig. 2.13. The best known of the anionic stacks are the tetracyano-complexes of Ni(II), Pd(II) and Pt(II), which crystallize with a wide variety of Group IA, IIA and IIIA cations, including numerous lanthanides. The ternary halides such as K_2PtCl_4 have similar structures. Cationic stacks are exemplified by $Pt(NH_3)_4Cl_2$. In Fig. 2.13 stacks of neutral complexes are exemplified by dimethylglyoximate (dmg), a ligand especially adapted to forming planar complexes because of the favourable arrangement of hydrogen-bonding when two mono-anions are placed on either side of a metal ion. No doubt this is one reason why $Ni(dmg)_2$ is much less soluble than dimethyl-glyoximates of other metals, to the extent that it is used as a gravimetric reagent for Ni. Alternating square planar cations and anions are found in the compounds $[M(RNH_2)_4][M'X_4]$, of which the prototype is Magnus' Green Salt. Note that the name of this salt already tells us that interaction between the complexes affects the optical spectrum because the constituent ions, taken separately, are respectively colourless and red.

Turning to mixed valency chain compounds with directly bonded metal atoms, three main categories of substance can be identified. First are ternary oxides of Pt containing interlocking chains of planar PtO_4 groups. These have formulae $M_xPt_3O_4$ (M = Na, Mg, Cd, Ni), and chains of Pt atoms along each of three orthogonal directions in the crystal. Within the chains the Pt–Pt spacing is only 2.80–2.85 Å, and between chains 3.4–3.5 Å, so they are probably not a very good approximation to one-dimensional. The second, and much larger category are what have been called 'metal-rich' compounds, because many of them are made by reacting a 'normal' oxidation state compound with excess metal. The majority are halides, though one of the most thoroughly studied is the salt of a polyatomic anion, $Hg_{2.86}(AsF_6)$. In this compound there are infinite parallel linear chains of closed spaced Hg atoms forming two-dimensional arrays, with the directions of the chains in successive layers being at right angles. Between the chains are holes occupied by AsF_6^- anions. A pecular feature of the structure is the small difference between the repeat distances of Hg atoms and the anions, i.e. the lattices formed by the cations and anions are not commensurate with one another. Metal-rich halides containing chains of metal atoms occur principally among the Group IIIA elements, including the lanthanides. In M_5X_8 (M = Sc, Gd, Tb; X = Cl, Br) we find single chains of edge-sharing M_6 octahedra, edge-bridged by halogen, and separated by chains of

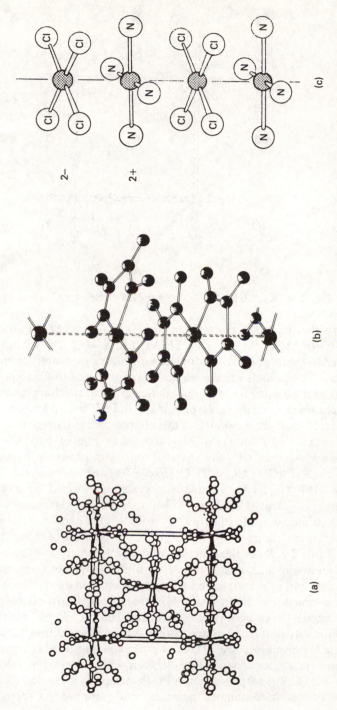

Fig. 2.13. Chains of square planar d^8 complexes, (a) $K_2Pt(CN)_4 4H_2O$, (b) bis(dimethylglyoximato)Ni(II), (c) Magnus' Green Salt.

Fig. 2.14. $K_2Pt(CN)_4Cl_{0.32}3H_2O$. Filled circles are Cl, large open circles K.

edge-sharing MX_6 octahedra in which the central M is in the $+3$ oxidation state. In M_2X_3 (M = Y, Gd, Tb, Er, Ln; X = Cl, Br) we again have chains of edge-sharing M_6 octahedra but without the separately metal halide chains. These compounds are discussed in greater detail in Chapter 5.

The third category of mixed valence metal atom chain compounds consists of coordination complexes, principally of Pt. The best known is KCP, already referred to at the beginning of this chapter. Mixed valency is introduced into the stacks of planar $Pt(CN)_4$ groups in one of two ways: either extra anions such as Cl^- are inserted into the channels between the stacks as in $K_2Pt(CN)_4Cl_{0.32}3H_2O$ (Fig. 2.14), or cations are removed, as in $K_{1.75}Pt(CN)_42H_2O$. Similar complexes are formed by oxalate or maleonitrilebithiolate ligands, by divalent cations like Mg^{2+} and Co^{2+}, and other anions including HF_2^-, $H(SO_4)_2^{3-}$ and N_3^-. We can write the general formula of such complexes as $M_mPtL_nX_p(H_2O)_q$ and most of those known are listed in Table 2.2. Since the chemical non-stoichiometry is established by adding or removing ions outside the stacks of Pt complexes themselves the number of ions can in principle be varied to accommodate the steric and electronic requirements of the Pt chain. For example it is possible to generate a level of partial oxidation which does not bear a simple relationship to the periodicity of the underlying lattice. As noted in Section 2.2 such lattices are called *incommensurate*. Table 2.2 shows that the average Pt oxidation numbers cluster around $+2.3$, though with a spread of values from $+2.49$ to $+2.14$. Not surprisingly the Pt–Pt distances along the chains decrease as the Pt oxidation number increases, and quite a good correlation has been

Table 2.2 *The formulae of conducting metal chain compounds* $A_m ML_n X_p (H_2 O)_q$

Anion non-stoichiometry

$$M = Pt, \qquad L = CN, \qquad n = 4$$

A:	K	K	NH_4	Rb	Rb	Cs	Cs	$C(NH_2)_3$
X:	$Br_{0.30}$	$Cl_{0.32}$	$Cl_{0.30}$	$Cl_{0.30}$	$(HF_2)_{0.26-0.4}$	$F_{0.19}$	$(N_3)_{0.25}$	$Cl_{0.25}$
q:	3	3	3	3	0–1.67	0		1

$$A = -, \qquad M = Ni, \qquad L = \text{planar macrocycle}, \qquad X = I_3, \qquad q = O$$

L:	OMTBP	TBP	Pc	DPG	OMTBP
p:	0.35	0.33	0.33	0.33	0.97

Cation non-stoichiometry

$$M = Pt, \qquad L = CN, \qquad X = -$$

A:	K	Rb	Cs
m:	1.75	1.73	1.72
q:	1.5	x	x

$$M = Pt, \qquad L = \text{oxalate}, \qquad X = -$$

A:	K	Rb	Co	Mg
m:	1.6	1.67	0.83	0.82
q:	1.2	1.5	6	6

---

found with the simple equation

$$d_x = d_0 + \alpha \log_{10} x \qquad (2.8)$$

where d_x is the Pt–Pt distance in a complex with Pt oxidation state $(2 + x)$, and d_0 is the single distance (Fig. 2.15). A similar expression was originally used by Pauling to fit the metal–metal distances in simple three-dimensional metals and alloys. Only in a few of these Pt chain compounds are the Pt atoms in special crystallographic positions requiring all the Pt–Pt spacings to be identically equivalent, but in the majority the spacings are almost equal (e.g. in $K_{1.75}Pt(CN)_4 1.5H_2O$ they alternate between 2.965 and 2.961 Å). No differences have ever been detected in the Pt–C or C–N bond lengths of different complexes in the chains so we must conclude that the effective Pt oxidation states are genuinely non-integral.

Another set of mixed valency chain compounds containing planar metal complexes stacked face to face is formed by partly oxidizing Ni(II) phthalocyanine (Pc), porphyrins (DMTP, TBP), or diphenylglyoximate (DPG) with iodine. The structure of one of these compounds is shown in Fig. 2.16 and the formulae of several are included in Table 2.2. Though formulae such as

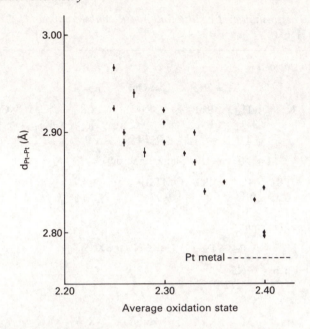

Fig. 2.15. Variation of the distance *d* between neighbouring Pt atoms with average Pt oxidation state in mixed valency chain compounds.

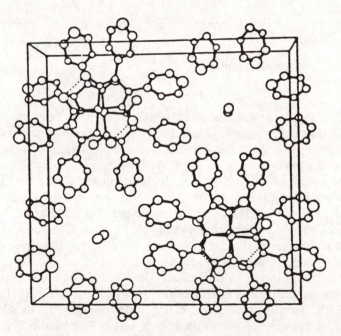

Fig. 2.16. Bis(diphenylglyoximato)Ni(I$_3$)$_{0.33}$.

$Ni(Pc)I_{1.0}$ suggest that these compounds might be integral oxidation state compounds of Ni(III), the iodine is actually present as I_3^-, inserted between the stacks, so the correct formulation is $Ni(Pc)(I_3)_{0.33}$ with an average Ni oxidation state of $+2.33$.

2.4 CORRELATION OF STRUCTURES AND PROPERTIES OF CHAIN COMPOUNDS

Among the chain compounds whose structures have been surveyed in the previous section are metals, semiconductors, insulators, superconductors, ferromagnetics, antiferromagnets, and so on. It is therefore important to make a correlation between the structures and physical properties. In general the occurrence of metallic or insulating behaviour in crystals can be rationalized in terms of two quantities, described in more detail in Chapter 1. The first of these is the resonance integral V_{AB}, proportional to the overlap integral, which measures the attraction of an electron on atom A towards atom B, and hence allows the energy of the system to be reduced by delocalizing electrons between sites. The second is the electron repulsion energy U which raises the total energy of the system when two electrons are placed on the same atom. When $V_{AB} \gg U$ the electrons become delocalized into a band. If the valence shells on each atom which go to make up the band are partly occupied the band, in turn, is partly occupied and the result is a metal. On the other hand, when $V_{AB} \ll U$ the electrons remain as far from one another as possible, there is no electron exchange between atoms and the result is an insulator. The parameter V_{AB} depends on the overlap between orbitals on adjacent centres. The latter in turn depends, not only on the distance between the atoms, but on the symmetry of the orbitals concerned. What distinguishes chain compounds is the simplicity of the topological connections between sites. This enables us to ask the simple question: are there bridging groups separating the metal ions with partly filled d-shells, or is there direct overlap between metal orbitals on neighbouring sites?

The parameter U depends on the radial extension of the valence shell orbitals since electron repulsion varies with r^{-1}. In a lattice, it is measured by the energy expended in changing the occupation of orbitals around two sites A and B:

$$M_A^{n+} + M_B^{n+} \rightarrow M_A^{(n+1)+} + M_B^{(n-1)+} \qquad (2.9)$$

that is, by the difference between the $(n-1)$th and the nth ionization potentials. For the ionization of d-electrons in transition metal ions this difference is very large, 18 eV for Ni^{2+}/Ni^{3+} for example, so the activation energy to transfer electrons between the degenerate d-shells in NiO is huge, and the compound is an insulator. Alternatively put, the valence bond configuration $Ni^{2+}O^{2-}Ni^{2+}$ ... is a good approximation to the ground state. Introducing mixed valency has a dramatic effect on the situation because the charge fluctuation is then built into the lattice chemically, and there is no change in the intra-site electron repulsion energy when an electron is transferred from A to B:

$$M_A^{n+} + M_B^{(n+1)+} \rightarrow M_A^{(n+1)+} + M_B^{n+}. \tag{2.10}$$

Of course, the energy of the process in eqn (2.10) will only be zero if the sites A and B have identical crystalline environments, a remark which forms the basis of the Robin–Day classification of mixed valency compounds. As we shall see, the presence of bridging ligands in mixed valency chain compounds always leads to trapping of the oxidation states, and hence to semiconducting rather than metallic behaviour.

Thus we have isolated two important features to look for in the structure of metal chain compounds: the presence or absence of bridging ligands and integral or mixed oxidation states. Surveying a large number of such compounds one concludes that there is indeed a characteristic pattern of physical behaviour associated with each of the four combinations of these features. For instance single valence chain compounds in which the interaction between neighbouring metal ions occurs through bridging anions, such as the hexagonal perovskites, are overwhelmingly insulators with quite weak ferro- or antiferromagnetic exchange. Single valence chains with directly interacting metal ions, such as the square planar d^8 compounds, are also insulating but show strong interaction between the molecular units in their excited states. Mixed valency chains with bridging anions are likewise semiconductors, because in all known cases displacement of the anions creates a difference in potential around the metal ion sites or, in other words, pins the phase of the charge density wave so that it cannot move along the chain. Only in the fourth category, where there is mixed valency combined with direct interaction between adjacent metal orbitals, do we find a metallic ground state. This is not just a question of enhanced electronic bandwidth (equal to $4V_{AB}$ in a one-electron model), or even of reduced electron repulsion arising from the mixed valency, but also that the average oxidation state is genuinely non-integral so that any Peierls distortion, with its resulting charge density wave, is incommensurate with the lattice. Table 2.3 summarizes our correlation in terms of the major types of physical property and gives examples of each category. In the following section we shall describe the properties of each of these in more detail.

Table 2.3 *Physical properties of metal atom chains*

SINGLE VALENCE

(1) *Bridged*

	Examples
(a) Local (d → d) transitions of metal ions only (no low-lying collective transitions)	Hexagonal perovskites ABX_3
(b) Magnetic exchange in one dimension (spins couple to crystal field transitions)	e.g. $CsFeCl_3$, $CsNiF_3$ $[N(CH_3)_4]MnCl_3$
(c) Insulators	

(2) *Direct metal–metal contact*

(a) Crystal field transitions (if any) strongly perturbed by charge transfer states in the ultraviolet. Electric dipole allowed transitions have large shifts from isolated molecule energies (dipolar interactions)	Magnus' Green Salt $Pt(NH_3)_4PtCl_4$ $BaPt(CN)_4 4H_2O$
(b) Diamagnetic	
(c) Insulators, becoming conducting under pressure	$MgPt(CN)_4 4H_2O$

MIXED VALENCE

(1) *Bridged*

(a) Lowest excited state is M → M charge transfer (local excitations such as d → d transitions only weakly perturbed)	Wolfram's Red Salt $[Pt(C_2H_5NH_2)_4]$– $[Pt(C_2H_5NH_2)_4Cl_2]Cl_4$
(b) Diamagnetic semiconductors (become conducting under pressure)	
(c) Mixed valence Robin–Day class II (distinct metal ion sites)	

(2) *Direct metal–metal contact*

(a) Metallic optical properties (plasma edge in visible)	$K_2Pt(CN)_4Br_{0.30} 3H_2O$
(b) Metallic conductivity at room temperature	
(c) Mixed valence Robin–Day class IIIB (identical metal ion sites)	

2.5 PHYSICAL PROPERTIES OF CHAIN COMPOUNDS: EXAMPLES

2.5.1 Single valence bridged chains

The hexagonal perovskite salts ABX_3 are typical one-dimensional magnetic insulators. Near-neighbour exchange constants are in the range 5–30 cm^{-1} and the ratio of intra- to inter-chain exchange is at least 10^3 in several cases. A specific example is $[N(CH_3)_4]MnCl_3$ (TMMC) in which the $(MnCl_3)_\infty$ chains are separated by bulky organic cations. Magnetic exchange interaction between the nearest neighbours along the chain is antiferromagnetic in this case, and is best described by the Heisenberg model (see Chapter 4 in Volume 1) with an exchange constant J estimated as 6.7 cm^{-1} from the temperature dependence of the susceptibility. The simple molecular field theory of cooperative magnetism (Chapter 4 of Volume 1) predicts a critical temperature of 73 K for this value if J if the lattice were three-dimensional. Experimentally, it is 0.83 K. On the other hand, inserting a ratio $J/J' = 10^3$ between the intra- and inter-chain exchange constants into eqn (2.5), the predicted value of T_c drops to 1.35 K. That there exist long regions of ordered spins at temperatures far above T_c is demonstrated by neutron diffraction. Where there is three-dimensional antiferromagnetic order the magnetic unit cell is a multiple of the chemical cell and magnetic Bragg reflections are found with Miller indices which are fractions of those arising from nuclear scattering. If the spins are only ordered along one direction in the crystal however, the magnetic scattering intensity develops well-defined peaks only when scanned along that direction. This is equivalent to saying that the intensity forms sheets in reciprocal lattice space. The width of the sheet, when scanned parallel to the direction of the correlated spins, is a direct measure of the correlation length ξ. Equation (2.4) gave the temperature dependence of ξ in exponential form but when $kT \gg J$ we can approximate it by the first linear term. The excellent accord between this theory and the experiment for TMMC is shown in Fig. 2.17.

Because of the very small overlap between orbitals even on neighbouring metal ions in this structure we are firmly within the $U \gg V_{AB}$ regime and hence in the realm of validity of the Heitler–London valence bond description. Therefore the excited states of lowest energy are ones which do not

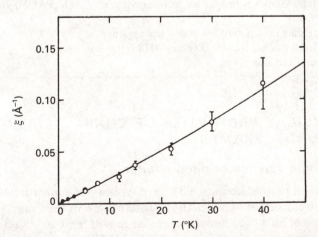

Fig. 2.17. Temperature dependence of the inverse correlation length ξ^{-1} in $[N(CH_3)_4]MnCl_3$, determined by neutron diffraction. The solid line is calculated from eq. 2.4 for $kT \gg J$.

change the number of electrons on each centre. Best known excitations of that kind are the ligand-field transitions, where electrons are promoted from one metal-localized orbital to another. At first sight the visible-near ultraviolet spectrum of TMMC looks like that of any other Mn(II) salt with octahedrally coordinated ligands. Only closer inspection reveals the effect of the one-dimensional magnetic exchange. As is well known, all the ligand-field transitions in Mn(II) compounds are spin forbidden because the ground term is the only sextet in the d^5 manifold, but they are more intense in those compounds where there is magnetic exchange between nearest neighbours because the decrease in spin projection (sextet to quartet) accompanying the electronic excitation on one ion can be compensated by making an equal and opposite 'spin flip' within the ground state of a neighbouring ion. Since the correlations between the spins vary with temperature the absorption band intensities also vary, in a way that is easily distinguishable from the more familiar vibronic intensity mechanism of ligand-field bands.

2.5.2 Single valence chains with direct metal–metal overlap

The examples of this kind of chain (Section 2.3.4) are all composed of square planar d^8 complexes. Being low spin they have no magnetic properties of interest, but peculiarities in their optical properties (often noticeable at once to the eye) indicate strong interactions between the molecular units in their excited states. A typical example is bis(dimethylglyoximato)Ni(II) which forms deep red crystals, though only pale yellow in solution. More striking are the tetracyanoplatinate(II) salts with Group IA, IIA and IIIA cations, whose unusual colours were first commented on more than 100 years ago. In dilute solution the ion $Pt(CN)_4^{2-}$ has no absorption bands in the visible or near ultraviolet but its solid salts all have a single broad absorption band in this region, which is only observed when the electric vector of the incident light is parallel to the chains of metal ions. So intense is the band that the crystals have metallic reflectivity over that frequency range (Fig. 2.18).

In their electrical properties all the d^8 metal chain compounds are insulators and the crystals contain well-defined molecular units. Hence the most approproate description of their excited states appears to be in the $U \gg V_{AB}$ limit, as in the bridged single valence chains. However, the transition giving the visible band in the tetracyanoplatinates is clearly electric dipole allowed; so, if it does not result in electron transfer from one molecule to another, it must originate from either an intramolecular charge transfer or a transition localized on the metal atom. Being polarized perpendicular to the molecular planes, the most plausible alternative of the former type is $5dz^2 \rightarrow CN\pi^*$, and of the latter $5dz^2 \rightarrow 6pz$. In either case the transition dipoles within each chain are all parallel. Interaction of the excited molecule with the rest of the crystal alters the transition energy from what it would be if the molecule were completely isolated. The simplest way of parametrizing

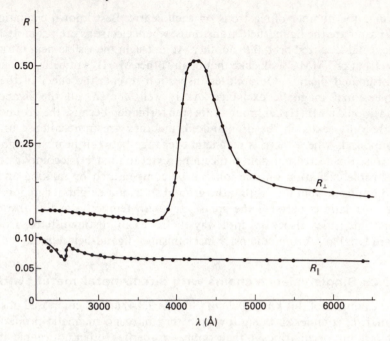

Fig. 2.18. Polarized reflectivity of $BaPt(CN)_4 4H_2O$.

such an interaction is an expansion in terms of multipoles, among which the dipole–dipole varies as the inverse cube of their separation, and the Pt–Pt distances in the tetracyanoplatinate salts vary from 3.7 to 3.1 Å, thus permitting a simple test of the hypothesis by plotting the observed frequency of the intense absorption band against $R(Pt-Pt)^{-3}$. Figure 2.19 shows that the fit is excellent.

2.5.3 Mixed valence bridged chains

A typical example of a mixed valency chain compound with ligands separating the metal ions is the Pt(II,IV) halide called Wolfram's Red Salt (WRS) (Fig. 2.11), whose structure was described in Section 2.3.3. Note that, like many compounds prepared in the last century, its trivial name characterizes its colour. The colour at once indicates an unusual excited state because neither of the constituent ions $[Pt(C_2H_5NH_2)_4]^{2+}$ nor $[Pt(C_2H_5NH_2)_4Cl_2]^{2+}$ has any intramolecular transitions, ligand field, or charge transfer, in the visible. The dark colour arises from an absorption polarized parallel to the chain which, as in the tetracyanoplatinates, is intense enough to give 'metallic' reflectivity over a large region of the visible (Fig. 2.20 shows the reflectivity of a closely related compound). Such compounds are diamagnetic,

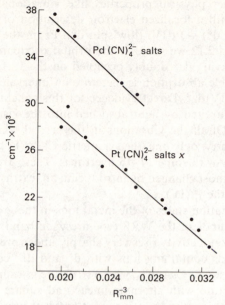

Fig. 2.19. Frequency of the intense absorption band in tetracyanopalladate and -platinate crystals with metal–metal spacing.

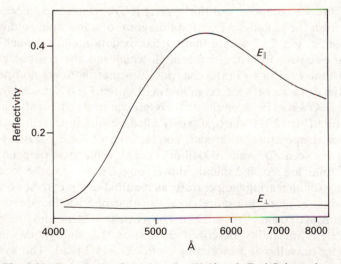

Fig. 2.20. Polarized reflectivity of a Wolfram's Red Salt analogue.

and all the other physical properties like vibrational and photoelectron spectra agree with a localized electron description of the ground state as Pt(II) (low spin d^8) + Pt(IV) (low spin d^6). From the ligand-field splitting diagrams of Fig. 2.12 we see that the orbital occupancies at the two sites differ only in having dz^2 doubly occupied on the Pt(II) and empty on the Pt(IV). The visible absorption is therefore an intervalence electron transfer Pt(II)dz^2 → Pt(IV)dz^2. Direct evidence for this conclusion comes from the Raman spectra excited by light absorbed into the visible band (resonance Raman effect). Of all the vibrations in the crystal which could be Raman active by symmetry, only one, the symmetric Cl–Pt(IV)–Cl stretching mode, appears strongly with some 14–15 overtones. These are precisely the bonds which could be most changed by introducing an 'extra' dz^2 electron into the excited state of the Pt(IV).

Since the oxidation states of the metal ions in the ground state are firmly trapped, compounds of the WRS type are wide band gap semiconductors, although their conductivity rises very sharply under pressure. Mixed valency chain compounds containing ions with d^8 and d^{10} configurations behave similarly. For instance $CsAuCl_3$, whose structure is a superlattice of the cubic perovskite structure with alternate linear and square planar coordination around the metal ions, is an insulator at ambient pressure, but increases in conductivity by 10^9 up to 60 kbar, when it is metallic. X-ray diffraction shows that at this pressure the difference between the two Au sites has disappeared so the compound is then effectively one of Au(II). It is assumed that pressure has a similar effect on WRS.

2.5.4 Mixed valence chains with direct metal–metal overlap

The best known compound of this type is KCP, referred to in Section 2.1. Starting from the ligand-field splitting diagram of a low spin square planar d^8 complex in Fig. 2.12 we see that partial oxidation could be achieved by removing electrons from the dz^2 orbital, which has the greatest extension parallel to the Pt–Pt axis in the chains. Given that there are no intervening ligands, the distance between neighbouring Pt atoms is not much larger than in Pt metal. Overlap between the dz^2 orbitals can lead to a highly directional energy band (Fig. 2.21) which, if partly filled, results in metallic properties.

At room temperature the specific conductivity of KCP parallel to the Pt chains is between 500 and 1000 $ohm^{-1} cm^{-1}$ while that perpendicular is some 10^4 times less. At, and slightly above, room temperature the conductivity decreases with increasing temperature, as required of a metal, but on cooling it approximates more and more closely to that of a semiconductor with a band gap Δ of 580 cm^{-1} (Fig. 2.22). This is clear evidence for the existence of the Peierls instability described in Section 2.2, though the transition between the metallic and insulating states is very broad. The average Pt oxidation number of $+2.30$ defines the wavevector k_F of the Fermi surface

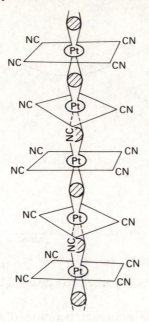

Fig. 2.21. Overlap of dz^2 orbitals in tetracyanoplatinate chains.

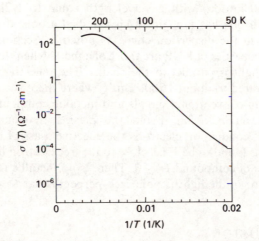

Fig. 2.22. Temperature dependence of the conductivity of KCP.

in KCP as $0.85\pi/a$, from eqn (2.6). X-ray and neutron diffraction show that in the low temperature Peierls-distorted phase the Pt atoms are displaced from their mean positions by an amount which varies sinusoidally along the chain with a repeat distance of $6.67a$, and a maximum amplitude of 0.025 Å. The static distortion is accompanied by the onset of three-dimensional order

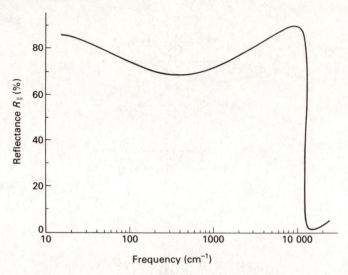

Fig. 2.23. Polarized reflectivity of KCP.

but above the Peierls transition temperature of 100 K the distortion is already present in dynamic form, though uncorrelated from chain to chain. In those conditions a longitudinal acoustic phonon (see Chapter 10 in Volume 1 for definition) with a wavelength equal to $1/2\mathbf{k}_F$ has a lower frequency than if one has a chain of closed shell atoms, thus producing a Kohn anomaly in the dispersion curve as noted in Section 2.2 (Fig. 2.3).

The optical properties of KCP are those of a metal when the electric vector of the incident light is parallel to the chains. The reflectivity remains high from the far infrared to about $16\,000\ \mathrm{cm}^{-1}$, where there is a plasma edge of the kind found in conventional metals and metallic compounds such as the tungsten bronzes (Fig. 2.23). From the plasma frequency and known concentration of conduction electrons the effective mass of the carriers can be estimated. It is found to be 1, i.e. close to the free electron limit, suggesting that in this class of compound $U \sim 0$. There is no metallic reflectivity when the electric vector of the light is polarized perpendicular to the chains.

2.6   CONCLUSIONS

We have seen that there are many inorganic and metal-organic compounds containing chains of metal ions, and that their physical properties span the whole range from magnetic insulators to semiconductors and metals. As in other solids this variety is the result of an interplay between the interaction of neighbouring metal ions and the repulsion between electrons on individual metal ions. Because the topology of connected polyhedra is particularly simple in one dimension, the structures of metal chain compounds are neatly characterized according to the number of bridging groups between cations. Their properties, too, divide neatly into categories according to the degree of metal–metal interaction, exemplified by the presence or absence of bridging ligands and whether the metal oxidation state is single or mixed. Finally, one-dimensional compounds exhibit a number of properties, such as Peierls distortions, not found in other classes of inorganic solid.

2.7 BIBLIOGRAPHY

No references are given to original publications, which can be found in the review articles and books listed below:

Keller, H. J. (ed.), *Low-dimensional cooperative phenomena*, Plenum Press, New York (1975).

Keller, H. J. (ed.), *Chemistry and physics of one-dimensional metals*, Plenum Press, New York (1977).

Miller, J. S. (ed.), *Extended linear chain compounds* (3 Vols.), Plenum Press, New York (1982).

Miller, J. S. and Epstein, A. J., *Prog. Inorg. Chem.*, **20**, 1 (1976).

Robin, M. B. and Day, P., *Adv. Inorg. Chem. Radiochem.*, **10**, 247 (1966).

Underhill, A. E. and Watkins, D. M., *Chem. Soc. Rev.*, **9**, 429 (1980).

The most comprehensive account of the structure of inorganic solids, is:

Wells, A. F., *Structural inorganic chemistry* 4th edn, Oxford University Press (1975).

3 Superconducting materials

J. Etourneau

3.1 INTRODUCTION

In 1908 the experiments of Kammerlingh-Onnes[1] on the liquefaction of helium led to the study of a number of phenomena in the temperature range 1 K to 14 K. One of the first experiments that he performed was the measurement of the resistivity of numerous materials as a function of temperature. Thus he found (1911) that the electrical resistance of a purified mercury sample dropped sharply from $0.08\ \Omega$ at above 4 K to less than $3 \times 10^{-6}\ \Omega$ at about 4.2 K. This drop occurred over a temperature interval of 0.01 K (Fig. 3.1). This phenomenon was called *superconductivity*. The most striking feature of all superconducting materials is that below a critical temperature, labelled T_{cr}, the electrical resistance is zero.

Since the discovery of superconductivity it has been found that a large number of metallic elements and alloys, and even some heavily doped semiconductors, are superconducting. In metals and alloys T_{cr} usually ranges from less than 1 K to about 18 K. After the discovery of the Nb_3Sn superconductor with A15-type structure, the efforts were mainly concentrated on intermetallic compounds. With a T_{cr} of 23.3 K in Nb_3Ge, however, a limit seemed to have been approached.[2,3] In spite of great efforts to increase this limit further, it stood as the record until 1986 when J. G. Bednorz and K. A. Müller observed that lanthanum barium copper oxide began its super-conducting transition as it was cooled below 35 K.[4]

The first part of this chapter is devoted to the fundamental properties which distinguish a superconductor from other solids. Electrical and magnetic properties will be described briefly as well as the thermodynamics of the transition. The main features of the microscopic theory of superconductivity formulated by Bardeen, Cooper, and Schrieffer (BCS) will be given.[5] In the second part, different classes of superconducting materials are reviewed, including elements and relatively high critical-temperature compounds. The study of elements is essential because it leads to a conceptual understanding of the occurrence of superconductivity, whilst the study of compounds with high transition temperatures and high critical fields is of interest for technological applications like magnets, motors, and transmission lines.

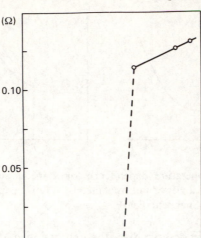

Fig. 3.1. Variation of the resistance with temperature for a mercury sample.[1]

A profound weakness of the descriptions and explanations of superconductivity in the framework of BCS theory is that they cannot predict whether any given material will exhibit superconductivity or not. Furthermore the theory does not permit the calculation of T_{cr} in any given case because the model neglects the complexity of real materials. Accordingly it is only by establishing empirical conditions between the superconducting transition temperature T_{cr} and chemical composition, atomic and structural properties that the search for high temperature superconductors has progressed. Thus in reviewing the different families of materials we shall emphasize the influence of crystal symmetry, atomic disorder, cluster formation, lattice instabilities, and atomic properties, such as the valence electron concentration (VEC), on their superconducting properties.

3.2 GENERAL PROPERTIES OF SUPERCONDUCTORS

3.2.1 Critical temperature

The critical temperature T_{cr} is that at which the metal–superconductor transition occurs in the absence of an applied magnetic field. As will be shown in Section 3.2.2.2, the transition occurs at a lower temperature when a magnetic field is present and accordingly the nature of the thermodynamic transition changes from second order to first order, i.e. there is a latent heat in a non-zero field.

As indicated in Fig. 3.2, the thermal variation of the resistivity for a normal

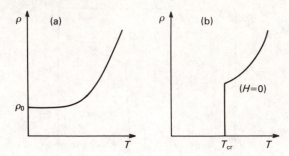

Fig. 3.2. Resistivity temperature dependence for a metal and a superconductor. (a) Low temperature resistivity of a normal metal. (b) Low temperature resistivity of a superconductor (in zero magnetic field).

metal above T_{cr} can be described by the expression $\rho_{(T)} = \rho_0 + BT^5$ where the constant term ρ_0 arises from impurities and defects, and the term in T^5 arises from the interaction of charge-carriers with vibrations. Below T_{cr}, such scattering mechanisms disappear and the resistivity drops abruptly to zero..

3.2.2 Magnetic properties—Meissner effect and critical field

Meissner and Ochsenfeld were the first to exemplify the difference in behaviour between a perfect metallic conductor and a superconductor, in the presence of a magnetic field.[6] A magnetic field, if not too strong, cannot penetrate inside a superconductor. Figures 3.3 and 3.4 show what happens when an external magnetic field H_e is applied parallel to the axis of a long metallic cylinder exhibiting a superconducting transition at T_{cr}. The manner in which penetration occurs with increasing field strength depends on the geometry of the sample. However, for the simplest geometry (long, thin, cylindrically shaped samples with their axes parallel to the applied magnetic field) two kinds of superconductors can be distinguished, called types I and II.

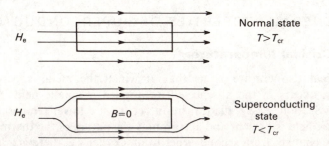

Fig. 3.3. The Meissner effect; the magnetic field is expelled from the sample when the superconductivity occurs.

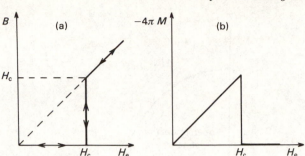

Fig. 3.4. External magnetic field effects in type I superconductors (a) Induction versus field. (b) Magnetization versus field.

3.2.2.1 Type I superconductors

Below T_{cr}, when the field is increased from zero up to a critical value H_c, surface currents prevent the penetration of the field and the induction B is zero within the sample (Fig. 3.4).

$$B = B_e + \mu_0 M = 0 \quad \text{(SI)} \tag{3.1}$$

$$B = H_e + 4\pi M = 0 \quad \text{(CGS)} \tag{3.2}$$

where B_e or H_e is the external magnetic field, M is the magnetization per unit volume and μ_0 is the permeability of free space.†

At $H = H_c$ the field penetrates into the sample which then behaves as a normal metal in which the induction B equals H_c. This result shows that, at a given temperature $T < T_{cr}$, an ideal superconductor expels the field and consequently $B = 0$ for $0 < H_e < H_c$ (Fig. 3.4). Thus, the superconductor exhibits perfect diamagnetism, a phenomenon that is called the *Meissner effect*. Superconductors showing such behaviour are of type I and the applied field which destroys the superconductivity is called the critical field H_c. H_c depends on temperature as shown in Fig. 3.5. The thermal variation of H_c is reasonably well represented by the relation

$$H_c = H_0 \left(1 - \frac{T^2}{T_{cr}^2} \right) \tag{3.3}$$

where H_0 is the critical field at 0 K.

† It is worthwhile noting that textbooks tend to be written in SI units and original work in magnetism is mostly in electromagnetic CGS units. The application of SI to magnetism can be contradictory and confusing. Since a vast amount of older information is firmly established in the CGS system and many research papers are published currently in different countries in CGS units, the CGS system is used in this chapter. Thus in the SI system formulae dealing with force, energy, magnetic moment and so on are in terms of the induction B (measured in Tesla). In free space $B = \mu_0 H$. In the CGS system the formulae are in terms of the field strength H measured in œrsteds. The permeability of free space is $\mu = 1$ so that the induction B (measured in Gauss) in free space is numerically equal to the field; nevertheless, for the sake of clarity μ_0 will be maintained in the formulae.

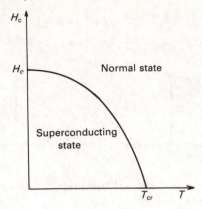

Fig. 3.5. The critical field versus temperature for a type I superconductor.

3.2.2.2   Type II superconductors

Type II superconductors behave differently from those of type I, as shown in Fig. 3.6. Below $H_{c1}(T)$ there is no penetration of the magnetic field; when the aplied field exceeds an upper critical field $H_{c2}(T) > H_{c1}(T)$, the whole sample is in the normal state and the field penetrates readily. When the applied field has a value between $H_{c1}(T)$ and $H_{c2}(T)$, there is a partial penetration of flux and the sample exhibits a complicated microscopic structure of both normal and superconducting regions, known as the *mixed state* or *vortex state*. Thus, for $H_e > H_{c1}$ the magnetization vanishes gradually as the field increases, rather than suddenly as for a type I superconductor. As in type I superconductors, the critical fields H_{c1} and H_{c2} vary with temperature as shown in Fig. 3.7. The curves $H_{c1}(T)$ and $H_{c2}(T)$ are the boundaries of three domains within which $B = 0$ (Meissner effect, perfect diamagnetism), $B < H$ (mixed state), and $B = H$ (normal state).

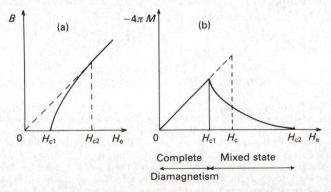

Fig. 3.6. External magnetic field effects in type II superconductors. (a) Induction versus field. (b) Magnetization versus field.

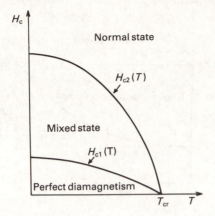

Fig. 3.7. Variation of the critical fields H_{c1} and H_{c2} with temperature for a type II superconductor.

3.2.3   Persistent and critical currents

Figure 3.2(b) shows the variation of the resistivity versus temperature for a superconductor. In the superconducting state, currents can flow with no discernible energy dissipation. For instance, closed currents set up in a ring of superconducting material have flowed for up to 2 years without showing any trace of decay. There are, however, some limitations since the super-conductivity can be destroyed by applying either a sufficiently large magnetic field (see above) or a sufficiently large current (Silsbee effect). In the latter case the critical current, for which superconductivity disappears, depends on the nature and geometry of the specimen.

3.2.4   Thermodynamics of the transition

The implication, from the Meissner effect, that the superconducting–normal transition is reversible, in both type I and type II superconductors, permits us to apply thermodynamics to the transition. The transition is reversible with respect to temperature (T), field (H), and pressure (P), but the superconductor undergoes only very small volume changes and the pressure dependence can be neglected. The equilibrium state of a superconductor in a uniform magnetic field is determined by the temperature T and the magnitude of the field H. We shall assume that the pressure P is fixed and that the superconductor is a long cylinder parallel to the field so that demagnetization effects are negligible.

The Gibbs free energy will be considered because in any system the stable state is that with the lowest free energy. The aim is to compare the difference in the magnetic contribution to the free energy of two phases, superconducting and normal, when they are in the same applied magnetic field. When the

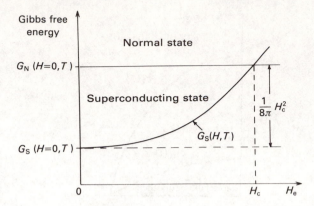

Fig. 3.8. Gibbs free energy versus field of normal and superconducting states.

sample is cooled below its transition temperature it becomes superconducting and accordingly the free energy of the superconducting state must be less than that of the normal state (Fig. 3.8).

The differential of the internal energy of a system in an external magnetic field H_e is

$$dU = T\,dS - P\,dV + \mu_0 V H_e\,dM + \sum_{i=1}^{m} \mu_i\,dn_i \tag{3.4}$$

where M is the magnetization per unit volume, μ_0 is the permeability of the free space, μ_i is the chemical potential of component i, and n_i the number of moles of component i in the volume V.

Upon integration using Euler's theorem:

$$U = TS - PV + \mu_0 V H_e M + \sum_{i=1}^{m} \mu_i n_i.$$

Complete differentiation leads to

$$dU = T\,dS + S\,dT - P\,dV - V\,dP + \mu_0 V H_e\,dM$$

$$+ \mu_0 V M\,dH_e + \sum_{i=1}^{m} \mu_i\,dn_i + \sum_{i=1}^{m} n_i\,d\mu_i. \tag{3.5}$$

Subtracting eqn (3.4) from (3.5) yields the condition

$$S\,dT - V\,dP + \mu_0 V M\,dH_e + \sum_{i=1}^{m} n_i\,d\mu_i = 0.$$

At both constant temperature and pressure

$$\sum_{i=1}^{m} n_i\,d\mu_i = -\mu_0 V M\,dH_e.$$

In a one-component system such as a metal or a compound the term $n_i \mu_i$ is identical to the Gibbs free energy of the material of volume V.

$$dG = -\mu_0 V M \, dH_e. \tag{3.6}$$

In the superconducting state one may write

$$G_S(H_c, T) - G_S(H = 0, T) = -\mu_0 V \int_0^{H_c} M \, dH \tag{3.7}$$

In the case of type-I superconductors the external magnetic field H_e produces a negative magnetization such that the induction $B = \mu_0 H_e + 4\pi M$ vanishes. If the field penetration is neglected (see Section 3.2.6) the magnetization per unit volume M is given by

$$M = -\frac{\mu_0 H_e}{4\pi}.$$

The free energy is thus increased when a magnetic field is applied to a superconductor:

$$G_S(H_c, T) = G_S(H = 0, T) + \frac{\mu_0^2 V}{4\pi} \int_0^{H_c} H \, dH$$

$$G_S(H_c, T) = G_S(H = 0, T) + \frac{\mu_0^2 V}{8\pi} H_c^2. \tag{3.8}$$

For most superconductors the normal state is paramagnetic and the magnetization is small compared with that of the superconducting state shown in Fig. 3.4(b) or Fig. 3.6(b) for fields below H_c or H_{c2}. Therefore considering that magnetization is unimportant in the normal state, we have

$$G_N(H_c, T) - G_N(H = 0, T) = 0 \tag{3.9}$$

where G_N is the Gibbs free energy of the normal state. At the critical field H_c for which there is an equilibrium between the superconducting state and the normal state, the Gibbs free energy of these two states must be equal for two-phase equilibrium:

$$G_N(H_c, T) = G_S(H_c, T). \tag{3.10}$$

Taking into account eqns (3.8), (3.9) and (3.10), the free energy difference ΔG between the two zero-field states is

$$\Delta G = G_N(H = 0, T) - G_S(H = 0, T) = \frac{\mu_0^2 V}{8\pi} H_c^2. \tag{3.11}$$

Physically the quantity $(\mu_0^2/8\pi)H_c^2$ represents the magnetic energy per unit volume received by the superconductor as the field is raised from zero to a value which completely eliminates superconductivity. This field is H_c for type-I conductors (see Fig. 3.4(b) and H_{c2} for type-II superconductors (see Fig. 3.6(b)). However for type-II superconductors one defines a '*thermodynamic critical field H_c*' calculated from the relation $(\mu_0/8\pi)H_c^2 = \int_0^{H_c} M \, dH$ which corresponds to the area under the type-II magnetization curve..

3.2.5 Specific heat

The specific heat C_N of a normal metal is usually regarded as being composed additively of contributions from the lattice and the conduction electrons, and at very low temperature it may be expressed as

$$C_N = \gamma T + \beta T^3 \tag{3.12}$$

where the first term γT is due to the electrons and the second is due to the lattice. From numerous experiments we can reasonably assume that the lattice properties are unchanged between the normal and the superconducting states. Thus, only the electronic contribution to the specific heat will be considered, C_{eN} and C_{eS} being respectively the contribution in the normal and the superconducting states. As shown in Fig. 3.9, when the transition occurs in a zero magnetic field, the electronic specific heat shows two striking features: there is a discontinuous jump in the specific heat at T_{cr} with approximately $C_{eS} \simeq 3C_{eN}$ as predicted by the Bardeen–Cooper–Schrieffer theory of superconductivity, and C_{eS} drops rapidly and non-linearly for $T < T_{cr}$, going to zero at $T = 0$ K as required by the third law of thermodynamics.

Two further points concerning the electronic specific heat of a superconductor merit attention. The fact that below the transition temperature the specific heat is greater in the superconducting state than in the normal state

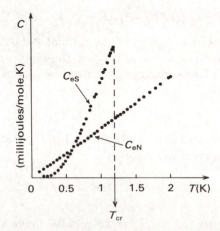

Fig. 3.9. Low temperature specific heat of normal and superconducting aluminium. The normal state below T_{cr} is restored by applying a magnetic field of 300 gauss.[91.]

means that, when a metal in the superconducting state is cooled through this region, the entropy of its conduction electrons decreases more rapidly with temperature than if they were in the normal state. In the superconducting state the electrons can be considered to be in a highly ordered state. Further, the electronic specific heat of a metal in the superconducting state varies exponentially with temperature. This is the variation to be expected if there is a gap in the range of energies available for electrons. Thus the number of electrons excited in energy levels above the gap at a temperature T is proportional to $e^{-\Delta E/2k_B T}$ where k_B is Boltzmann's constant and ΔE the energy gap.

It should be noted that some superconductors do not exhibit an energy gap and are subsequently called *gapless* superconductors. These materials are not typical and we do not give any details about them in this chapter.

3.2.6 Penetration depth

In a magnetic field a superconductor exhibits a persistent shielding current which leads to the absence of field within samples of macroscopic dimensions. This current must be confined within a region very close to the surface. Expressed in terms of the field, one may conclude that H drops to a vanishingly small value over a characteristic penetration depth λ_L whose value is of the order of 10^{-5} cm; λ_L is called the London penetration depth. Although this thickness is very small, it plays an important role since it corresponds to the region where the magnetic contribution to the free energy density changes from its normal value to that appropriate for the super-conductor.

According to the London equation the applied field H_e does not suddenly drop to zero at the surface of a type I superconductor. For a sample thicker than λ_L the magnetic field decays exponentially as it penetrates into the material, i.e.

$$H(x) = H_e \exp\left(-\frac{x}{\lambda_L}\right) \tag{3.13}$$

where H_e is the applied magnetic field at the surface and x is the distance in from the surface (Fig. 3.10).

In fact for a number of calculations, it is enough to use the approximation that the magnetic field remains constant to a distance λ within the metal and then abruptly falls to zero.

At $T = 0$ K, for a superconducting pure metal containing n_s conduction electrons per unit volume, the London penetration depth is defined by

$$\lambda_L(0) = \left[\frac{mc^2}{4\pi n_s e^2}\right]^{1/2}. \tag{3.14}$$

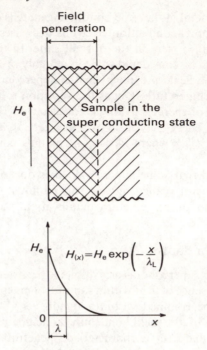

Fig. 3.10. Field penetration in a superconductor.

In the quasi free electron model n_s is given by

$$n_s = \frac{2m^* V_F^2}{3} N(0) \tag{3.15}$$

where m^* is the effective mass of the electrons, V_F is the Fermi velocity, and $N(0)$ is the density of states at the Fermi energy (for one direction of spin).

Then, using eqns (3.14) and (3.15), we find

$$\lambda_L(0) = \left[\frac{3c^2}{8\pi N(0) V_F^2 \, e^2} \right]^{1/2}. \tag{3.16}$$

For a pure metal in the free electron approximation, the penetration is temperature dependent. Near the critical temperature T_{cr}, one obtains within the Landau–Ginzburg model

$$\lambda(T) \propto \lambda_L(0) \left[\frac{T_{cr}}{T_{cr} - T} \right]^{1/2}. \tag{3.17}$$

3.2.7 Qualitative features of the microscopic theory of superconductivity

What has long been the accepted microscopic theory of superconductivity was proposed by Bardeen, Cooper, and Schrieffer (BCS) in 1957.[5] It is outside the scope of the present chapter to develop the formalism needed for an adequate description of their theory. We shall only describe the basic principles and major theoretical predictions in a qualitative way.

The main assumptions and results of the BCS theory are as follows. In a metal there exists an attractive force between pairs of conduction electrons caused by virtual phonon exchange and a repulsive force due to the screened Coulomb interaction. When their combined effect is attractive, the metal may become superconducting. Further, in a superconductor, electrons are condensed into pairs called Cooper pairs. The electrons are paired in such a way that each pair, in the lowest energy state, shows zero total wavevector (i.e. zero total momentum) and zero total spin (i.e. singlet state).

3.2.7.1 Basic ideas of the BCS theory

Before introducing the main features of the BCS microscopic theory we shall recall several points concerning the quasi free-electron model for metals. According to this model, which deals with the electrons belonging to the outer atomic shells, the effect of the lattice is taken into account by replacing the electron mass by its effective value m^*. Furthermore Coulomb and exchange electron–electron interactions are not considered.

Thus, the unperturbed collective electronic states may be described in terms of the one-electron state occupation numbers. The one-electron states are spin-orbitals $|\mathbf{k}, \sigma\rangle$ characterized by the wavevector $\mathbf{k}$ defined in Chapter 2 and the spin index $\sigma = \uparrow$ or $\downarrow$; they form the conduction band. Their energy $E_{\mathbf{k}}$ which is spin-independent is given by the relation

$$E_{\mathbf{k}} = \frac{\hbar^2 \mathbf{k}^2}{2m^*}. \tag{3.18}$$

The lowest energy collective states are obtained when the N electrons available for the conduction band occupy the smallest-k spin-orbitals taking into account the Pauli exclusion principle. These states are localized within the Fermi sphere of radius k_F (Fermi radius). This sphere is centred at the origin of the reciprocal space. All the electronic states at the surface of the sphere (Fermi surface) have the same energy (Fermi level) given by the formula

$$E_F = \frac{\hbar^2 \mu_F^2}{2m^*}. \tag{3.19}$$

In this ground state the probability for a one-electron state to be occupied by an electron is thus exactly 1 for $k < k_F$ and 0 for $k > k_F$ (Fig. 3.11)

When the temperature is raised above 0 K, higher energy collective states are available, i.e. electrons can occupy individual states with energy larger than E_F, leaving empty an equal number of states with energy lower than

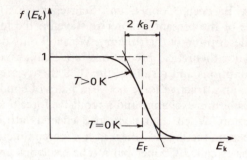

Fig. 3.11. Probability $f(E_k)$ vs. E_k that a quantum state of kinetic energy is occupied by an electron for normal metal.

E_F. The probability for a one-electron state to be occupied by an electron is then given by the Fermi–Dirac distribution (Chapter 1, Section 1.2)

$$f(E_k) = \frac{1}{1 + e^{(E_k - E_F)/k_B T}}.$$ (3.20)

In fact the probability $f(E_k)$ no longer shows a discontinuity at E_F, but decreases continuously from 1 to 0 within a small energy range ($\sim 2k_B T$) near E_F (Fig. 3.11). Such an individual state description cannot be used in the theory of superconductivity since, as was guessed very early on, the phenomenon has to be related to some kind of attractive electron–electron interaction.

The discovery of the isotope effect on the critical temperature of superconducting elements by Reynolds et al.[7] and by Maxwell[8] gave a strong hint that the explanation for superconductivity had to be sought in a coupling between the conduction electrons and the lattice vibrations. These authors measured the transition temperatures of separated mercury isotopes and found a positive result that could be interpreted as $T_{cr} M^{1/2} \simeq$ constant, where M is the isotopic mass which can be involved only through the lattice vibrations (phonons).

Frölich[9] and later on Cooper[10] developed a model for this coupling leading to an electron–electron interaction which may be attractive under some very restrictive conditions. This may be stated in a simplified form as follows: any electron may be significantly scattered from a state defined by $\mathbf{k}_0$ to a state defined by $\mathbf{k}_1$ only if the energy difference $|E_{\mathbf{k}_0} - E_{\mathbf{k}_1}|$ is less than $\hbar\omega_D$ where ω_D is the Debye frequency. The matrix element of the scattering process is usually denoted by $V_{\mathbf{k}_0 \mathbf{k}_1}$. Since the scattering must occur between an occupied state and an empty one, one may conclude that only the individual states whose kinetic energy lies within $\hbar\omega_D$ near E_F are involved in the scattering process.

The physical meaning of this model can be schematized as follows: as an electron moves across the lattice, it attracts the nearby cations and slightly

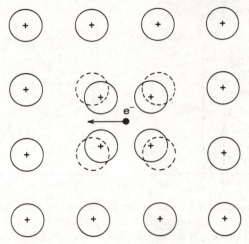

Fig. 3.12. Polarization of lattice near electron.

displaces them towards itself (Fig. 3.12). Near this moving electron, such a displacement creates a small increase in positive charge which has to be compensated for by a corresponding decrease at larger distances in order to maintain the total charge constant. Such a local charge perturbation in the lattice produces a weak short-range attractive potential capable of capturing another electron. This potential leads to the formation of electron pairs (Cooper pairs). The weakness of the perturbing potential is reflected in the severe conditions required for scattering as indicated above.

The BCS model starts with the basic idea of electron pairing and describes a collective state in terms of electronic pairs rather than individual states as is the case for normal metals. Since each electron pair involves two spin-orbitals $|\mathbf{k}, \sigma\rangle$ and $|\mathbf{k}', \sigma'\rangle$, it then appears that *the lowest energy collective state* is built up from electron pairs of spin-orbitals having both opposite spins and wavevectors. This electronic collective state is called the BCS ground state. It may be described as follows: the two spin-orbitals forming a pair, $|\mathbf{k}, \uparrow\rangle$ and $|-\mathbf{k}, \downarrow\rangle$, are *simultaneously* empty or occupied by two electrons. Such a situation leads to the concept of average occupation probability g_k for the pair $(|\mathbf{k}, \uparrow\rangle|-\mathbf{k}, \downarrow\rangle)$ (Fig. 3.13(b)). This is the origin of the name *correlated state* given to the superconducting state. The g_k values are determined through a variational process in order to minimize the energy.

The excited states are obtained by pair-breaking so that the electrons no longer have equal and opposite momenta and behave almost like free electrons and for this reason are referred to as *quasi-particles*. It must be understood by *quasi-particle* an electron in the state $|\mathbf{k}, \uparrow\rangle$ while the complementary state $|-\mathbf{k}, \downarrow\rangle$ is *empty*. If, as a result of splitting up a pair, e.g. by radiation, there is a quasi-particle in the state $|-\mathbf{k}, \uparrow\rangle$, then this state

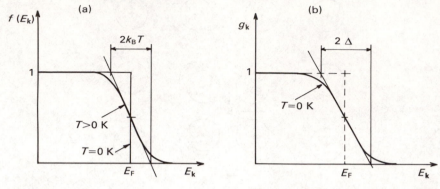

Fig. 3.13. (a) Probability $f(E_k)$ vs. $|k|$ for a quantum state with energy E_k to be occupied by an electron for normal metal. (b) Probability g_k vs. $|k|$ for the two electron states ($|k, \uparrow\rangle |-k, \downarrow\rangle$) to be occupied as a whole in the BCS ground state.

is now definitely occupied whereas previously the probability of its being occupied was g_k (Fig. 3.13(b)).

We may therefore ask how much energy is needed to break up a pair so as to produce two electrons or more properly two quasi-particles. There is a minimum energy, named Δ, required to create a quasi-particle. This energy which corresponds to single-particle excitations is calculated from the expression

$$\Delta = 2\hbar\omega_m \exp\left(-\frac{1}{N(0)V}\right) \tag{3.21}$$

where $N(0)$ is the density of states (spin $\uparrow$ or $\downarrow$ only) at the Fermi level the parameter V is a suitable average of $V_{k_0 k_1}$ defined previously and $\hbar\omega_m$ is an average vibrational energy near E_F.

Since the excitations are created in pairs, the minimum energy required to create excitations from the ground state, at 0 K is

$$\Delta E = 2\Delta. \tag{3.22}$$

It is useful at this point to emphasize the distinction between the average occupation probability g_k and the Fermi function $f(E_k)$ since they are both occupation numbers and their variation versus E_k may be somewhat similar, as shown in Figs 3.13(a) and 3.13(b). Strictly speaking g_k is a quantum mean value which describes the low-lying state of the superconducting phase. As a result it has no real meaning at non-vanishing temperatures, except close to absolute zero when one considers elementary excitations of the Cooper pairs. g_k decreases steeply, from unity to zero over an energy range 2Δ near the Fermi level E_F (defined for the normal phase). In contrast to g_k, $f(E_k)$ is essentially a statistical or thermodynamical function describing non-interacting fermion systems. $f(E_k)$ also shows a sharp decrease from 1 to 0 near E_F at temperatures above absolute zero and at 0 K the change at E_F becomes discontinuous. The energy interval over which the functions fall to zero is temperature dependant ($2k_BT$) for $f(E_k)$, whereas it takes the unique value 2Δ for g_k.

The number of quasi-particles increases as the temperature is raised above absolute zero. The creation of a quasi-particle, i.e. an electron in the state $|\mathbf{k}, \uparrow\rangle$ without a partner in the state $|-\mathbf{k}, \downarrow\rangle$, prevents the formation of the pair state $(|\mathbf{k}, \uparrow\rangle, |-\mathbf{k}, \downarrow\rangle$. Consequently the pair interaction energy is diminished because the number of scattering events in which the Cooper pair may participate is lessened. Accordingly the energy gap ΔE decreases as the number of quasi-particles increases, i.e. as the temperature rises. Thus ΔE appears to be temperature dependent according to the expression

$$\Delta E(T) = \Delta E(0)\left(1 - \frac{T}{T_{cr}}\right)^{1/2}. \tag{3.23}$$

Consequently $\Delta E(T)$ vanishes at the critical temperature T_{cr}, where a second-order phase transition from the superconducting to the normal state is observed. It can be shown that the critical temperature T_{cr} is proportional to the gap at 0 K:

$$T_{cr} = \frac{1.14}{k_B} \hbar\omega_m \exp\left(-\frac{1}{N(0)V}\right) \tag{3.24}$$

$$T_{cr} = \Delta E(0)/3.5k_B. \tag{3.25}$$

As can be seen from these expressions, high T_{cr} values are related, through V, to strong electron–phonon coupling which results in a rather low electrical conductivity in the normal state.

The BCS theory disregards two important factors: the Coulomb repulsion of electrons and the effects of retardation in the electron–phonon interaction. For these reasons it cannot lead to quantitative agreement with experimental data. However it provides a convenient basis for explaining numerous observed phenomena. Taking account of these two factors, the work of Bogolyubov *et al.*,[11] Eliashberg,[12] and McMillan[13] led to the following new formula for the superconducting transition temperature:

$$T_{cr} = \frac{\Theta_D}{1.45} \exp\left(-\frac{1.04(1 + \lambda)}{\lambda - \mu^*(1 + 0.62\lambda)}\right). \tag{3.26}$$

The product $N(0)V$ is so important that a term is attached to it, namely the electron–phonon coupling constant and labelled λ in many text books, reviews and papers. Equation (3.26) is known as the McMillan formula for strong coupling superconductors, i.e. for $\lambda \sim 1$. The Debye temperature $\theta_D = \hbar\omega_D/k_B$ is used as the typical phonon frequency.

The factor μ^* is the effective Coulomb repulsion between Cooper pair electrons. It is expressed in terms of the matrix element V_c of the Coulomb repulsion at the Fermi surface as

$$\mu^* = \frac{V_c(0)}{1 + V_c N(0) \dfrac{\ln E_F}{\omega_D}}. \tag{3.27}$$

The λ factor can itself be expressed as a function of the phonon density of states $F(\omega)$ and of $\alpha^2(\omega)$ which is a convenient average value of the square of the electron–phonon interaction matrix element

$$\lambda = 2 \int_0^{\omega_D} \frac{\alpha^2(\omega)F(\omega)}{\omega} \, d\omega. \tag{3.28}$$

In a rigorous way McMillan showed that in the case of a one-atom lattice with an atomic mass M, λ could be written in the form

$$\lambda = \frac{N(0)\langle J^2 \rangle}{M \langle \omega^2 \rangle} \tag{3.29}$$

where $\langle J^2 \rangle$ is a suitable average over the Fermi surface of the electron–ion potential gradient matrix element and $\langle \omega^2 \rangle$ is the mean square phonon frequency.

Superconductors can be classified into groups of weak, intermediate and strong coupling ones according to the value of the product $N(0)V = \lambda$. In the weak coupling case, $N(0)V \ll 1$ and eqn (3.26) reduces to the BCS result with $N(0)V$ replaced by $N(0)V - \mu^*$.

3.2.7.2   Coherence length in a superconductor

The coherence length, ξ, in a superconductor can be understood considering the probability, $P(r) \, dv_2$ of finding an electron with momentum $-p$ and spin down in a volume element dv_2 at a distance r from a volume element dv_1 which contains an electron with momentum p and spin up.

In a normal metal the probability is generally considered as independent of r since there are no correlations. In the superconducting state the probability for very large values of r is the same as in the normal state. However for small values of r the probability is greater, indicating that paired electrons are more likely to be close together than far apart as shown in Fig. 3.14. The distance over which the probability is found to be greater than that corresponding to large values of r is called the coherence length. In practical terms the coherence length ξ can be identified with the spatial extent of the pair wavefunction.

It may be noted that $k_B T_{cr}$ is the energy range of the Bloch states which are significantly affected by the electron pairing. From this and the uncertainty principle it can be deduced that for a *pure material*, the spacial extent ξ_0 of a pair in the quasi free electron model is

$$\xi_0 \simeq \frac{\hbar V_F}{k_B T_{cr}} \tag{3.30}$$

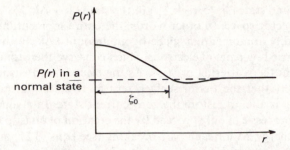

Fig. 3.14. Probability $P(r)$ for an electron with momentum p and spin down to be at a distance r from another electron with momentum p and spin up.

where V_F is the Fermi velocity. In the Ginzburg–Landau framework, the coherence length ξ is temperature dependent according to

$$\xi(T) = 0.74 \left[\frac{T_{cr}}{T_{cr} - T} \right]^{1/2} \xi_0 \tag{3.31}$$

3.2.7.3 Classification of superconductors by means of the Ginzburg–Landau constant

In the previous sections two characteristic lengths $\lambda_L(T)$ and $\xi(T)$ have been defined. Since both vary as $(T_{cr} - T)^{-1/2}$, for $T \to T_{cr}$, their ratio

$$K = \frac{\lambda(T)}{\xi(T)} \tag{3.32}$$

is practically temperature independent and remains finite at T_{cr}. K is known as the Ginzburg–Landau constant of the material. Its value permits us to classify the superconductors into two types: for $K < \sqrt{2}/2$ the superconductor is type-I, while for $K > \sqrt{2}/2$ is type-II.

3.2.7.4 The zero resistance phenomenon in a superconductor

The most striking feature of superconductors is that these materials offer absolutely no resistance to the flow of an electrical current below their critical temperature T_{cr}.

   In the normal state the effect of an electrical field is to displace the Fermi surface continuously as a whole, in a direction opposite to the field. Due to the presence of impurities, lattice defects and phonons, the electrons are scattered with a change in momentum $\mathbf{p} = \hbar \mathbf{k}$ and consequently the material offers a resistance. When the electrical field is maintained constant, the scattering leads to a steady current. When the electric field is switched off, one observes the restoration of the zero current state.

   In the superconducting state, the presence of an electric field displaces the centre of gravity of each pair by an amount $\Delta \mathbf{k}$. Thus, each pair changes from the ground state ($|\mathbf{k}, \uparrow\rangle, |-\mathbf{k}, \downarrow\rangle$) to the state ($|\mathbf{k} + \Delta\mathbf{k}, \uparrow\rangle, |-\mathbf{k} + \Delta\mathbf{k}, \downarrow\rangle$) in the wavevector space. In other words, the entire momentum distribution is shifted bodily in momentum space by an amount $\hbar\Delta\mathbf{k}$. In this picture the current is carried by pairs of electrons which conserve their total momentum ($\mathbf{p} = 2\hbar\,\Delta\mathbf{k}$) in the scattering process. As the electric field is suppressed it is highly unlikely that the Fermi surface returns to its initial position. Only pair-breaking is able to restore the zero current state. In a superconducting state such a process can only proceed by the creation of quasi-particles which requires an energy 2Δ for each pair-breaking (see eqns (3.21) and (3.22)), an amount of energy which is not available from a single phonon. Thus the non-zero flow state is a metastable one. On the other hand, as the current density increases, the momentum distribution becomes more and more displaced until it becomes energetically possible for a Cooper pair to split into two quasi-particles.

3.2.8 Superconductivity and magnetism

Many experiments have shown that the presence of magnetic impurities in materials tends to cancel their superconducting properties. As indicated previously (Section 3.2.7) the BCS theory of superconductivity depends strongly on the degeneracy of the spin orbitals $|\mathbf{k}, \uparrow\rangle, |-\mathbf{k}, \downarrow\rangle$ which form a Cooper pair. The presence of unpaired spins (spin density) in superconducting materials creates internal magnetic fields able to split the energies of these spin orbitals, since they have opposite spin numbers. As one might expect, it can be shown that this effect is rather similar to that of an external magnetic field, and consequently it lowers the superconducting transition temperature, or even suppresses the superconductivity.

At this point a clear distinction has to be made between the electrons responsible for magnetism and those able to form the Cooper pairs. For example, in the pure transition metals magnetism is of the collective type. Hence conduction electrons participate in both magnetism and superconductivity. Generally, in these materials, magnetism and superconductivity exclude one another. On the other hand, in pure rare earth metals and rare earth alloys the electrons in the incomplete 4f shells form localized magnetic moments. In these materials the conduction electrons, which are in energy bands build up from outer 5d–6s orbitals, interact magnetically with the 4f local moments. Such an indirect magnetic interaction is called Ruderman–Kittel–Kasuya–Yosida indirect exchange.[14, 15] Therefore electrons of different types contribute to both magnetism and superconductivity. The strength of the 4f–5d or 4f–6s interaction will depend on the overlap of the corresponding orbitals. Consequently, coexistence of magnetism and superconductivity should be sought in materials in which conduction and magnetic orbitals only overlap weakly. The ternary rare earth compounds $REMo_6X_8$ (Chevrel phases, Section 3.3.2.3.2) and $RERh_4B_4$ (Section 3.3.2.3.3) appear to belong to this category. They offer an interesting experimental and theoretical basis for studying the interplay between long-range ordering and superconductivity.

Experimental evidence shows that superconductivity may coexist with antiferromagnetism. When the Néel temperature T_N is lower than T_{cr} two types of behaviour may be observed. As the temperature is decreased through T_N the value of H_{c2} for type II superconductors either increases (e.g. $ErMo_6S_8$ and $SmRh_4B_4$) or abruptly decreases (e.g. $REMo_6S_8$ with RE = Tb, Dy and $NdRh_4B_4$).

A number of models have been developed, involving several mechanisms in which antiferromagnetism and pair-breaking are interrelated. Thus the change in the mean magnetization as well as the spin fluctuations near T_N act to break the pairs through exchange and Zeeman coupling. Also spin waves, which create local spin deviations, result in an effective repulsive interaction between electrons in different spin states. Besides its sign, this interaction is similar to the electron–electron interaction via the phonons, which is responsible for the occurrence of superconductivity owing to its attractiveness. Another influence of antiferromagnetism on superconductivity comes from the change in lattice periodicity often associated with antiferromagnetic ordering. This may involve a change in the Brillouin zone which, of course, would affect the Fermi surface.

Superconductivity and ferromagnetism are antagonistic: in a superconductor the electrons tend to group into pairs with anti-parallel spins, whereas in a ferromagnetic there is a tendency for all spins to be parallel. From a thermodynamic point of view, the ferromagnetic state is normally favoured over the superconducting one. Thus, relative to the paramagnetic normal state, the free energy of the ferromagnetic state is lowered by roughly cNk_BT_c, where c is the atomic fraction of magnetic ions, N is the total number of atoms and T_c the magnetic ordering temperature. On the other hand, in the superconducting state the free energy is diminished by approximately $(k_BT_{cr}/E_F)Nk_BT_{cr}$, because k_BT_{cr}/E_F corresponds to the proportion of the conduction electrons which participates in the superconductivity. Generally the factor c is larger than k_BT_{cr}/E_F.

In certain materials the superconducting transition temperature T_{cr} is above the magnetic ordering temperature. In most cases the onset of ferromagnetic ordering destroys the superconducting state. Such a behaviour is called *re-entrant superconductivity* (e.g. $HoMo_6S_8$ and $ErRh_4B_4$). However, it seems that, in certain special conditions, superconductivity and ferromagnetism may coexist.[16,17] For example, it has been shown that for $HoMo_6S_8$ and $ErRh_4B_4$, a sinusoidally modulated magnetic state coexists with superconductivity in a narrow range of temperatures below the magnetic ordering transition T_c. However the question is not yet settled, mainly because of an incomplete knowledge of the nature of the sinusoidal magnetic structure.

3.3   SUPERCONDUCTIVITY IN PURE ELEMENTS, ALLOYS, AND COMPOUNDS

A problem for chemists is to answer the following question: how can we improve the superconducting properties of a given material (e.g. raising the critical temperature and critical field) or what kinds of new compounds might exhibit superconductivity? In principle it should be possible to answer the first question knowing the parameters which influence the superconductivity (Debye frequency, density of electron states at the Fermi surface, Coulomb interaction, and electron–phonon interaction). However, in practice the establishment of empirical rules and correlations between superconductivity and various crystallochemical and physical properties within a series of materials can also give a partial answer to these questions.

In this section we give first a brief description of the main materials which have been found so far to be superconducting, and secondly we focus attention on the correlations, indicated above, which are 'universal' for all types of compound.

3.3.1   Superconductivity and the periodic table

Many elements have been found to be superconducting, as indicated in Fig. 3.15, although some of them only exhibit a superconducting transition under special conditions. Thus most of the non-transition elements (Si, P, Ge, As, Se, Sb, Te, Bi) and metals (Y, Cs, Br, V, Ce) only become superconducting under pressure.[18] Several elements are superconducting when they are either prepared in thin films (Li, Cr, Si, Eu) or irradiated by α-particles (Pd).[19] Recently, temperatures down to 50 μK were used to investigate the superconducting properties of rhodium, which shows a superconducting transition at 325 μK.[20]

Most of the metallic elements are superconductors at sufficiently low temperature and the discovery of further ones may result from advances in low-temperature measurements. It is also clear that metals which order ferromagnetically, such as transition metals with incomplete d and f shells (Co, Fe, Ni, Gd, etc.), do not exhibit any superconducting transition. This indicates that superconductivity and ferromagnetism are mutually exclusive phenomena. In pure elements T_{cr} does not exceed 10 K under normal conditions, the highest superconducting transition temperatures being observed for Pb ($T_{cr} = 7.19$ K) and for Nb ($T_{cr} = 9.25$ K). It is worth noting that the application of high pressure to pure elements may, in certain cases, bring about an increase of T_{cr}. For instance T_{cr} goes from 5.5 K at 1 bar up to 12.9 K at 200 bar in the case of lanthanum.[21]

According to the value of the Ginzburg–Landau constant $K = \lambda(T)/\xi(T)$ given in Section 3.2.7.3, the superconductors can be classified into two

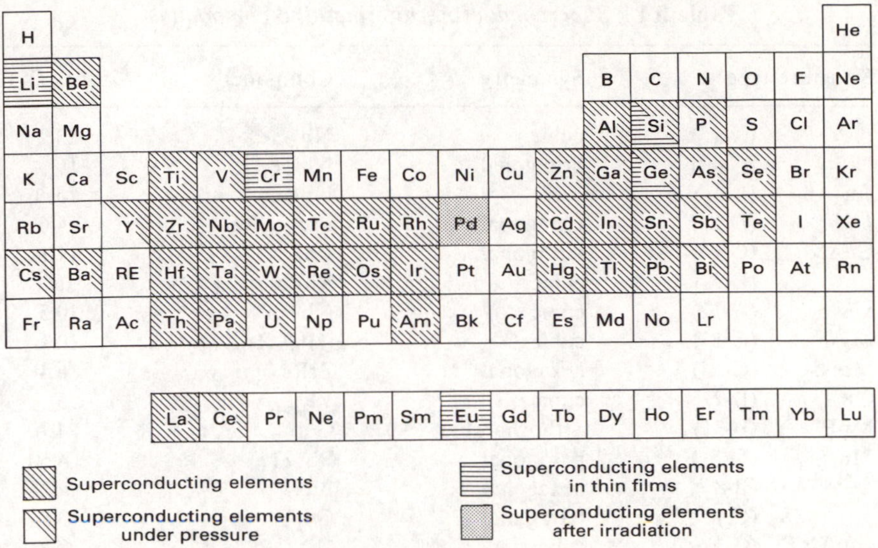

Fig. 3.15. Superconducting elements in the periodic system of the elements.

groups: type-I ($K < \sqrt{2}/2$) and type-II ($K > \sqrt{2}/2$). Pure metals are usually type-II. It is, however, possible for even pure metals to be type-II superconductors (e.g. Nb, V, Tc).

3.3.2 Superconductivity in binary and ternary compounds

Before giving the main characteristics of these superconducting compounds we have to define what a binary or a ternary compound is.

If two metals are able to form a binary compound, two possibilities exist: either the two metals may form a solid solution over a certain range of composition, or the alloying of the two metals may lead to the formation of intermetallic compounds which are characterized by well-defined stoichiometric compositions. A ternary compound is the combination of three elements in a crystal structure which is different from that of any binary combination of those three elements.

In this section, we distinguish the compounds in which the bonds are essentially metallic (e.g. metallic solid solutions and intermetallic compounds) from those which possess a considerable share of ionic and covalent bonds along with the metallic type of bond. This later type of compound, which we call for simplicity *chemical compound*, is formed from both metallic and non-metallic atoms, or from non-metallic atoms only.

Tables 3.1, 3.2, and 3.3 list numerous compounds corresponding to the different types mentioned above. For each structural type the material exhibiting the highest superconducting transition temperature is given.

Table 3.1   *Superconducting intermetallic compounds*[a]

Structure type		Symmetry	Compounds	T_{cr} (K)
β-W	(A-15)	cubic	Nb_3Ge	23.2
$CrSi_2$	(C-40)	hexagonal	$NbGe_2$	16
Cu_3Au	(L-1_2)	cubic	$NbRu_3$	15–16
β-U	(D-8_b)	tetragonal	$Mo_{0.38}Re_{0.62}$	14.6
$CuAl_2$	(C-16)	tetragonal	$RhZr_2$	11.1
	(E-9_3)	cubic	$RhZr_3$	11
α-Mn	(A-12)	cubic	$NbTc_3$	10.5
$MgZn_2$	(C-15)	cubic	$(Hf_{0.5}Zr_{0.5})V_2$	10.1
$MgNi_2$	(C-14)	hexagonal	$ZrRe_2$	6.4
CsCl	(B-2)	cubic	VRu	5
MnP	(B-31)	orthorhombic	GeIr	4.70
Mn_5Si_3	(D-8_8)	hexagonal	Pb_3Zr_5	4.60
	(B-8_1)	hexagonal	BiNi	4.24
	(C$_c$)	tetragonal	Ge_2Y	3.8
anti-CaF_2	(C-1)	cubic	$Ga_{0.7}Pt_{0.3}$	2.9
FeSi	(B-20)	cubic	AuBe	2.64
	(D-1_c)	orthorhombic	$AuSn_4$	2.38
	(L-1_0)	tetragonal	NaBi	2.25
	(D-2_d)	hexagonal	Au_5Ba	0.7

[a] The notations in brackets are those used in 'Strukturbericht' and Landolt–Börnstein Zahlenwerte und Funktionen—4 teil: Kristalle; Springer-Verlag, Berlin (1955).

Table 3.2   *Binary superconducting chemical compounds*[a]

Structure type		Symmetry	Compounds	T_{cr} (K)
NaCl	(B-1)	cubic	NbN	17.3
Pu_2C_3	(D-5_c)	cubic	$(Y_{0.7}Th_{0.3})_2C_{3.1}$	17
		hexagonal	MoN	13–14.8
	(C-49)	orthorhombic	$ZrGe_2$	8
	(C-27)	hexagonal	$NbSe_2$	7.2
CaB_6	(D-2_1)	cubic	YB_6	6.5–7.1
	(C-2)	cubic	$Rh_{0.53}Se_{0.47}$	6
UB_{12}	(D-2_b)	cubic	ZrB_{12}	5.85
	(D-0_e)	tetragonal	Mo_3P	5.31
		hexagonal	Nb_3S_4	4
	(C-6)	trigonal	PdTe	1.53

[a] See footnote to Table 3.1.

Table 3.3 *Ternary superconducting chemical compounds*[a]

Structure type	Symmetry	Compounds	T_{cr} (K)
Mo_3Se_4	rhombohedral	$PbMo_6S_8$	15.2
$MgAl_2O_4$ (H-I$_1$)	cubic	$LiTi_2O_4$	13.7
Fe_2P	hexagonal	$ZrRuP$	13.02
$CeCo_4B_4$	tetragonal	$Lu_{0.75}Th_{0.25}Rh_4B_4$	11.9
Ti_3S_4	hexagonal	$Li_{0.3}Ti_{1.1}S_2$	13
$LuRuB_2$	orthorhombic	$LuRuB_2$	9.9
	cubic	$Yb_6Rh_8Sn_{26}$	8.6
$Sc_5Co_4Si_{10}$	tetragonal	$Sc_5Rh_4Si_{10}$	8.50
$LaRu_3Si_2$	hexagonal	$LaRu_3Si_2$	7
$LuRu_4B_4$	tetragonal	$ScRu_4B_4$	7.2
YOs_3B_2	orthorhombic	YOs_3B_2	5.5
$U_2Co_3Si_5$	orthorhombic	$LaRh_3Si_5$	4.45
$ZrOs$ or $SrSi_2$	cubic	$LaRhSi$	4.35
$ThCr_2Si_2$	tetragonal	YRh_2Si_2	3.11
$CeCo_3B_2$	hexagonal	$LaRh_3B_2$	2.70

[a] High-T_{cr} oxides are listed in Table 3.4.

3.3.2.1 *Superconducting solid solutions of transition metals*

As a rule, disordered solid solutions are formed when alloying takes place between neighbouring metals in the periodic table. It is among these materials that the superconductors now used in the production of wire for superconducting coils were discovered. Thus the solid solutions $Nb_{0.75}Zr_{0.25}$ (T_{cr} = 11 K) and $Nb_{0.75}Ti_{0.25}$ (T_{cr} = 10 K) have become technologically the most important superconducting materials among ductile solid solutions. In contrast to intermetallic compounds, they are quite ductile, and wire made from them is therefore much easier to use. The critical current densities are between 10^5 and 10^6 amp/cm^2 and the critical fields are around 180–190 kOe. $Nb_{0.75}Ti_{0.25}$ is advantageous for high current densities, whilst $Nb_{0.75}Zr_{0.25}$ withstands higher critical fields.[22] Mo–Tc compounds with $T_{cr} \simeq$ 14 K exhibit the highest transition temperature among the transition metal solid solutions. Unfortunately technetium is expensive and difficult to use. The Mo–Re alloys with a transition temperature exceeding 11 K are among the most ductile superconductors known, but they lose their superconducting properties in fields above 20 kOe.[22].

3.3.2.2 *Superconducting intermetallic compounds*

The number of intermetallic compounds, which are metal–metal or metal–non-metal combinations, is very large. They crystallize in many different crystal structures, and a number of them are superconducting as shown in Table 3.1.

It is worth noting that the structural types of intermetallic compounds that have received the most attention are A-15, C-15, C-14, B-2, A-12 and D-8$_b$ (Table 3.1). However the most important and interesting intermetallic compounds are those with the A-15 or β-W structure. Since the discovery of the superconductivity of V$_3$Si ($T_{cr} = 17$ K, A-15 type structure) by Hardy and Hulm in 1953,[23] a lot of compounds with this structure have been prepared with $T_{cr} \geqslant 15$ K: Mo$_3$Re ($T_{cr} = 15$ K), Nb$_3$Sn ($T_{cr} = 18$ K), Nb$_3$Al ($T_{cr} = 18.8$ K), Nb$_3$Si ($T_{cr} = 19$ K), Nb$_3$Ga ($T_{cr} = 20.3$ K) and Nb$_3$Ge ($T_{cr} = 23.2$ K). Nb$_3$Ge shows, so far, the highest known superconducting transition temperature apart from the copper oxides.[4] These reasons lead us to describe the A-15 compounds in somewhat more detail than the others.

The structure of the A-15 compounds is shown in Fig. 3.16. The stoichiomeric formula is A$_3$B. The A atom is always a transition element of Groups IV, V, or VI of the periodic table (Ti, V, Cr, Zr, Nb, Mo, Ta, W); the B atom can be either a non-transition element (Al, Si, P, Ga, Ge, As, In, Sn, Sb, Pb, Bi) or a transition element (Co, Ni, Ru, Rh, Pd, Re, Os, Ir, Pt, Au). The B atoms form a body-centred cubic lattice. On each face there are equidistant A atoms which form continuous chains through the system. The three chains do not intersect. Deviations from stoichiometry can occur in both A and B components. Thus, for A$_{3+x}$B$_{1-x}$ the linear chains remain intact and the excess of A atoms occupies the B sites, whereas for A$_{3-x}$B$_{1+x}$ there is a partial breakdown of the A-atom chains.

The superconducting properties of the A-15 type compounds are strongly influenced by the composition and the degree of atomic order. Two main classes of substance crystallizing with this structure type may be distinguished, called *typical* and *atypical*.

The *typical* compounds must fulfill two conditions: no atoms with d-levels near the Fermi energy may occupy the cubic sites of the unit cell, and the Fermi energy E_F must lie close to the edge of a d-band built up from the orbitals of the components on the linear chain sites. Such compounds (e.g. V$_3$Si, V$_3$Ga, V$_3$Ge, V$_3$Au, Nb$_3$Ga, Nb$_3$Ge, etc.) exhibit a pronounced

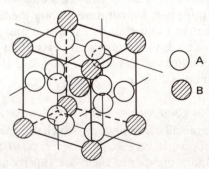

Fig. 3.16. The structure of the A$_3$B compounds (A-15) (β-tungsten type).

maximum of the superconducting transition temperature T_{cr} as well as the density of states, at the stoichiometric 3:1 composition. For example, in V_3Au the transition from a disordered state to a highly ordered state is accompanied with an increase of T_{cr} of almost a factor of 300 (from $T_{cr} \simeq 0.0012$ K to $T_{cr} = 2.97$ K).[24] Conversely in the *atypical* A_3B compounds (Mo_3Ir, Mo_3Pt, Cr_3Os, Cr_3Ir, V_3Ir, etc.) the B atoms contribute to the common d-band at E_F. This explains both why the maximum T_{cr} does not correspond to a stoichiometric composition and why the degree of atomic order influences weakly the value of T_{cr}.

With regard to structural transformations, Batterman and Barrett[25] were the first to demonstrate that V_3Si and several other superconductors of β-W structure undergo a second-order transformation from cubic to tetragonal at a temperature T_m invariably larger than T_{cr}. Such a transformation is called *martensitic* in the sense that no diffusion is involved. The structural transformation temperature may vary from sample to sample: for V_3Si between 18 and 30 K, for Nb_3Sn between 35 and 50 K. The single-crystal nature of the material is not conserved when it is transformed from a cubic to a tetragonal structure. A domain structure twin has been detected in the tetragonal state.[26] However, Batterman and Barrett[25,27] prepared a V_3Si single crystal which was superconducting below 17 K and for which no structural distortion was detected down to 1.9 K.

3.3.2.3 *Superconducting chemical compounds*

In this section we deal with materials that are formed from both metallic and non-metallic atoms or with the participation of non-metallic atoms only. Hydrides, borides, carbides, nitrides, chalcogenides, silicides, phosphides, intercalation compounds, and polymers belong to these types of material. Some of them are listed in Tables 3.2 and 3.3.

It is worth noting that a high T_{cr} is observed in a large number of materials for which the constituents themselves show low T_{cr} or are not superconductors at all. Thus we can give as examples ZrN ($T_{cr} = 10.7$ K), MoC ($T_{cr} = 14.3$ K), $Li_xTi_{1.1}S_2$ ($T_{cr} = 10\text{--}13$ K), $C_{8.8}K$ ($T_{cr} = 0.125$ K) and $(SN)_x$ ($T_{cr} = 0.26$ K). The pure elements Zr, Mo, Ti have $T_{cr} < 1$ K and Li, K, C, N, and S are not superconductors. On the other hand, theoretical calculations by Cohen have indicated that highly doped semiconductors and semimetals might become superconducting at temperatures around 0.21 K.[28] According to him, if superconductivity does occur among semiconductors and semimetals, the transition temperature should be a function of the number of carriers present in the sample. Such behaviour has been found in germanium tellurides, $Ge_{1-x}Te$ (e.g. $T_{cr} \simeq 0.3$ K for $Ge_{0.937}Te$ with the NaCl structure).[29]

Our aim is not to describe all the superconducting families of chemical compounds listed in Tables 3.2 and 3.3, but to focus our attention on several classes of material considered as important from theoretical and application points of view.

3.3.2.3.1   Compounds having the NaCl-type structure   A large number
of hydrides, carbides, nitrides, phosphides, and chalcogenides of transition
metals and their solid solutions crystallize with the NaCl-type structure (B-1).
A NaCl-type structure can be considered as two face-centred cubic lattices
which are enmeshed into each other as shown in Fig. 3.17. In other words
metallic A atoms form a cubic face-centred cubic lattice, and non-metallic
B atoms occupy all the octahedral sites. In practice, all the B-1 compounds
are prone to exhibit deviations from stoichiometry. The compounds can be
metal or non-metal deficient, or both simultaneously (e.g. $Ti_{0.8}\square_{0.2}O$,
$TiO_{0.7}\square_{0.3}$, $Ti_{0.85}\square_{0.15}O_{0.85}\square_{0.15}$ in which $\square$ represents vacancies). At
certain compositions the vacancies tend to order and give rise to a
superstructure.

Among the compounds having the B-1 structure, the carbides and nitrides
of 4d and 5d transition elements show the highest superconducting transitions.
Whereas the substitution of a 5d-metal by a 4d-metal does not significantly
alter the transition temperature, the replacement of a 4d-metal by a 3d-metal
leads to an abrupt decrease of T_{cr}. The highest T_{cr} values have been obtained
for NbN ($T_{cr} = 17.3$ K) and $(NbN)_{0.75}(NbC)_{0.25}$ ($T_{cr} = 17.8$ K).[30] As with
the A-15 intermetallic compounds, the B-1 materials turn out to be very
sensitive to deviations from stoichiometry. T_{cr} decreases as the number of
metallic or non-metallic vacancies increases.

The compounds of palladium with hydrogen and deuterium merit special
attention since PdH and PdD with a NaCl-type structure exhibit high
critical temperatures (9 K and 11 K respectively).[31,32] For the hydrides it
is $Pd_{0.55}Cu_{0.45}H$ which shows the highest superconducting transition
($T_{cr} \simeq 16.6$ K).[33,34]

Palladium metal is not superconducting because the electrons present in
a narrow 4d band at the Fermi level are strongly correlated and the electronic

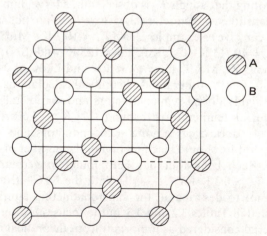

A

B

Fig. 3.17. The NaCl-type structure (B-1).

system is close to the critical condition for the occurrence of localized magnetic moments. There are spin density fluctuations which result in repulsion of electrons of opposite spins so that the Cooper pairs cannot exist. The palladium hydrides are superconducting because the 1s electrons of hydrogen atoms occupy the 4d-band of the palladium in such a way that they destroy the spin fluctuations.

3.3.2.3.2 Chalcogenides Among the chemical compounds that we are considering, the chalcogenides merit special attention, on the one hand since superconductivity was found to exist in several materials exhibiting a very low conductivity above T_{cr}, and on the other hand because some of them possess both high critical temperature and magnetic field.

Oxides: Generally the metallic oxides of the transition metals are not superconducting or do not exhibit high T_{cr} (e.g. $T_{cr} \sim 1$ K in TiO_x with the rocksalt structure). However, three ternary oxides have already been found to be superconducting at relatively high temperature before the discovery of warm ceramic superconductors: (*i*) the hexagonal tungsten bronzes Rb_xWO_3 ($0.16 \leqslant x \leqslant 0.33$) for which the maximum transition temperature ($T_{cr} \sim 7$ K) has been observed around $x = 0.25$,[35] (*ii*) the $Li_xTi_{3-x}O_4$ compounds with the $MgAl_2O_4$ spinel structure for which the highest T_{cr} is 13.7 K for $x = 1$,[36,37] and (*iii*) the perovskite $Ba(PbBi)O_3$ with a T_{cr} of 13 K.[38]

So far, high-T_{cr} behaviour has been observed in about fifteen prototype compounds listed in Table 3.4 in order of increasing structural complexity.[39–51] The critical temperatures indicated in parentheses are only approximate and vary as a function of chemical composition. For some structures, symmetry changes occur as a function of temperature, cation substitution and oxygen content.

The bismuth based oxide $Ba_{1-x}K_xBiO_3$, with $T_{cr} \sim 30$ K for $x = 0.4$, has the cubic perovskite structure shown in Fig. 3.18.[52,53] In contrast to all other high-T_{cr} superconductors it contains no copper and, furthermore, the superconductivity occurs within the framework of a three-dimensionally connected bismuth–oxygen array.

Copper-based oxides of composition $(La_{2-x}M_x)CuO_4$, which adopt the K_2NiF_4-type structure (Fig. 3.19), have been known since 1973 and derive from non-metallic La_2CuO_4 by partial replacement of La by M = Sr, Ba.[54,55] The highest critical temperatures occur at $T_{cr} = 40$ K for $x = 0.15$ in $La_{2-x}Sr_xCuO_4$. On the other hand the undoped La_2CuO_4 oxide, which crystallizes in an orthorhombically distorted modification, is semiconducting when it is strictly stoichiometric. However, a superconducting transition occurs in samples treated either under high oxygen pressure[56] or by electrochemical oxidation[57] to yield $La_2CuO_{4+\delta}$ ($T_{cr} \sim 40$ K). Fluorination of La_2CuO_4 lends to a superconducting oxyfluoride $La_2CuO_4F_y$[58] ($T_c \sim 40$ K).

The structures of $YBa_2Cu_3O_7$ and $YBa_2Cu_3O_6$, shown in Fig. 3.20, can be derived from the perovskite structure (Fig. 3.18) by tripling one cell axis

Table 3.4 *Main prototypes of high-T_{cr} superconductors*

Compounds	T_{cr} (K)
$(Ba_{1-x}K_x)BiO_3$	30
$(La_{2-x}Sr_x)CuO_4$	40
$(Nd, Ce, Sr)_2CuO_4$	28
$(Nd_{2-x}Ce_x)CuO_4$	24
$YBa_2Cu_3O_7$	90
$YBa_2Cu_4O_8$	80
$Y_2Ba_4Cu_7O_{14}$	40
$Pb_2Sr_2NdCu_3O_8$	70
$TlBa_2CaCu_2O_7$	103
$TlBa_2Ca_2Cu_3O_9$	120
$Tl_2Ba_2CuO_6$	90
$Tl_2Ba_2CaCu_2O_8$	112
$Tl_2Ba_2Ca_2Cu_3O_{10}$	125
$Pb_2Sr_2Ca_{1-x}Y_xCu_3O_8$	46
$Pb_{0.5}Sr_{2.5}Y_{0.5}Ca_{0.5}Cu_2O_{7-\delta}$	100

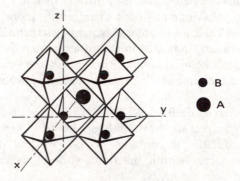

Fig. 3.18. Cubic perovskite structure ABO3 ($Ba_{1-x}K_xBiO_3$). The basic structural unit is a cube in which metallic atom (A) lies at the centre, eight smaller metallic atoms (B) occupy the corners and 12 non-metallic atoms (oxygen) are at the midpoints of the edges.

to accommodate cation ordering $-Y-Ba-Ba-Y-Ba-Ba-$ on the A site, and by removing up to two (out of nine) oxygen atoms in the case of $YBa_2Cu_3O_7$. The oxygen defect concentration, δ, in $YBa_2Cu_3O_{7-\delta}$ varies between 0 and 1. For $0 < \delta < 0.5$ the compound is orthorhombic and superconducting, and for $0.5 < \delta < 1$ it is tetragonal and insulating. Two plateaux are observed for T_{cr} as δ varies: $\delta < 0.15$ ($T_{cr} \sim 90$ K) and $0.25 < \delta < 0.45$ ($T_{cr} = 50-60$ K). The superconducting oxides $YBa_2Cu_4O_8$ ($T_{cr} = 80$ K) and $Y_2Ba_4Cu_7O_{14}$

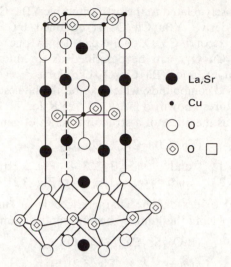

Fig. 3.19. Tetragonal K_2NiF_4 type structure of $La_{2-x}M_xCuO_4$ (M = Sr, Ba). Flat sheets of corner-sharing CuO_6 octahedra are separated by lanthanum–strontium–oxygen rocksalt layers.

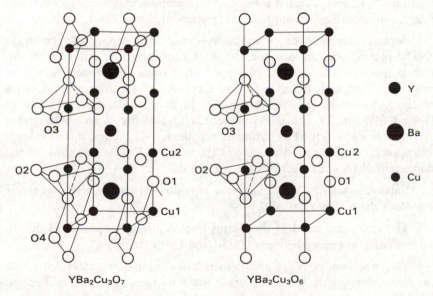

Fig. 3.20. Structure of the $YBa_2Cu_3O_{7-\delta}$ oxides for the two limit cases: $\delta = 0$ and $\delta = 1$.

($T_{cr} = 40$ K) are closely related to the orthorhombic $YBa_2Cu_3O_7$ compound from which they derive. $YBa_2Cu_4O_8$ is obtained by an insertion into $YBa_2Cu_3O_7$ of a second Cu–O chain parallel to one of the short cell dimensions. $Y_2Ba_4Cu_7O_{14}$ can be described as an intergrowth between tetragonal $YBa_2Cu_3O_6$ and $YBa_2Cu_4O_8$-like slabs.[59]

The thallium-based compounds, which exhibit the highest superconducting transition temperatures known so far ($T_{cr} \sim 125$ K for $Tl_2Ba_2Ca_2Cu_3O_{10}$), can be considered as members of a structural series of general formula

$$(TlO)_n Ba_2 Ca_{m-1} Cu_m O_{2+2m}$$

$$
\begin{array}{lll}
n = 1 & \text{and} & m = 1, 2, 3, 4 \quad \text{Fig. 3.21(a)} \\
n = 2 & \text{and} & m = 1, 2, 3 \quad\quad \text{Fig. 3.21(b)}
\end{array}
$$

The bismuth-based oxides can be characterized by a similar formula, in which $n = 2$, by replacing thallium by bismuth and barium by strontium,

$$(BiO)_2 Sr_2 Ca_{m-1} Cu_m O_{2+2m}$$

$$(m = 1, 2, 3 \quad \text{Fig. 3.21(b)})$$

The thallium-based compounds are not strictly isostructural with the bismuth ones.

All the high-T_{cr} superconductors have orthorhombic or tetragonal symmetry, except $Ba_{1-x}K_x BiO_3$ which is cubic. Certain structural, chemical, and electronic generalizations can be made about the copper-oxide based superconductors.

1. Superconductivity occurs within two-dimensional copper–oxygen arrays (CuO_2 planes) and in compounds exhibiting mixed valent copper. An average formal oxidation state of copper larger than $+2$ can be achieved by cation substitutions (e.g. La^{3+}–Sr^{2+} or Ba^{2+}–K^+), by introducing vacancies on the cation sites (e.g. the Sr site in $Bi_2Sr_{2-x}CaCu_2O_8$,[60] Tl site in $Tl_2Ba_2CuO_6$ or Ca site in $Tl_2Ba_2CaCu_2O_8$) or by introducing excess oxygen. It is also worthwhile noting that Y–Ba–Cu–O exhibits mixed valence compounds belonging to both classes III: with $N/(N-1)$ electrons for the planes and $N/(N-2)$ electrons for the chains.[61,62]

2. Superconducting CuO_2 planes are separated by charge reservoir layers (e.g. CuO chains for YBaCuO).

3. The electronic band at the Fermi level is strongly hybridized due to the similarity in energy between Cu:3d and O:2p states.

4. The superconductivity appears in all cases at the doping level for which the antiferromagnetic insulating state loses its local moment to become metallic.

5. The charge transport occurs via electron holes belonging to the Cu:3d and O:2p bands, and probably originating in the CuO_2 layers. It is worthwhile noting that electrons as charge carriers are found only in $(Nd_{2-x}Ce_x)CuO_4$.

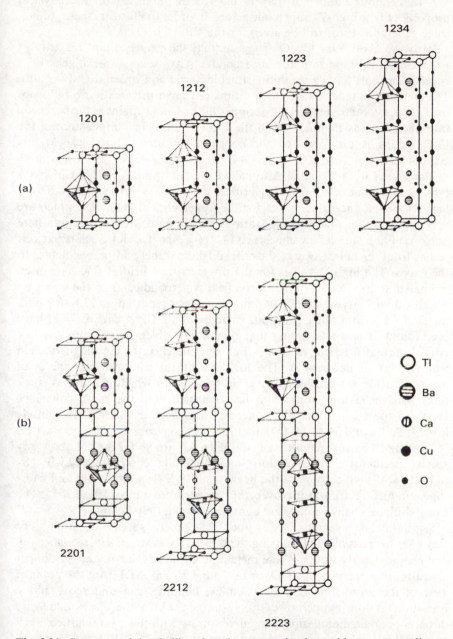

Fig. 3.21. Structure of the thallium-based superconducting oxides corresponding to the general formula $(TlO)_n Ba_2 Ca_{m-1} Cu_m O_{2+2m}$: (a) $n = 1$ and $m = 1, 2, 3, 4$; (b) $n = 2$ and $m = 1, 2, 3$.

A tremendous number of papers have been published on the physical properties of the high-T_{cr} superconductors. In order to illustrate these studies, some general features will be given for the $YBa_2Cu_3O_{7-\delta}$ system.

Early work on $YBa_2Cu_3O_7$ suggests that the presence and integrity of CuO chains is important for superconductivity because the highest T_{cr}s occur in crystals having the most stoichiometric and ordered CuO chains. The fact that magnetic rare earth atoms (R) have practically no influence on T_{cr} supports this view because copper in the Cu–O chains is further away from the R atoms than copper in the CuO_2 planes. The importance of the CuO chains is also stressed for copper-substituted $YBa_2(Cu_{3-x}M_x)O_7$ compounds.[63]

The $YBa_2Cu_3O_{7-\delta}$ oxide, also called the '123 phase' or YBaCuO, is a type II superconductor. The H_{c1} critical field is very low, but H_{c2} is very high. However, the values given for the upper critical field, H_{c2}, which are very important for practical applications, do not agree well, especially where polycrystalline samples are concerned (34–80 Tesla at 77 K). Such behaviour results from the anisotropy of the critical field parallel and perpendicular to the c axis. The highest values for the upper critical field at 0 K have been estimated at 200–300 Tesla when the field is perpendicular to the c axis.

Critical currents of 40–3000 A/cm^2 have been reported at 77 K for bulk materials. For thin films, however, currents of $10–10^5$ A/cm^2 at 77 K have been found. It is worth noting that the main problem is not only the low critical current for bulk materials, but also the effect of the magnetic field which tends to decrease it. The low critical currents and the effects of magnetic fields are attributed to grain boundaries which can act as weak links. This behaviour might also be connected with the small coherence length of the 1 2 3 oxide. The coherence length is an order of magnitude lower (7 Å $\perp$ **c** and 34 Å $//$ **c**) than that found in conventional superconductors.

Sulphides: The superconducting sulphides known so far do not show any special peculiarity. We mention here only the compounds $Li_xT_{1.1}S_2$ ($0.1 \leqslant x \leqslant 0.3$) which possess the hexagonal Ti_3S_4-type structure and show superconducting transitions over the temperature range 10–13 K.[64] The Chevrel-phase chalcogenides are considered separately below.

Selenides: The selenides $NbSe_2$ and $NbSe_3$ merit special attention since they offer the possibility of studying the coexistence between superconductivity and charge-density-waves. These metallic compounds, having a low dimensionality (one-dimensional:1-D, or two dimensional:2-D) from the point of view of the anisotropy of their chemical bonding, can undergo a phase transition at low temperatures to a state in which the electron density displays periodic modulations incommensurate with the crystal lattice. Such structural instabilities are called charge-density-waves (CDW) (see Chapter 2).

$NbSe_2$ (2H-polymorph) and $NbSe_3$, which can be considered as having 2-D and 1-D dimensionality, respectively, show incommensurate charge-density-waves. $NbSe_3$ exhibits a superconducting transition at very low

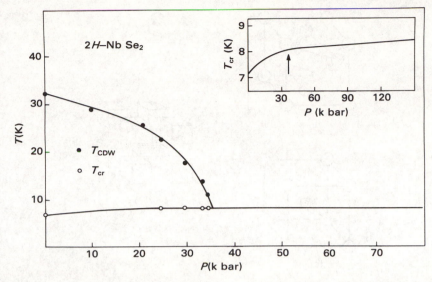

Fig. 3.22. Pressure dependence of the CDW transition temperature (T_{CDW}) and of the superconducting transition temperature (T_{cr}) for NbSe$_2$.[61]

temperatures only under pressure.[65] NbSe$_2$ on the other hand is supercon-ducting at normal pressure ($T_{cr} \simeq 7.2$ K) and T_{cr} increases with pressure. Since the pressure is known to suppress the CDW, superconductivity occurs as the CDW disappears. Such behaviour is illustrated in Fig. 3.22.[66,67]

Chevrel phases: The ternary molybdenum chalcogenides with the general formula M$_x$Mo$_6$X$_8$, first synthesized by Chevrel *et al.*,[68] are of great interest owing to their extraordinary superconducting properties, discovered in 1972 by Matthias *et al.*[69] Such compounds exist for a broad variety of elements, M = alkaline, alkaline earths, 3d elements, Ag, Cd, In, Sn, Pb, Sc, Y, lanthanides, actinides and X = S, Se, Te. x in M$_x$Mo$_6$X$_8$ may be between 0 and 4, depending on the size of the M atom.

The Chevrel phases show three striking features: the superconducting transition temperatures are high (e.g. $T_{cr} \simeq 15$ K for PbMo$_6$S$_8$); the critical fields H_{c2} are the highest found so far (e.g. $H_{c2} \simeq 500$ kOe for PbMo$_6$S$_8$ at 4.2 K); and although the M atoms can be magnetic rare earths, in most cases such a magnetic sublattice does not destroy the superconductivity. Alongside the ternary borides reported in Section 3.3.2.3.3, these materials provide a remarkable opportunity for studying the coexistence of superconductivity and long-range magnetic order.

The structures of the Chevrel phases are derived from that of the binary Mo$_6$X$_8$. Most of the M$_x$Mo$_6$X$_8$ compounds crystallize in a rhombohedral structure with $\alpha_{rh} \simeq 90°$ and space group R$\bar{3}$. As shown in Fig. 3.23, the structure can be visualized as a slight rhombohedral distortion of the CsCl

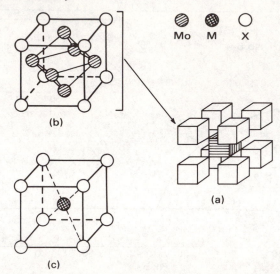

Fig. 3.23. Structure of the Chevrel phase $M_x Mo_6 X_8$ (M = large cations: Ag, In, Sn, Pb ... and X = S, Se, Te). (a) Arrangement of $Mo_6 X_8$ blocks and M atoms in a rhombohedral distortion of the CsCl-type structure. (b) $Mo_6 X_8$ unit. (c) Environment of the M atom.

structure, where the caesium atoms are replaced by M atoms and the chlorine atoms by $Mo_6 X_8$ units (Fig. 3.23(a)). These $Mo_6 X_8$ blocks can be considered as deformed cubes with the 8 X atoms at the vertices of the cubes and the 6 Mo atoms at the centres of the cube faces (Fig. 3.23(b)). The M atoms are located in the three dimensional channels between the $Mo_6 X_8$ units. The M atoms having a large size (e.g. Sn, Pb, ...) are surrounded by eight X chalcogen atoms belonging to eight different $Mo_6 X_8$ units (Figs 3.23(a) and 3.23(c)). The occupation of this site is more complex for the smaller M atoms (e.g. Cu, Ni, Fe ...) and the structure, which remains rhombohedral for small values of x, becomes triclinic as x increases.

It must be noted that the size of the Mo_6 octahedral cluster is approximately independent of the M and X elements and that the Mo–Mo intercluster distances depend strongly on the M and X elements and are generally 20 per cent larger than those found within the Mo_6 cluster. Furthermore, the Mo atoms are separated from the M atoms by the chalcogen atoms, leading to large Mo–M distances.

Numerous experiments involving various substitutions of the M elements for lead or tin, or chalcogen or halogen for sulphur, or rhenium for molybdenum, show that only the electrons coming from the octahedral Mo_6 clusters are responsible for the occurrence of superconductivity. These electrons, which are confined within the $Mo_6 X_8$ units, belong to a narrow Mo 4d conduction band whose width depends on the intercluster interactions.

The Fermi level lies in this narrow band and the corresponding density of states is strongly dependent upon the stoichiometry and the nature of the element M; M provides electrons to the Mo_6X_8 cluster, as shown by band-structure calculations.[70]

The indirect exchange interaction between localized magnetic moments of 4f rare earth ions occurs via the electrons of the 5d-6s bands because of overlap between the 5d-6s and 4f orbitals. The 4d-electrons of the molybdenum atoms also interact with the localized moments, but this interaction is weak due to a very small overlap between 4d and 4f orbitals. Thus the contribution of the Mo 4d superconducting electrons to the indirect magnetic exchange turns out to be small in these chalcogenides. This explains, for instance, the coexistence of antiferromagnetism and superconductivity in $REMo_6S_8$ compounds (RE = Gd, Tb, Dy, Er).

3.3.2.3.3 Borides

A relatively low transition temperature is the main feature of the superconducting binary borides. For example, the cubic hexaboride YB_6 and dodecaboride ZrB_{12} show a transition at 6.5–7.1 K and 5.85 K respectively.[71] At present many investigators are focusing their attention on ternary borides since they have high superconducting transition temperatures which exceed 11 K and they permit the study of the coexistence of long-range magnetic order and superconductivity. At present, the ternary borides most intensively investigated are the MT_4B_4 compounds (M = Sc, Y, lanthanides, Th, U and T = Ru, Os, Co, Rh, Ir). Four polytypes occur with closely related structures: (*i*) $CeCo_4B_4$-type (tetragonal with $P4_2/nmc$ space group); (*ii*) $LuRu_4B_4$-type (body-centred tetragonal with $I4_1/acd$ space group); (*iii*) $LaRh_4B_4$-type (orthorhombic with *Ccca* space group); (*iv*) $NdCo_4B_4$-type (tetragonal with $P4_2/n$ space group).[72]

All the polytypes are characterized by the presence of T_4 tetrahedra and pairs of boron atoms in their structure. The $CeCo_4B_4$-, $LuRu_4B_4$-, and $LuRh_4B_4$-type compounds form predominantly with M elements of medium and relatively small atomic radius ($1.75 < r_M < 1.08$ Å), whereas $NdCo_4B_4$-type compounds form mainly with M metals of relatively large atomic radius ($r_M > 1.8$ Å). Among the polytypes indicated above, the first three are favourable for superconductivity. However, only two of them exhibit high superconducting transition temperatures (e.g. YRh_4B_4 with $CeCo_4B_4$-type, $T_{cr} = 11.3$ K, and $Y(Rh_{0.85}Ru_{0.15})_4B_4$ with $LuRu_4B_4$-type, $T_{cr} = 9.4$ K). In both $CeCo_4B_4$- and $LuRu_4B_4$-type structures, the T_4 tetrahedra and M atoms are arranged in a slightly distorted NaCl lattice (Fig. 3.24).

However, whereas in the $CeCo_4B_4$-type phases the T_4 tetrahedra of the same orientation are arranged in sheets perpendicular to the *c* axis (Fig. 3.24(a)), they are distributed equally in each phase in an ordered way in the $LuRu_4B_4$-type structure (Fig. 3.24(b)).[73]

In general, the fact that the $CeCo_4B_4$-type compounds are metastable and form only as high-temperature phases is consistent, for example, with the

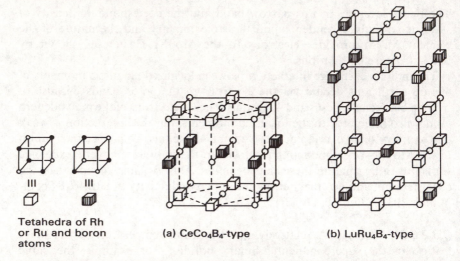

Tetahedra of Rh or Ru and boron atoms

(a) CeCo$_4$B$_4$-type

(b) LuRu$_4$B$_4$-type

Fig. 3.24. Structure of the MT$_4$B$_4$ borides. (The cubes representing Rh$_4$B$_4$ or Ru$_4$B$_4$ units are now drawn to scale for sake of clarity.) (a) Schematic structure of CeCo$_4$B$_4$-type compounds. The dashed lines outline the primitive tetragonal cell. (b) Schematic structure of LuRu$_4$B$_4$-type compounds.

large value of the electron–phonon coupling parameter λ calculated for ErRh$_4$B$_4$.[74] On the other hand, the relative confinement of the Rh 4d electrons within the Rh$_4$B$_4$ cluster may be responsible for peaks in the conduction electron density at the Fermi level and consequently for the relatively high value of T_{cr}.

One of the most interesting borides is ErRh$_4$B$_4$ since it exhibits a superconducting transition at 8.7 K and orders ferromagnetically at lower temperature ($T_C \simeq 0.93$ K) with the simultaneous disappearance of the superconducting state. However, it has been shown that over a narrow temperature range ErRh$_4$B$_4$ (CeCo$_4$B$_4$-type) exhibits both superconductivity and long-range ferromagnetic order. This phenomenon is explained by the coexistence of ferromagnetic and superconducting regions.[75] Recently the coexistence of ferromagnetism and superconductivity has been shown down to very low temperatures in a body-centred tetragonal phase of ErRh$_4$B$_4$ (LuRu$_4$B$_4$-type). In this sample, obtained by carbon doping of ErRh$_4$B$_4$, the critical temperature for the onset of superconductivity is 7.5 K. The ferromagnetic ordering of Er spins, which appears at 0.7 K, does not destroy the superconductivity.[76]

The MT$_4$B$_4$ ternary borides resemble the chalcogenides, described in Section 3.3.2.3.2, in the sense that they are cluster compounds. For example, in the M$_x$Mo$_6$X$_8$ and MRh$_4$B$_4$ compounds the 4d electrons of Mo and Rh are confined within the Mo$_6$X$_8$ and Rh$_4$B$_4$ clusters, producing sharp structure in the electron densities of states at the Fermi level. This could account for superconductivity with high values of T_{cr}. Since the conduction electrons at the Fermi level are mainly of Mo or Rh 4d character, the exchange interaction between the conduction electron spins and the rare earth magnetic moments is very weak. This could explain the possibility for coexistence of superconductivity and long range magnetic ordering in the ternary borides.

3.3.2.3.4 Silicides Except for the A-15 compounds V_3Si ($T_{cr} = 17.1$ K) and Nb_3Si ($T_{cr} = 18-19$ K) reported in Section 3.3.2.2, the other supercon-ducting binary silicides known to date have relatively low critical temperatures (e.g. α-$ThSi_2$, $T_{cr} = 3.16$ K; β-$ThSi_2$, $T_{cr} = 2.41$ K; Mo_3Si, $T_{cr} = 1.7$ K; W_3Si_2, $T_{cr} = 2.84$ K; $LaSi_2$, $T_{cr} = 2.5$ K; La_5Si_3, $T_{cr} = 1.6$ K). However Lejay *et al.*[77] have shown that it is possible to increase the superconducting transition temperature of α-$ThSi_2$ when silicon atoms are partially replaced by noble metals such as Ir or Rh. Such a substitution gives rise to the compounds α-$ThIr_xSi_{2-x}$ ($0 \leqslant x \leqslant 1$) and α-$ThRh_xSi_{2-x}$ ($0 \leqslant x \leqslant 0.96$).

The structure of α-$ThSi_2$ is tetragonal and can be described on the basis of a silicon-centred trigonal prism of the Th atoms as shown in Fig. 3.25. These Th_6 trigonal prisms are packed alternately up and down to result in infinite rows of prisms which are connected with infinite zig-zag chains of either Si atoms in $ThSi_2$ or both Si and Rh or Ir atoms in the α-ThM_xSi_{2-x} (M = Ir or Rh) compounds. The drawing of the α-$ThSi_2$ structure shows that half of the Th_6 prism rows run parallel to the xOz plane and half of them are parallel to the yOz plane. As shown in Figs 3.26 and 3.27, the sharp

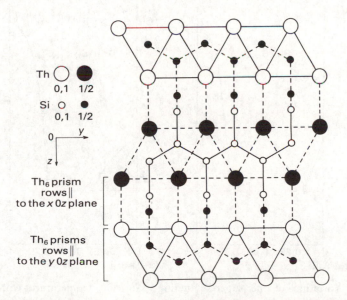

Fig. 3.25. Projection of the structure of α-$ThSi_2$ onto the yOz plane.

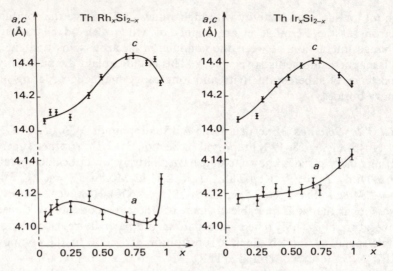

Fig. 3.26. Variation of the lattice parameters with x for $ThRh_xSi_{2-x}$ and $ThIr_xSi_{2-x}$.

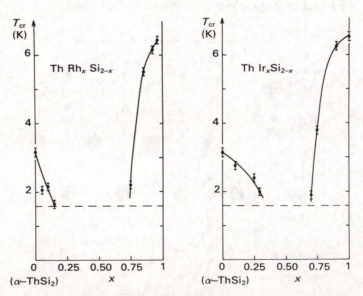

Fig. 3.27. Variation of the superconducting transition temperature with x for $ThRh_xSi_{2-x}$ and $ThIr_xSi_{2-x}$.

Table 3.5   *Ternary superconducting silicides*

Structure type	Symmetry	Compounds	T_{cr} (K)	Reference
$Sc_5Co_4Si_{10}$	tetragonal	$Sc_5Rh_4Si_{10}$	8.50	81
		$Sc_5Ir_4Si_{10}$	8.40	
		$Sc_5Co_4Si_{10}$	5	
		$Lu_5Ir_4Si_{10}$	3.70	
		Y_5IrSi_{10}	2.60	
$LaRu_3Si_2$	hexagonal	$LaRu_3Si_2$	7	82
$Sc_2Fe_3Si_5$	tetragonal	$Lu_2Fe_3Si_5$	6	81
$U_2Co_3Si_5$	orthorhombic	$La_2Rh_3Si_5$	4.45	78
		$Y_2Rh_3Si_5$	2.70	
ZrOs or $SrSi_2$	cubic	LaRhSi	4.35	79
		LaIrSi	3.30	
$ThCr_2Si_2$	tetragonal	YRh_2Si_2	3.11	80
$CeNiSi_2$	orthorhombic	$LaRhSi_2$	3.42	80
		$LaIrSi_2$	2.03	

increase of T_{cr} observed for $x > 0.75$ corresponds to a depression of the c parameter due to the formation of short Rh–Si or Ir–Si bonds along the c direction.

The search for new ternary superconductors has increasingly focused on silicides such as those shown in Table 3.5. Chevalier *et al.*[78–80] recently prepared new superconducting ternary silicides in the rare-earth–noble-metal (Rh or Ir)–silicon systems. It is worth noting that in the reported ternary silicides, the 3d, 4d, or 5d transition metals are not connected to an independent three-dimensional network via homonuclear bonds. In contrast to the molybdenum chalcogenides, e.g. the Chevrel phases described in Section 3.3.2.3.2, these silicides only become superconducting if they contain no magnetic rare earths. (Note that Fe and Co are non-magnetic in $RE_2Fe_3Si_5$ and $Sc_5Co_4Si_{10}$.[81])

3.3.3   Correlations between superconducting transition temperature and structural and atomic properties of the materials

From the study of transition metals and their alloys as well as that of binary compounds it has become apparent that correlations exist between the superconducting transition temperature and structural properties such as crystal symmetry, atomic disorder, clustering, lattice instabilities, and atomic properties such as, for example, the average valence electron concentration.

In this section we develop some ideas that should allow the chemist to discuss and criticize his results in the search for new superconducting materials.

3.3.3.1   *Superconductivity and crystal symmetry*

In binary systems the highest T_{cr} occurs with the most symmetric cells such as the cubic β-W, NaCl-type, Pu_2C_3-type structures, the tetragonal α-Mn type structure and the hexagonal $MgCu_2$-type structure as shown in Tables 3.1, 3.2, and 3.3. However, the highest T_{cr} has been found in Nb_3Ge having the cubic β-W-type structure (A-15).[3] Few superconductors have low symmetry and none are triclinic.

The situation appears to be different in the ternary materials where a triclinic compound $Cu_{1.8}Mo_6S_8$ has been found to be superconducting with a relatively high superconducting transition ($T_{cr} \simeq 11$ K).[83] Furthermore it is worth emphasizing that few ternary superconductors have cubic symmetry, the last ones discovered being the silicides LaRhSi, and LaIrSi.[79] It follows from these observations that symmetry does not play a prominent role for the occurrence of superconducting properties in transition metal compounds.

3.3.3.2   *Correlation between superconducting transition temperature T_{cr} and other physical properties of the materials*

Matthias *et al.* have pointed out that there are certain conditions which apparently favour the appearance of superconductivity.[71] Conductors in which these conditions are satisfied have relatively high transition temperatures. They pointed out the existence of a correlation between the occurrence of superconductivity and the valence electron concentration (VEC) in numerous metals and binary compounds. To calculate VEC, one counts as valence electrons all electrons outside the last filled shell of the elements appearing in the formulae of the compounds. To show superconductivity a metal or alloy must have a mean number of valence electrons per atom between 4 and 8. In the case of the transition metals and their alloys, two pronounced maxima are observed near 4.7 and 6.5 valence electrons per atom in the curves $T_{cr} = f(\text{VEC})$ (Fig. 3.28); VEC = 4 and 5.6 do not favour superconductivity. The same electron-concentration dependence of T_{cr} established for the transition metals and their alloys can be found in other series, as for example the A-15 compounds (Fig. 3.29). However in materials having the NaCl-type structure, the larger T_{cr} occur only within the range of electron-concentration 4.5 to 5. It is also worth noting that for other structure types such as Pu_2C_3, MoB_2, or $ThSi_2$ only a new range of electron-concentration of 3.5–4 is favourable for superconductivity. In this context it appears remarkable that the VEC values of 4.5 and 6.5 are particularly favourable for the occurrence of superconductivity in most of the metals and binary compounds.

An examination of the superconducting properties of well-known ternary compounds shows that the correlation between T_{cr} and VEC is not so clear

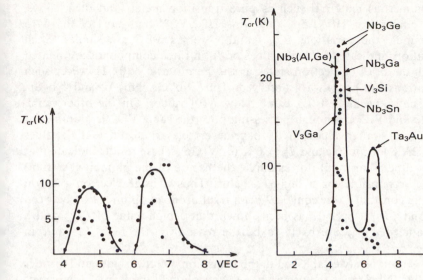

Fig. 3.28. Variation of T_{cr} of different solid solutions of transition metals vs. VEC.

Fig 3.29. Variation of T_{cr} of A-15 compounds vs. VEC.

as in the metals and binary compounds. However, except for a few chalcogenides, the highest values of T_{cr} are found essentially in two ranges of VEC values, 4–5 and 5–6.[80] Such correlations between T_{cr} and VEC are observed independently of any assumption as to the redistribution of electrons between the electronic states of different symmetry. However in order to understand more precisely the superconducting properties of the materials we must refer to the BCS theory and its improvements. Thus, we consider the electron and phonon spectra.

As indicated previously in Section 3.2.7.1, eqns (3.24), (3.26), and (3.27) show that T_{cr} is closely related to the electron–phonon coupling λ: the larger the values of ω_D and λ, the greater the value of T_{cr}. Therefore a high density of states at the Fermi level $N(0)$ as well as a high density of low frequency phonons modes $F(\omega)$ are favourable for the occurrence of high superconducting transition temperatures.

The electron density of states at the Fermi level, $N(0)$, can be deduced from the electronic specific heat coefficient γ or from the Pauli paramagnetic susceptibility χ, since in most cases both are proportional to $N(0)$ which itself depends on the density of conduction electrons per cubic centimeter and on the bandwidth. However, the bandwidth cannot be reduced too much since, in this case, Coulomb and exchange interactions may become dominant and give rise to magnetism or to an insulating state. The phonon density of states $F(\omega)$ can be obtained from the inelastic neutron scattering measurements.

In numerous materials such as pure transition metals and their alloys, Laves phases (e.g. HfV_2, ZrV_2 and $Zr_{0.5}Hf_{0.5}V_2$)[84] and most of the A-15 compounds, the variations T_{cr}, γ and χ versus VEC are similar. Such correlations indicate that high values of T_{cr} in these compounds are related to high densities of electron states at the Fermi level $N(0)$. However such correlations are not always observed. Thus Nb_3Ge, Nb_3Al, and Nb_3Sn, which have the highest T_{cr}, exhibit low $N(0)$ values. On the other hand, Nb_3Ge and V_3Ge, although possessing both the same VEC favourable for high T_{cr} and similar values of $N(0)$, have nevertheless quite different T_{cr} values ($T_{cr} \simeq 23$ K for Nb_3Ge and $T_{cr} = 6$ K for V_3Ge). These results indicate that factors other than $N(0)$ (for example, the nature of the phonon spectrum) can be responsible for a high T_{cr} value. Thus a large electron–phonon coupling constant λ (see eqn (3.29)) can result from an optimization between $N(0)$ and $\langle J^2 \rangle / \langle \omega^2 \rangle$. It has been shown that a high value of $\langle J^2 \rangle$ and a low value of $\langle \omega^2 \rangle$ are likely to be the reason for the high T_{cr} value in Nb_3Ge.[85]

It has been firmly established in the transition metal carbides and nitrides (e.g. NbC, NbN) that the occurrence of soft phonon modes (i.e. low values of $\langle \omega^2 \rangle$) is responsible for their high T_{cr}. The existence of such soft modes is explained by the scattering of the metal d electrons into empty non-metal p states at the Fermi level.[86,87] Thus one can explain why, in NbC, carbon vacancies destroy superconductivity; empty carbon p-like states are less and less available as the number of carbon vacancies increases and consequently the p–d scattering mechanism can no longer play a part.

3.3.3.3　*Superconductivity and structural instabilities*

The coincidence of a high T_{cr} value and lattice instabilities is a common feature of binary superconductors. These structural instabilities can manifest themselves either as a modification of the crystal structure between room temperature and T_{cr} or by a softening of the lattice moduli or a softening of the short-wave branches of the phonon dispersion curves. However, so far, an insufficient number of investigations have been carried out to ascertain whether there is a universal correlation between lattice instability and superconductivity.

Thus, for example, the cubic–tetragonal transformations in V_3Si and Nb_3Sn, although known for rather a long time, are still not thoroughly understood. It seems that there is no direct relationship between superconducting behaviour and the tetragonal distortion. For instance, Batterman and Barrett prepared a crystal of V_3Si for which no structural transformation could be observed down to 1.9 K whereas it became superconducting below 17 K.[25,27] It should be noted that the structural transformation always occurs at temperatures higher than T_{cr}.

On the other hand, stoichiometric high temperature superconductors are generally difficult to prepare (e.g. Nb_3Ge, Nb_3Ga, MoC, NbN). In the vicinity of stoichiometric compositions of Nb_3Si, MoC, and NbN, different structural phases can be observed.

The existence of high density of states in a narrow band at the Fermi level is certainly favourable for obtaining a high T_{cr} value but at the same time reflects a certain instability of the material. Thus we can say that a strong electron lattice coupling in a material favours both the occurrence of superconductivity and a lattice instability. However such a lattice instability is not in itself responsible for the high T_{cr} value.

3.3.3.4 Superconductivity and crystalline disorder

The effect of crystalline disorder on superconductivity has been investigated in numerous materials such as, for example, the A-15 compounds and the $RERh_4B_4$ borides. Such disorder can be caused by variations in growth procedure, stoichiometry, or by damage induced by neutron, α-particle or electron irradiation. Disorder reduces T_{cr} drastically, as found experimentally. Anderson *et al.* propose a theory of the 'universal' degradation of T_{cr} in superconductors.[16] According to these authors, Anderson electronic localization in extremely disordered systems provides a natural explanation for a degradation of T_{cr}. They show that the Coulomb repulsion μ^* (see eqn (3.27)) between Cooper pair electrons increases with disorder and consequently leads to a decrease of T_{cr} (see eqn (3.26)).

3.3.3.5 Guidelines for seeking superconductors having high transition temperatures and high critical fields

In order to propose guidelines for seeking superconductors having high transition temperatures and high critical fields, it appears necessary to make the distinction between the conventional superconductors (metals, alloys, intermetallics), whose behaviour can be described in the framework of the BCS theory, and the new high-T_{cr} superconducting oxides for which the critical temperatures clearly exceed 23 K and the pairing mechanism is not yet identified.

3.3.3.5.1 Conventional superconductors *The search for high critical temperature superconductors*: According to eqns (3.24) and (3.26) high critical temperatures are observed in materials showing a large electron–phonon interaction λ. Large λ values can result from the existence of either high electronic densities of states $N(0)$ at the Fermi level or low frequency phonons (see eqns (3.28) and (3.29)). High $N(0)$ values could be sought in materials exhibiting narrow bands at the Fermi level. Compounds in which clusters form, for instance the Chevrel phases, satisfy this criterion. However, according to eqn (3.29) it does not require the existence of very large $N(0)$ to give high-T_{cr} superconductivity if the ratio $\langle J^2\rangle/\langle\omega^2\rangle$ is large. On the other hand, it is believed that a very strong electron–phonon interaction as characterized by a large $\langle J^2\rangle/\langle\omega^2\rangle$ may lead to structural instability or

phase decomposition. Such a possibility agrees for example with the systematic trend observed in going from the stable compounds V_3Si ($T_{cr} = 17$ K) and Nb_3Sn ($T_{cr} = 18$ K) to the nearly stable Nb_3Al ($T_{cr} = 18.8$ K) and on to the metastable Nb_3Ge ($T_{cr} = 23.2$ K).

It is also worth mentioning that according to a theory of Hanke *et al.* an enhancement in $\langle J^2 \rangle$, accompanied by a lattice softening, i.e. a decrease in $\langle \omega^2 \rangle$, can result from a hybridization of the relatively localized electronic states (e.g. the Nb 4d states) and the conduction-band states (e.g. the s, p states from a non-metal element such as C, Ge, Si, etc.) at the Fermi level.[88] Such p–d hybridization has been indicated by XPS experiments carried out on Nb_3Sn and Nb_3Ge.[89,90]

The search for high critical magnetic field superconductors: High magnetic field superconductivity occurs in a few high-T_{cr} type-II superconducting materials. One must recall that the so-called high-field superconductors are those in which H_{c2} happens to be particularly high. The origin of such high values may be intrinsic, impurity-dominated, or include both effects acting in unison.

The microscopic theory indicates that H_{c2} is related to the temperature dependent coherence length $\xi(T)$ by

$$H_{c2} = \frac{\phi_0}{2\pi \xi^2(T)} \tag{3.33}$$

where ϕ_0 is the flux quantum $hc/2e$. Coherence lengths for typical high field materials are of the order of 100 Å or less, whereas considerably larger values are found in pure superconductors such as for instance Al (16000 Å), Sn (2300 Å) and Pb (830 Å). It is convenient to introduce a distinction first proposed by Anderson between 'clean' superconductors in which the electron mean free path l is much larger than the coherence length ξ_0 (see eqn (3.30)), and 'dirty' superconductors in which $l \ll \xi_0$.[91] In the class of 'clean' superconductors one finds pure metals (type-I), while the 'dirty' one contains alloys, metals with impurities, binary and ternary chemical compounds (type-II).

Due to the fact that the electron motion in 'dirty' superconductors is mainly governed by lattice disorder, impurities or defects diffusion processes, the thermal dependence of the coherence length $\xi(T)$ is no longer proportional to ξ_0 but to $\sqrt{\xi_0 l}$:

$$\xi(T) = 0.85(\xi_0 l)^{1/2} \left[\frac{T_{cr}}{T_{cr} - T} \right]^{1/2}. \tag{3.34}$$

Taking into account eqn (3.30) giving ξ_0, $\xi(T)$ is proportional to $(l \, V_F/T_{cr})^{1/2}$. Consequently, high critical fields should be obtained for dirty superconductors exhibiting high T_{cr} and small l and V_F values, i.e. for those having a low conductivity in the normal state.

Thus, for instance, high critical fields have been observed in Nb–Ti alloys due to their small l value and in numerous Chevrel phases $M_xMo_6X_8$ in which both V_F and l are very small (weak overlap between Mo_6S_8 cluster orbitals involves a narrow Mo 4d conduction band). Note that, so far, the highest critical field compounds are Chevrel phases (e.g. $H_{c2} = 600$ kOe at $T = 0$ K for $PbMo_6S_8$). We should also draw attention to the case of the A-15 compounds, and particularly that of Nb_3Ge ($T_{cr} = 23.2$ K), for which the high H_{c2} critical field observed (380 kOe) results from the combination of the small V_F and l and the high T_{cr}.

3.3.3.5.2 High-T_{cr} superconductors It is too early to propose guidelines for the search for new high-T_{cr} superconductors, since it is apparent that superconductivity is not linked to the presence of particular metal elements or certain structural features and symmetries. For example, there exists no universal correlation between T_{cr} and the oxygen content (charge carrier concentration) or the Cu(III)/Cu(II) ratio. Superconductivity can be found in oxides having an average formal oxidation state either higher or lower than $+2$. On the other hand, the theoretical problems are immense and efforts are being made to know how the BCS theory should be modified to account for the properties of these new materials. It is possible that this theory will survive essentially intact, but with a more comprehensive electron interaction than the electron–phonon interaction alone.

So far the highest superconducting transition temperatures have been found in copper oxides exhibiting two-dimensional copper–oxygen arrays and mixed valence of copper. However, high critical temperatures can also be found in perovskites which do not contain copper, e.g. $Ba_{1-x}K_xBiO_3$ for which superconductivity occurs within the framework of the three dimensionally connected bismuth–oxygen array. The superconducting transition temperature of around 30 K in $BaBiO_3$-based compounds is remarkable as it exceeds the T_{cr} of all conventional superconductors. However, most of the cuprate-based superconductors have T_{cr} well above 77 K and the obvious question arises whether the same microscopic mechanism is responsible for the pairing.

3.3.4 Metallurgical aspect of superconducting materials

The sensitivity of superconducting properties to the state of a material is most pronounced. Annealing, cold working, homogeneity, dislocation densities, and the presence of impurity atoms and other scattering centres are all liable to modify the critical temperature and magnetic properties of a specific material. Sample configuration is now becoming an important parameter governing the properties of a superconductor. Thus new superconducting properties have been found with multifilamentary and thin film materials. Such sample configurations are developed to provide solutions to the requirements of advanced engineering projects such as nuclear fusion, magnetohydrodynamics, magnetic energy storage, D.C. and A.C. motors and generators, D.C. and A.C. power transmission, etc.

The major practical application for type-II superconductors is as a winding for high-field magnets; obviously this would be extremely limited if the superconducting state was destroyed by the passage of a modest current, i.e. for materials exhibiting a low critical current. Fortunately, whatever the superconductors, whether conventional or high-T_{cr} type, high critical currents can be observed thanks to the 'flux pinning' phenomenon. Thus in type-II superconductors, magnetic flux can penetrate into the superconducting state leading to a 'mixed structure' (Section 3.2.2.2) showing superconducting areas and normal areas. When an electric current passes through a type-II superconductor in the 'mixed state', it takes the path of least resistance and only crosses the superconducting regions. The magnetic flux lines, or vortex, which penetrate the normal regions, can move in the presence of an electric current due to the Lorentz force and accordingly the flux is expelled from the sample. Even when no external field is applied the current generates its own magnetic field. If such a situation happens the mixed state is destroyed and the sample is no longer superconducting. Thus the superconducting state can disappear, in a zero applied field, for a critical value of the current, J_{cr}. Fortunately the flux lines can be pinned by different types of defects (precipitates, grain boundaries, dislocations). Therefore the art of producing a good superconductor is to be able to control the microstructure of the material in order to maximize the flux pinning, since a decrease of the flux pinning diminishes the value of J_{cr}.

Three basic criteria have to be satisfied before superconductors can be used in high-field applications: a high critical temperature T_{cr}, a high critical field H_{c2}, and a high critical current density J_{cr}. So far, practical superconducting materials which have these properties can be divided into two groups: the ductile solid solution alloys such as Nb–Ti and the brittle intermetallic compounds mainly having the A-15 structure (e.g. V_3Ga and Nb_3Sn). Applications using magnetic fields between 90 kOe and 140 kOe seem to favour Nb_3Sn and those requiring less than 90 kOe use Nb–Ti alloys. At present, most superconducting magnets in use or under development utilize some form of copper–(Nb–Ti) or copper–Nb_3Sn multifilamentary composites. Further development work is needed on superconducting composite conductors suitable for use above 180 kOe. There are several promising materials for this purpose, including the A-15 compounds Nb_3Ge and $Nb_3Al_{0.7}Ge_{0.3}$ and the Chevrel phase compounds.[92, 93]

Concerning the high-T_{cr} superconducting oxides, an empirical approach and a very simple method of preparation are the main explanations for the successes achieved by many laboratories in the rush to find new superconductors by substituting other elements for the barium and lanthanum used in Bednorz and Müller's material.[4] Ceramic superconductors are prepared by careful mixing and heating of simple oxides, just as the alloys were prepared by mixing of molten metals. Conventional solid state chemistry routes are used: oxides are fused together by furnace heating, although this stage is preceded either by grinding oxide powders together or by precipitating oxide precursors from solution. These solid state reactions should be made under controlled atmosphere. The main drawbacks of the superconducting oxides result from the fact that they are not ductile and cannot therefore be deformed plastically. It is not easy to produce superconducting wires from materials of this type. Their complex crystal structures impede the sliding of atomic planes and the propagation of distortions. However, thin films can be used in a large number of applications, typically with a thickness in the micrometre range and deposited on appropriate supports. Over and above the metallurgical problems, the $YBa_2Cu_3O_{7-\delta}$ oxides, for example, undergo degradation in moist air. An efficient protection against atmospheric moisture and gas exchange has been achieved by fluorination of the oxide at low temperatures.[94,95]

ACKNOWLEDGEMENTS

The author expresses his sincere thanks to Prof. R. Georges of the University of Bordeaux I and Prof. J. M. D. Coey of the University of Dublin for fruitful discussions.

3.4 REFERENCES

1. Kamerlingh Onnes, H., *Akad. Van Wetenschappen* (Amsterdam), **14**, 113 and *Leiden Comm.*, **122b**. 124c (1911).
2. Gavaler, J. R., *Appl. Phys. Lett.*, **23**, 480 (1973).
3. Testardi, L. R., Wernick, J. H., and Royer, W. A., *Solid State Commun.*, **15**, 1 (1974).
4. Bednorz, J. G. and Müller, K. A., *Z. Phys. B. Condensed Matter*, **64**, 189 (1986).
5. Bardeen, J., Cooper, L. N., and Schrieffer, J. R., *Phys. Rev.*, **108**, 1175 (1957).
6. Meissner, W. and Ochsenfeld, R., *Naturwiss*, **21**, 787 (1933).
7. Reynolds, C. A., Serin, B., Wright, W. H., and Nesbitt, L. B., *Phys. Rev.*, **78**, 487 (1950).
8. Maxwell, E., *Phys. Rev.*, **78**, 477 (1950).
9. Frölich, H., *Phys. Rev.*, **79**, 845 (1950).
10. Cooper, L. N., *Phys. Rev.*, **104**, 1189 (1956).
11. Bogolyubov, N. N., Tolmatchev, V. V., and Shirkov, D. V., In *Novgj mjetod v tjeorii svjerkhprovodimosti* (ed. Gurov, K. P.) (*Izv. Akad. Nauk. SSSR*, Moscow (1958)). (A new method in the theory of superconductivity, Academy of Science, Moscow (1958); Consultants Bureau, New York (1959).)
12. Eliashberg, G. M., *Zh. Eksperim. i Teor. Fiz.*, **38**, 966 (1960) (*Soviety Phys. JETP* **11**, 696 (1960)).

13. McMillan, W. L., *Phys. Rev.*, **167**, 331 (1968).
14. Ruderman, M. H. and Kittel, C., *Phys. Rev.*, **96**, 99 (1954).
15. Yosida, K., *Phys. Rev.*, **106**, 893 (1957).
16. Anderson, P. W., Muttalib, K. A., and Ramakrishnan, T. V., *Phys. Rev.*, **28**, 117 (1983).
17. Gor'kov, L. P. and Rusinov, A. I., *Sov. Phys. JETP*, **19**, 922 (1964).
18. Roberts, B. W., *Nat. Bur. Stand. U.S. Techn.* Note 983 Suppl. (1978).
19. Vonsovsky, S. V., Izyumov, Y. A., and Kurmaev, E. Z., *Superconductivity of transition metals*, Springer Series in Solid State Sciences, vol. 27, Springer-Verlag, Berlin, Heidelberg, New York (1982).
20. Buchal, C. and Welter, J. M., *Plat. Met. Rev.*, **27**, 170 (1982).
21. Balster, H. H., *Ber. Kernforschungsanlage Julich*, **1126**, 639 (1974).
22. Matthias, B. T., *Physics Today.* August, 23 (1971).
23. Hardy, G. F. and Hulm, J. K., *Phys. Rev.*, **89**, 884 (1953).
24. Van Reuth, E. C. and Poulis, N. J., *Phys. Lett.*, **A 25**, 390 (1967).
25. Batterman, B. W. and Barrett, C. S., *Phys. Rev.*, **13**, 390 (1964).
26. Madar, R., Senateur, J. P., and Fruchart, R., *J. Solid State Chem.*, **18**, 59 (1979).
27. Batterman, B. W. and Barrett, C. S., *Phys. Rev.*, **145**, 297 (1966).
28. Cohen, M. L., *Phys. Rev.*, **134**, A 511 (1964).
29. Hein, R. A., Gibson, J. W., Mazelsky, R., Miller, R. C., and Hulm, J. K., *Phys. Rev. Lett.*, **12**, 320 (1964).
30. Matthias, B. T., *Phys. Rev.*, **92**, 874 (1953).
31. Skowskiewicz, T., *Phys. Status Solidi*, **A 11**, K 123 (1972).
32. Stritzker, B. and Buckel, W., *Z. Phys.*, **257**, 1 (1972).
33. Heim, G. and Stritzker, B., *Appl. Phys.*, **7**, 239 (1975).
34. Stritzker, B., *Z, Phys.*, **268**, 261 (1974).
35. Remeika, J. P., Geballe, T. H., Matthias, B. T., Cooper, A. S., Hull, G. W., and Kelly, E. M., *Phys. Lett.*, **A 24**, 565 (1967).
36. Johnston, D. C., Prakash, H., Zachariasen, W. H., and Viswanathan, R., *Mat. Res. Bull.*, **8**, 777 (1973).
37. Johnston, D. C., *J. Low Temp. Phys.*, **25**, 145 (1976).
38. Sleight, A. W., Gilson, J. L., and Bierstedt, P. E., *Solid State Commun.*, **7**, 27 (1975).
39. Jorgensen, J. D., *Jpn. J. Appl. Phys.*, **26**, (Suppl. 26-3), 2017 (1987).
40. Hewat, A. W., Capponi, J. J., Chaillout, C., Marezio, M., and Hewat, E. A., *Solid State Comm.*, **64**, 301 (1987).
41. Rao, C. N. R., *Mod. Phys. Lett.*, **B2**, 1217 (1988).
42. Wilson, J. A., *J. Phys.*, **C20**, L 911 (1987).
43. Wilson, J. A., *J. Phys.*, **21**, 2067 (1988).
44. Cava, R. J., in *Proceedings of the European Workshop on High T_{cr} Superconductors and Potential Applications.* Commission of the European Communities, Geneva (July 1987).
45. Cava, R. J., Battlogg, B., Krajewski, J. J., Rupp, L. W., Schneemayer, L. F., Siegrist, T., Van Dover, R. B., Marsh, P., Deck, W. F., Gallagher, P. K., Glarum, S. H., Marshall, J. H., Farrow, R. C., Wadzczak, J. V., and Trevor, P., *Nature*, **336**, 211 (1988).
46. Santoro, A., Beech, F., Marezio, M., and Cava, R. J., *Physica*, **C 156**, 693 (1988).
47. Day, P. in *High temperature superconductors.* Proceedings of the First Latin

American Conference (ed. Nicolsky, R.), Singapore: World Scientific Publ. Co., preprint (1988).

48. Sleight, A. W., *Science*, **242**, 1519 (1988).
49. Rouillon, T., Provost, J., Hervieu, M., Groult, D., Michel, C., and Raveau, B., *Physica*, **C 159**, 201 (1989).
50. Yvon, K. and François, M., *Z. Phys. B Condensed Matter*, **76**, 413 (1990).
51. Raveau, B., Michel, C., Hervieu, M., and Provost J., *Physica*, **C153–155**, 3 (1988).
52. Mattheiss, L. F., Gyorgy, E. M., and Johnson, D. W. Jr. *Phys. Rev.*, **B37**, 3745 (1988).
53. Cava, R. J., Batlogg, B., Krajewski, J. J., Farrow, R., Rupp, L. W. Jr, White, A. E., Short, K., Deck, W. F., and Kometani, T., *Nature*, **332**, 814 (1988).
54. Goodenough, J. B., Demazeau, G., Pouchard, M., and Hagenmuller, P., *J. Solid State Chem.*, **8**, 325 (1973).
55. Michel, C. and Raveau, B., *Rev. Chim. Minér.*, **21**, 407 (1984).
56. Beille, J., Cobonel, R., Chaillout, C., Chevalier, B., Demazeau, G., Deslandes, F., Etourneau, J., Lejay, P., Michel, C., Provost, J., Raveau, B., Sulpice, A., Tholence, J. L., and Tournier, R., *C.R. Acad. Sci.*, **304**, 1097 (1987).
57. Grenier, J. C., Wattiaux, A., Lagueyte, N., and Park, J. C., *Physica C*, **173**, 139 (1991).
58. Chevalier, B., Tressaud, A., Lepine, B., Amine, K., Dance, J. M., Mozano, L., Hickey, E., and Etourneau, J., *Physica*, **C167**, 97 (1990).
59. Yvon, K. and François, M., *Z. Phys. B Condensed Matter*, **76**, 413 (1989).
60. Cheetham, A. K., Chippindale, A. M., and Hibble, S. J., *Nature*, **333**, 21 (1988).
61. Pouchard, M., Grenier, J. C., and Doumerc, J. P., *CR Acad. Sci.*, **305**, 571 (1987).
62. Pouchard, M., Grenier, J. C., Doumerc, J. P., Demazeau, G., Chaminade, J. P., Wattiaux, A., Dordor, P., and Hagenmuller, P., *Proceedings MRS Spring Meeting*, San Diego, CA, USA (1989).
63. Tarascon, J. M., Barboux, P., Miceli, P. F., Greene, L. H., Hull, G. W., Eibschutz, M., and Sunshine, S. A., *Phys. Rev.*, **B37**, 7458 (1988).
64. Barz, H. E., Cooper, A. S., Corenzwit, E., Marezio, M., Matthias, B. T., and Schmidt, P. H., *Science*, **175**, 884 (1972).
65. Monceau, P., Peyrard, J., Richard, J., and Molinie, P., *Phys. Rev. Lett.*, **39**, 161 (1977).
66. Berthier, C., Molinie, P., and Jerome, D., *Solid State Comm.*, **18**, 1393 (1976).
67. Berthier, C., Jerome, D., Molinie, P., and Rouxel, J., *Solid State Comm.*, **19**, 131 (1976).
68. Chevrel, R., Sergent, M., and Prigent, J., *J. Solid State Chem.*, **3**, 515 (1971).
69. Matthias, B. T., Marezio, M., Corenzwit, E., Cooper, A. S., and Barz, H. E., *Science*, **175**, 1465 (1972).
70. Nohl, H., Klose, W., and Anderson, O. K., in *Superconductivity in ternary compounds I* (eds Fischer, Ø. and Maple, M. B.), Springer-Verlag, Berlin, Heidelberg, New York, p. 165 (1982).
71. Matthias, B. T., Geballe, T. H., Andres, K., Conrenzwit, E., Hull, G., and Maita, J. P., *Science*, **159**, 530 (1968).
72. Yvon, K., in *Superconductivity in ternary compounds I* (eds Fisher, Ø. and Maple, M. B.), Springer-Verlag, Berlin, Heidelberg, New York, p. 87 (1982).
73. Johnston, D. C. and Braun, H. F., in *Superconductivity in ternary compounds: II Superconductivity and Magnetism* (eds Maple, M. B. and Fischer, Ø.), Springer-Verlag, Berlin, Heidelberg, New York, p. 11 (1982).

74. Freeman, A. J. and Jalborg, T., in *Superconductivity in ternary compounds: II Superconductivity and Magnetism* (eds Maple, M. B. and Fischer, Ø.), Springer-Verlag, Berlin, Heidelberg, New York, p. 167 (1982).

75. Sinha, S. K., Crabtree, G. W., Hinks, D. G., and Mook, H. *Phys. Rev. Lett.*, **48**, 950 (1982).

76. Genicon, J. L., Sulpice, A., Tournier, R., Chevalier, B., and Etourneau, J., *J. Physique-Lettres*, **44**, L 725 (1983).

77. Lejay, P., Chevalier, B., Etourneau. J., Hagenmuller, P., and Tarascon, J. M., *Mat. Res. Bull.*, **18**, 67 (1982).

78. Chevalier, B., Lejay, P., Etourneau, J., Vlasse, M., and Hagenmuller, P., *Mat. Res. Bull.*, **17**, 1211 (1982).

79. Chevalier, B., Lejay, P., Cole, A., Vlasse, M., and Etourneau, J., *Solid State Comm.*, **41**, 801 (1982).

80. Chevalier, B., Lejay, P., Etourneau, J., and Hagenmuller, P., *Mat. Res. Bull.*, **18**, 315 (1983).

81. Braun, H. F. in *Ternary superconductors* (eds Shenoy, G. K., Dunlap, B. D., and Fradin, F. Y.), North Holland, New York, Amsterdam, Oxford, p. 225 and p. 239(1981).

82. Barz, H., *Mat. Res. Bull.*, **15**, 1489 (1980).

83. Yvon, K., Baillif, R., and Flükiger, R., *Acta Cryst.*, **B 25**, 2859 (1979).

84. Inoue, K. and Tachikawa, K., *J. Jpn Inst. Metals*, **39**, 1266 (1975).

85. Tsuei, C. C., *Bull. Amer. Phys. Soc.*, **23**, 406 (1978).

86. Schwarz, K. and Rösch, N., *J. Phys. C, Solid State Phys.*, **9**, L 433 (1976).

87. Schwarz, K., *J. Phys. C, Solid State Phys.*, **10**, 195 (1976).

88. Hanke, W., Hagner, J., and Bilz, H., *Phys. Rev. Lett.*, **37**, 1560 (1976).

89. Höchst, H., Hüfner, S., and Goldman, A., *Solid State Comm.*, **19**, 899 (1976).

90. Pollak, R. A., Tsuei, C. C., and Johnson, R. W., *Solid State Comm.*, **23**, 879 (1977).

91. Anderson, P. W., *J. Phys. Chem. Solids*, **11**, 26 (1959).

92. Gregory, E., *Applied Polymer Symposium*, **29**, 1, John Wiley and Sons, (1976).

93. Hulm, J. K. and Matthias, B. T., *Science*, **208**, 881 (1980).

94. Tressaud, A., Chevalier, B., Lepine, B., Dance, K. M., Lozano, L., Grannec, J., Etourneau, J., Tournier, R., Sulpice, A., and Lejay, P., *Mod. Phys. Lett. B*, **2**, 1183 (1988).

95. Tressaud, A., Chevalier, B., Lepine, B., Amine, K., Dance, J. M., Lozano, L., Darriet, J., Etourneau, J., Tournier, R., Sulpice, A., Lejay, P., Diot, J. L., and Maestro, P., *Solid State Ionics*, **32/33**, 1188 (1989).

3.5   BIBLIOGRAPHY

Burger, J. P., *La supraconductivité des métaux, des alliages et des films minces*, Masson Ed., Paris (1974).

De Gennes, P. G., *Superconductivity of metals and alloys*, Benjamin Ed., New York (1966).

Ginsberg, D. M., *Physical properties of high temperature superconductors I*, World Scientific (1989).

Etourneau, J., *Les matériaux supraconducteurs: problèmes et perspectives d'utilisation. Nouvelles orientations des recherches. Revue Phys. Appl.*, **21**, 649–657 (1986).

Fischer, Ø. and Maple, M. B. (eds), *Topics in current physics—Superconductivity in ternary compounds in two volumes*. Springer-Verlag, Berlin, Heidelberg, New York (1982).

Park, R. D., *Superconductivity* (two volumes), Marcel Dekker, New York (1969).

Rickayzen, G., *Theory of superconductivity*. Monographs and texts in *Physics and Astronomy*, vol. XIV, Interscience, John Wiley and Sons, New York, London, Sydney (1965).

Rose-Innes, A. C. and Rhoderick, E. H., *Introduction to superconductivity*, International Series in Solid State Physics, volume 6, Pergamon International Library (1969).

Saint-James, D., Sarma, G., and Thomas, E. J., *Type II superconductivity*, Pergamon Press (1969).

Vonsovsky, S. V., Izyumov, Y. A., and Kurmaev, E. Z., *Superconductivity of transition metals*, Springer Series in *Solid state sciences*, Vol 27, Springer-Verlag, Berlin, Heidelberg, New York (1982).

4 Metal-rich compounds

Arndt Simon

4.1 INTRODUCTION

There are two main reasons why solid state chemistry presents difficulties
for many chemists. One is the quasi-infinity which we meet in a crystal.
Instead of having simple small molecules containing only a few atoms, large
units which contain approximately Avogadro's number of atoms have to be
surveyed. Fortunately, symmetry allows us to cut these large units into
smaller pieces and to deal only with a representative part. There are many
different ways of cutting down crystal structures. In this chapter one special
way will be chosen which tries to link molecular and solid state chemistry.
The second problem in understanding solid state compounds arises from the
frequent occurrence of unusual compositions. Forgetting about the residual
non-stoichiometry which is inherent to the solid state, even idealized
compositions like Hf_3P_2 or Ni_3S_2 seem difficult to understand on the basis
of a conventional valence concept with integral oxidation numbers. The
difficulty in accepting such compositions as nothing unusual is rather
surprising. No chemist is puzzled by a composition C_3O_2. Instead, he will
write down a formula like $O{=}C{=}C{=}C{=}O$ which tells him that all valence
electrons of the C atom which are not involved in C–O bonding are used
for homonuclear C–C bonds. The corresponding conditions also hold for
metal-rich compounds.

The last remark helps to define the term 'metal-rich compound' as it is
frequently used. The term as such might be misleading, because it is not the
proportion of metal atoms in a compound that counts but rather the number
of metal valence electrons which are not involved in metal–non-metal
bonding, nor trapped as non-bonding electrons, but which are used for
metal–metal bonding. This number follows from an application of the
generalized $(8 - N)$-rule.[1] The concentration of metal-centred electrons,
MCE,[2] belonging to a cluster or given formula unit is an important guideline
for discussing M–M bonding. The metal-centred electrons can be involved
in bonding, nonbonding and even antibonding interactions between M
atoms. Other synonyms instead of MCE are sometimes used.[1,3–5] The golden
metallic NbO (MCE = 3) will be discussed as a metal-rich compound, NaCl
(MCE = 0), of course, will not; $MoCl_2$ is a metal-rich compound, but even
Li_3N is not. Composition is important in a second sense since it determines

the nature of the M–M bonding in a critical way, influencing whether finite metal clusters are formed or whether M–M bonding extends in one, two, or three dimensions.

It is not intended in this chapter to give a survey of the vast number of known metal-rich compounds, but rather to discuss recent developments in the field and, using selected compounds, to outline ideas which link structure and bonding of seemingly different compound classes. More detailed information and references are found in a number of books and reviews.[6–27]

We shall enter the field of metal-rich compounds from discrete molecules or clusters containing only metal atoms. Non-metal atoms can occupy cavities or form bonds to the surface of such clusters. As long as there are still electrons left for M–M bonding, the entire unit of metal and non-metal atoms is called a cluster. In the solid state the relative amount of non-metal atoms on the surface determines how closely the metal atoms of adjacent clusters can approach. At a high enough metal content they become fused, and such 'condensation' results in extended M–M bonding.

4.2 METAL CLUSTERS IN THE GAS PHASE

In spite of the fact that this chapter is part of a book on solid state chemistry, it is helpful to start a discussion on metal-rich compounds with the gas phase. Clusters of atoms are frequently observed in metal vapours and much recent interest has focused on the kinds of species that exist and on their possible structures and chemical reactions. It is remarkable how many aspects of this work are of fundamental importance to solid state chemistry and vice versa. One example to illustrate this importance is the fascinating group of compounds containing polyatomic Zintl ions formed by the post-transition metals. These compounds have to be mentioned in an article on metal-rich compounds. Merely mentioning them, however, might be sufficient, because a recent review provides comprehensive information about the structures and bonding in these Zintl ions and the properties of compounds which contain them.[21] Naked clusters like Sn_9^{4-}, Sn_9^{3-}, Ge_9^{2-}, Pb_5^{2-}, Bi_4^{2-}, and Sb_7^{3-}, as well as a number of mixed-metal species, exist on the anionic side. Interestingly, the same kinds of clusters are found in co-evaporation experiments, e.g. with caesium and tin. Mass spectra show the ions $Cs_3Sn_2^+$, $Cs_3Sn_5^+$, $Cs_3Sn_9^+$, $Cs_5Sn_4^+$, and $Cs_5Sn_9^+$ to have the highest abundances.[28] Such gas phase experiments are likely to provide some hints for possible new compounds in the solid state. The second example is concerned with vapours of alkali metals, which are of particular interest, both theoretically and experimentally, due to the simple electronic structure of the atoms. Diatomic species have been known for a long time, but larger clusters are also now accessible by a range of methods including high-energy ion bombardment of the metals and seeded beam expansion.[29] In supersonic nozzle beams, clusters Na_x ($x \leqslant 16$) could be identified.[30] The relative abundances for Na_3 and Na_4 are nearly an order of magnitude smaller than for Na_2 and drop another order of magnitude for Na_5 and Na_6.

The thermodynamic stability of alkali metal clusters is low but can be greatly enhanced by the incorporation of non-metal atoms. This stabilization is obvious when the diatomic unit Na_2 is compared with the molecule Na_2O, but also holds for hypermetalated species like Na_3O, and Na_4O.[31] They are, at least as isolated gas phase species, thermodynamically stable against all possible dissociation processes. The binding of the metal atoms to such clusters is much stronger than to clusters containing metal atoms only, and there is a growing list of stable hypermetalated clusters like the above and Li_3O, Li_4O, Li_5N and Li_6C, which are experimentally detected or calculated to be stable.[32] Much of the interest in these species comes from the seemingly 'hypervalent' nature of the central non-metal atoms, suggesting a violation of the octet rule. But, of course, the bonding can be easily described in terms of normal valencies and strongly heteropolar bonds between the central non-metal atoms and the surrounding metal atoms. According to their valencies the non-metal atoms can take only a fraction of the electrons from the metal atoms. Electrons therefore stay with the metal atoms and provide additional metal–metal bonding according to simplified formulations, e.g. $Li_6C = (Li^+)_6C^{4-}(e^-)_2$. More precisely, the occupancy of the 10 valence electrons in the stable octahedral species Li_6C is $(3a_{1g})^2(2t_{1u})^6(4a_{1g})^2$ and the HOMO $(4a_{1g})$ is a Li–Li bonding orbital. What seems to be exotic with hypermetalated molecules is rather normal with metal-rich solids, as indicated by the few examples mentioned in the introduction and the many more examples still to be discussed.

4.3   ALKALI METAL SUBOXIDES

Clusters are found with metal-rich alkali metal oxides which closely resemble the hypermetalated species in the gas phase, but exist in solids.[15] The heavy alkali metals rubidium and caesium form a remarkable series of compounds with compositions M_xO ($x > 2$). Most of them melt or decompose below room temperature.

As far as their crystal structures are known, all of these suboxides follow the same unique building pattern. The principle of this pattern is evident from Fig. 4.1. The compounds Rb_9O_2 and $Cs_{11}O_3$ are composed of clusters with the same compositions, Rb_9O_2 and $Cs_{11}O_3$, respectively, which contain the oxygen atoms in an octahedral environment of metal atoms. Two or three such M_6O octahedra are condensed via common faces. The resulting clusters are only observed in the solid state. They are dissociated in the liquid suboxides and are not found in the gas phase. Reaction of caesium vapour with oxygen also yields hypermetalated species, but in mass spectra no ions

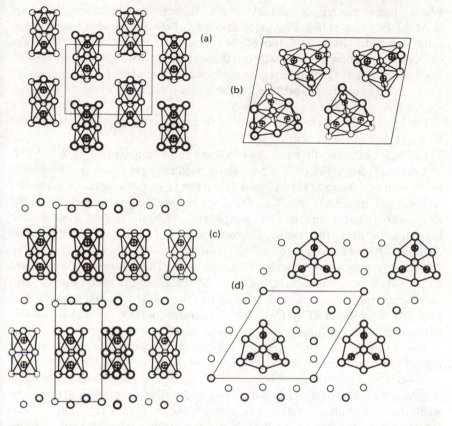

Fig. 4.1. Projections of the crystal structures of some alkali metal suboxides.[15] Oxygen atoms are labelled by crosses. The unit cells are outlined and strong connecting lines emphasize Rb_9O_2 and $Cs_{11}O_3$ clusters. (a) Rb_9O_2 (monoclinic, projected along [100]), (b) $Cs_{11}O_3$ (monoclinic, along [010]), (c) Rb_6O (hexagonal, along [10.0]), and (d) Cs_7O (hexagonal, along [00.1]).

are detectable which can be related to the M_9O_2 and $M_{11}O_3$ clusters that are found in the solid state.[33]

The Rb_9O_2 and $Cs_{11}O_3$ clusters are close-packed in the compounds of the corresponding compositions. The crystal structures rather resemble structures of molecular crystals. A discussion of the bonding closely follows the ideas outlined for the gaseous hypermetalated species. The strong bonds between oxygen and metal atoms in a covalent picture have essentially σ character using metal s and oxygen 2p states. Those metal valence electrons which are in excess are used for additional (weaker) M–M bonding. Due to the extreme differences in electronegativity between Rb, Cs, and O, respectively, the electron transfer to the oxygen atoms is large and the formulations $Rb_9O_2 = (Rb^+)_9(O^{2-})_2(e^-)_5$ and $Cs_{11}O_3 = (Cs^+)_{11}(O^{2-})_3(e^-)_5$ are allowed.

We are facing clusters of ions which are stabilized by additional electrons in M–M bonding states. The close approach between these clusters in the solids leads to overlap of the M–M bonding orbitals between different clusters with concomitant electronic delocalization throughout the crystal. Such suboxides may be referred to as 'complex metals' or 'cluster metals', where charged clusters substitute for the single ionic cores in an alkali metal, e.g. $Cs = Cs^+e^-$.

The above view of chemical bonding in Rb_9O_2 and $Cs_{11}O_3$ is supported by a number of experimental facts.

1. The interatomic distances $d(Rb-O)$ are in the range from 264 to 285 pm, and $d(Cs-O)$ from 268 to 298 pm. These values approximately correspond to the sum of ionic radii for M^+ and O^{2-}. The large distances $d(O-O) = 382$ and 404 pm in Rb_9O_2 and $Cs_{11}O_3$, together with significant shifts of the O^{2-} ions towards the cluster peripheries, indicate strong electrostatic repulsion between the highly charged anions. The range of intercluster distances $d(M-M)$ peak around 530 and 550 pm in Rb_9O_2 and $Cs_{11}O_3$, respectively, compared with the distances in metallic Rb (488–563 pm) and Cs (527–609 pm). These intercluster M–M distances are much longer than the intracluster distances, which lie in the ranges 354 to 403 pm (Rb_9O_2) and 367 to 416 pm ($Cs_{11}O_3$). The structural details are in qualitative agreement with the bond model for alkali metal suboxides.

The bond length/bond strength concept is frequently used to analyze interatomic distances on a quantitative level. Analyses of M–M distances in metal-rich compounds have been performed extensively[34] using Pauling's formula, $d = d(1) - 60\log(s)$, with s = bond order (strength), $d(1)(pm)$ = single bond distance.[35] Other expressions like $s = [(d/d(1)]^{-N}$ have been introduced more recently,[36] where N denotes specific constants for different element combinations. As such analyses will be made occasionally in the following text one should be aware of the principal limitations of the concept; these can be clearly demonstrated with the alkali metal suboxides.

All expressions for a bond length/bond strength relation have in common the fact that the shorter a distance the stronger is the bond assigned to this distance. One has to be very careful in applying this criterion to M–M bonds. According to Pauling's equation, the short distance $d(Cs-Cs) = 367$ pm between the atoms on the trigonal axis of the $Cs_{11}O_3$ cluster corresponds to bond orders in the range $33 \leqslant n \leqslant 58$ depending on the assignment of coordination numbers $14 \geqslant cn \geqslant 8$ to the reference state, bcc caesium. This result is, of course, nonsense in both cases. The short distance simply reflects the considerably smaller size of the Cs^+ ion compared to the uncharged atom and also reflects the constriction of the $(Cs^+)_2$ pair in the centre of the $Cs_{11}O_3$ cluster by three bridging O^{2-} ions. In general, M–M distances in metal-rich compounds are greatly influenced by metal–non-metal interactions. Another less severe problem with bond length/bond strength relations becomes obvious when the distances $d(Cs-O)$ are analysed. Using the expression $s = (d/233.5)^{-6.6}$, which has been derived as the best fit to the Cs–O distances in 27 different Cs salts of oxoacids,[36] the average bond order sum for the oxygen atoms in $Cs_{11}O_3$ is 1.58 ± 0.02 instead of 2. Nearly the same low value $\sum s = 1.57$ is calculated for Cs_2O, which crystallizes in the anti-$CdCl_2$ structure. Obviously, the bond length/bond strength relation is specific for the compound class and different for binary and ternary caesium oxygen compounds. But the nearly identical values for the insulator Cs_2O and the metallic $Cs_{11}O_3$, with its rather distorted coordination octahedra around the O^{2-} ions, clearly show the same kind of bonding to occur between Cs and O atoms in both compounds.

2. In agreement with the bond model, the copper-red compound Rb_9O_2 and the violet $Cs_{11}O_3$ are both metallic conductors. Their specific conductivities at room temperature are half the conductivities of the pure metals, Rb and Cs, and their variations as a function of temperature are very similar to those of the pure metals. Unfortunately, conductivity measurements do not allow a quantitative proof of the bond model in terms of the concentration of electrons in M–M bonding states, because both the concentration of (free) carriers, n, as well as their mobility, μ, determine the conductivity according to $\sigma = \eta \mu e$.

The free carrier concentration can be measured via the plasma edges. Reflectivity data show that the colours of Rb_9O_2 and $Cs_{11}O_3$ arise from plasma oscillations. The change of the plasma energies from 3.35 to 2.41 eV when we go from Rb to Rb_9O_2, and from 2.78 to 1.73 eV for Cs to $Cs_{11}O_3$, is a consequence of the lower free-electron concentrations in the compounds because electrons are localized at the O^{2-} ions. The compound spectra exhibit a number of features around the plasma edge and the numerical values for the concentrations of free electrons per formula unit, 3.0 and 2.7, derived from the plasma energies and normalized to the value 1.0 for the pure metals, are significantly lower than expected for Rb_9O_2 and $Cs_{11}O_3$ in a free-electron model.[37] The metal valence electrons are not simply free s electrons. The result is more exciting than disappointing. It is well known from the superconducting transition at 11 MPa, as well as from band structure calculations, that caesium becomes a d metal when it has approximately one-third of its volume at ambient pressure. The interatomic distances are then close to the shortest intracluster distances in $Cs_{11}O_3$. Last but not least, recent investigations of the metallization of CsI under pressure support the idea of a contribution of the Cs d electrons to chemical bonding in this ionic compound.[38,39] Does the cluster chemistry of Rb and Cs in their suboxides reflect some d metal chemistry of these elements?

Detailed information about the chemical bonding in alkali metal suboxides comes from UV photoelectron spectroscopy.[40] This surface method—the photoelectrons have an escape depth of only a few atomic layers—simultaneously informs us about the electronic band structure of the bulk as well

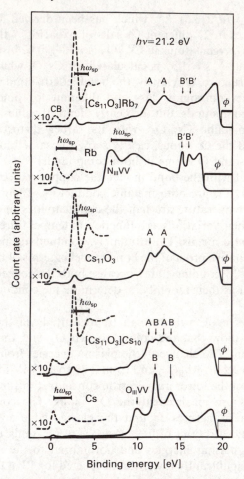

Fig. 4.2. He I photoelectron spectra of Rb, Cs, $Cs_{11}O_3$, $Cs_{11}O_3Cs_{10}$ ($= Cs_7O$) and $Cs_{11}O_3Rb_7$. Intensities are enhanced by a factor 10 with the dashed curves. Assignments are as following: Φ = workfunction; A = Cs $5p^{3/2}/5p^{1/2}$ core levels for the cluster atoms; B = Cs $5p^{3/2}/5p^{1/2}$ for caesium and atoms between clusters; B′ = Rb $4p^{3/2}/4p^{1/2}$ for Rb and atoms between clusters; $N_{III}VV$, $O_{III}VV$ = Auger peaks; structures in the dotted curves: conduction band (CB), O 2p level, energy losses due to surface plasmon excitation ($\hbar\omega_{sp}$).

as electronic surface effects (Fig. 4.2). The metallic character of $Cs_{11}O_3$ is verified as the density of states at the Fermi level ($E_b = 0$). From a comparison of the band widths for Cs and $Cs_{11}O_3$, the concentration of electrons per formula unit in the conduction band of $Cs_{11}O_3$ is calculated as MCE = 5.3. The narrow oxygen 2p band at $E_b = 2.7$ eV is consistent with the localized nature of the electrons in the O^{2-} ions. The energy loss structure associated with this narrow band, which is due to a surface plasmon excitation at $\hbar\omega_{sp} = 1.55\,eV$, yields an independent determination of the concentration of electrons in the conduction band of $Cs_{11}O_3$ which comes out as MCE = 5.1, relative to 1.0 for Cs. This value is in excellent agreement with the bond model. Obviously all metal valence electrons are involved in the energy loss process.

3. The chemical bonding in alkali metal suboxides is reflected in two distinct facets of their vibronic behaviour. The bonding within the clusters is strong due to M–O and some additional M–M bonding, but the coupling between clusters via M–M bonding is weak. This weak intercluster bonding shows up in an anomalous increase of the low temperature specific heat at approximately 0.7 K for compounds containing the $Cs_{11}O_3$ cluster.[41] Obviously, the replacement of Cs atoms in elemental caesium by $Cs_{11}O_3$ clusters only increases the mass of the vibrating unit, and the hump in the specific heat near 0.7 K corresponds to the lowest acoustic branch in Cs rescaled by the mass of the cluster. The intracluster vibrations are decoupled from the low-frequency bulk phonons and give rise to characteristic Raman spectra of the alkali metal suboxides. The most prominent bands can be assigned on the basis of fairly simple model calculations which minimize the total energies for the clusters with the assumption of ionic, two-body interactions. These calculations lead to the interesting result that the clusters $[(Rb^+)_9(O^{2-})_2]^{5+}$ and $[(Cs^+)_{11}(O^{2-})_3]^{5+}$ are not stable but loose positive ions. For the Rb_9O_2 cluster the energy minimum is obtained with a charge $0.7+$ for Rb, corresponding to a net charge $2.3+$ on the cluster. The corresponding values for the $Cs_{11}O_3$ cluster are $0.9+$ and $3.9+$, respectively. Additional electrons in M–M bonding states are necessary for the clusters to be stable; this is taken into account as partial shielding of the positive charges in the electrostatic model.

The arguments in 1.–3. support the view of Rb_9O_2 and $Cs_{11}O_3$ as examples of 'cluster metals'. Further support comes from the unusual chemistry of these compounds, including the formation of 'intermetallic compounds' with Rb and Cs. The binary phases $Rb_9O_2Rb_3 = Rb_6O$, $Cs_{11}O_3Cs = Cs_4O$ and $Cs_{11}O_3Cs_{10} = Cs_7O$, as well as some ternary compounds like $Cs_{11}O_3Rb$, $Cs_{11}O_3Rb_2$, and $Cs_{11}O_3Rb_7$, have been prepared and characterized. In each of these phases the cluster-to-alkali-metal ratio is strictly stoichiometric. Figures 4.1(c) and 4.1(d) illustrate the structural principle for the most metal-rich phases, Rb_6O and Cs_7O. In Rb_6O, layers of Rb_9O_2 clusters alternate with close-packed layers of Rb atoms and the arrangement of all the Rb atoms is close-packed as a consequence of the fact that the cluster itself represents a cutting from a 3-dimensional close-packed arrangement of Rb atoms. The $Cs_{11}O_3$ cluster differs in this respect, but it is remarkable that the matrix of Cs atoms between columns of $Cs_{11}O_3$ clusters in Cs_7O is again close-packed.

The properties of the clusters and the pure alkali metals are simply additive in the suboxides. For example, the molar volumes of $Cs_{11}O_3Cs$ and

$Cs_{11}O_3Cs_{10}$ exceed the volume of $Cs_{11}O_3$ by 69.9 and 696.5 cm^3 mol^{-3}, respectively, compared with the value for elemental caesium, 69.2 cm^3 mol^{-1}. The specific heat of, for example, $Cs_{11}O_3Cs_{10}$ closely corresponds to the appropriate sum of the specific heats of $Cs_{11}O_3$ and pure caesium. Finally, the photoelectron spectrum of $Cs_{11}O_3Cs_{10}$, with its characteristic Cs5p band region (Fig. 4.2), looks like a superposition of the spectra of $Cs_{11}O_3$ and Cs. Only the electronic properties of the surface, the work function Φ and the energy of the surface plasmon $\hbar\omega_{sp}$, indicate the homogeneous nature of the compound.

So far, the alkali metal suboxides have been discussed from a cluster viewpoint. Another viewpoint can be taken, namely that these compounds are essentially alkali metals which have been contaminated by oxygen in an ordered way. Clustering of impurity atoms in a metal is a common phenomenon and is generally explained by the argument that a joined deformation of the lattice by several impurity atoms is easier than several single deformations. The extent of the deformation critically depends upon the balance of M–M and M–X bonding. The alkali metals represent one extreme. They are very soft metals. M–M bonds are weak and the M–O bonds are so dominant that discrete clusters around the O atoms are formed. Transition metals like Nb and Ta represent the other extreme. M–M bonding is strong here and the known suboxides, e.g. Nb_6O, Nb_4O, Ta_6O and Ta_4O,[42] are characterized by clusters of interstitial O atoms in essentially unchanged *bcc* lattices of the metals. These extremes have model character for the behaviour of impurity atoms in metals.

Viewing the alkali metal suboxides as contaminated metals has an interesting physical consequence. The two or three O^{2-} ions incorporated in the Rb_9O_2 and $Cs_{11}O_3$ clusters, respectively, are highly repulsive for the conduction electrons. The alkali metal suboxides therefore have the properties of microscopically 'foamed' metals where the conduction electrons are excluded from the 'bubbles'. Instead of a continuous filling of the Fermi sphere as is the case with the pure alkali metals, certain electron wavevectors are forbidden in the case of the suboxides leading to a raised Fermi level[43] and therefore to a decreased work function, Φ. Figure 4.2 gives direct evidence for this effect. The gap between the energy (21.2 eV) of the incident light and the threshold of photoelectrons in the spectra is due to the work function and the gap is significantly smaller with $Cs_{11}O_3$ than with caesium metal.

As cluster compounds and as void metals the alkali metal suboxides represent a unique compound class. They seem to stand rather isolated. Yet, a closer inspection shows that they embody the principles that are also important with other metal-rich systems. In the alkali metal suboxides, M_6O octahedra are condensed into M–M bonded clusters. In the structures of the metal-rich alkaline earth nitrides Ca_2N, Sr_2N and Ba_2N,[44] which crystallize in the anti-CdCl$_2$ type structure, M_6N octahedra are condensed via edges to form layers of composition M_2N which are M–M bonded according to $(M^{2+})_2N^{3-}e^-$. Except for the different dimensionality of the structural units, the bonding in Ba_2N and $Cs_{11}O_3$ is very comparable. The distance $d(Ba–N) = 276$ pm closely corresponds to the sum of ionic radii; the environment of the N^{3-} ions is constricted to $d(Ba–Ba) = 378.8$ and 401.6 pm, whereas the interlayer distance $d(Ba–Ba) = 447.6$ pm is near to the shortest distance in elemental barium (435 pm). Again, these compounds and structures are not unique to the alkaline earth metals, since similar structures and closely related chemical bonding occur with transition metal compounds. For example, Ti_2O[45] and Ag_2F[46] both crystallize with the anti-CdI_2 structure and exhibit strong M–X and M–M bonding as in Ba_2N. The similarities between the beginning and the end of the transition metal series indicate a broad chemistry of metal-rich compounds in between.

4.4 ISOLATED METAL CLUSTERS

In the Born–Haber cycle only a part of the heat of atomization of the metal has to be raised if M–M bonds are present in a compound. Following Schäfer's argument,[7] M–M bonding should be common with those metals which have the highest heats of sublimation. In fact, a broad chemistry of metal-rich compounds and very stable metal clusters is centred in the periodic table around the metals Nb, Mo, Ta, W, and Re. In the following section, the principles of this chemistry will be discussed starting with the well-known octahedral metal clusters and, as before, the consequences of bringing these clusters into close contact in a solid.

Qualitatively speaking, transition metals like Nb or Mo have a large enough number of valence electrons to build strongly bonded M_6 units which need not be stabilized by interstitial atoms but have bonding capability for non-metal atoms on their surface. It is textbook knowledge that two characteristic cluster types exist, one with non-metal atoms X above all the octahedra faces (M_6X_8) and the other with X above all edges (M_6X_{12}). The bonding in the first can be simply described with 12 electron pairs on the (free) edges, the other with 8 pairs on the faces of the octahedron. The numbers of 12 and 8 M–M bonding states which can each be occupied by 2 electrons have been rationalized repeatedly in terms of MO arguments. If only d states of the metal atoms are taken into account for M–M bonding, the d_{xy} orbitals are used for bonding of the four X atoms around each M atom and a unitary transformation for the 12 bonding combinations of the remaining 6×4 d states yields 12 degenerate orbitals directed approximately along the edges of an octahedron. They point along the edges in the M_6X_8 cluster and into the centres of the octahedral faces in the M_6X_{12} cluster to form 2 centre/2 electron bonds and 3 centre/2 electron bonds, respectively.[47–50] The $Mo_6Cl_8^{4+}$ and $Nb_6Cl_{12}^{2+}$ ions, which represent closed-shell configurations

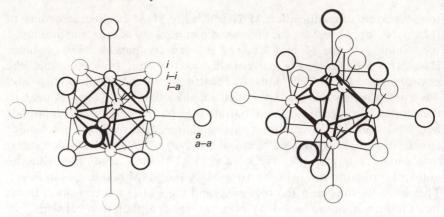

Fig. 4.3. $M_6X_8^iX_6^a$ and $M_6X_{12}^iX_6^a$ clusters (small circles are M atoms). The different functionalities of the X atoms correspond to Schäfer[7] (*i*, *a*, if belonging to one cluster, *i–i*, *i–a*, *a–a*, if interconnecting).

of the two cluster types, are effectively isoelectronic in the sense that each of them contains 20 electron pairs in M–M and M–X bonding states.[51]

Assigning the X atoms as X^i if they are located above octahedral faces or edges (Fig. 4.3), additional X atoms can be bonded to the apices, X^a, or X^{a-a} if they bridge adjacent clusters.[7] The X^i atoms can also have a bridging function (X^{i-i}, X^{i-a}). If the 'encapsulation' of the metal cores of such clusters by non-metal atoms is complete, the cores are well isolated in a solid compound. It is interesting to follow the structural consequences as well as the consequences in physical properties when this 'encapsulation' is cut open and a closer approach of the cores is made possible. In terms of the Herzfeld–Mott criterion metallization should ultimately occur.[52] Keeping in mind that this volume criterion essentially depends on distances, some intermediate steps in the approach of the cluster cores will involve metallic systems with low dimensionality.

The compound $K_4Nb_6Cl_{18} = K_4Nb_6Cl_{12}^iCl_6^a$ represents the limit of well separated cluster cores. The dark brownish-green compound is a semiconductor due to the fact that the valence band is entirely filled with 16 electrons per formula unit in the Nb–Nb bonding states. But $Ta_6Cl_{15} = Ta_6Cl_{12}^iCl_{6/2}^{a-a}$ and $Nb_6I_{11} = Nb_6I_8^iI_{6/2}^{a-a}$ are semiconductors too, although the missing numbers of electrons in M–M bonding states—MCE = 15 and 19 instead of 16 and 24, respectively—fulfill the first supposition of metallic conductivity, namely a partially filled valence band. But the second supposition, that the approach of the metal cores is close enough to yield a partially occupied conduction band, is not met. Chemical bonding in these compounds is well described in terms of discrete molecular orbitals within isolated clusters. Due to the localization of the d electrons in M–M bonding states in the clusters, interesting magnetic phenomena arise which have been investigated in detail for Nb_6I_{11}.[53-55] The compound undergoes a second order phase transition at 274 K, accompanied by a change from a doublet to quartet ground state for the cluster. Such spin crossover transitions are known to occur with mononuclear transition metal complexes as a consequence of changing ligand field strength.[56] In the cluster compound the geometry of the Nb_6 core itself changes and thereby the separation between occupied and empty states varies. Spin crossover transitions should be quite common with metal clusters due to the large number of, and small energy gaps between, electronic levels.

Closer approach of the clusters than in the preceding examples is possible by substituting the monovalent anions X by divalent atoms S, Se, Te, thus increasing the M/X ratio. MCE can be kept high enough for efficient M–M bonding by changing to the metals Mo or Re. Compounds like $Re_6Se_8Cl_3 = Re_6Si_4^i Se_4^{i-a} Cl_2^a$ and $Mo_6Br_6S_3 = Mo_6Br_4^i S_2^i S_{2/2}^{i-i} Br_{4/2}^{a-a}$ illustrate the stepwise closer approach between the metal atoms of adjacent clusters.[57,58] In the rhenium compound Re_2Se_8 clusters are arranged in layers and inclined in such a way that some of the Se atoms come into X^a positions for adjacent clusters. The Cl atoms are in between such layers. The molybdenum compound offers the interesting feature that inner atoms of the $Mo_6(Br, S)_8$ cluster are used to join neighbouring units (S^{i-i}). In both cases, the M–M bonding states are completely filled by 24 electrons per cluster and the compounds are semiconductors. The situation changes with $Mo_6S_6Br_2 = Mo_6S_6^{i-a} Br_2^i$, a compound composed of bare $Mo_6(S, Br)_8$ clusters which are densely packed in a rhombohedral structure and aligned in such a way as to bring all S atoms into X^a positions for adjacent clusters (S^{i-a}) with the Br atoms lying on the 3-fold axis.[59] The intercluster distances $d(Mo–Mo) = 322.5$ pm are still significantly longer than the intracluster distances, 271.9 and 273.2 pm, but the Mo–S bonds within (240.1 to 247.5 pm) and between clusters (248.5 pm) have nearly all the same lengths. The dense packing of the clusters as well as the fact that the M–M bonding states are only partially occupied by 22 electrons per cluster make the compound a metal and indeed a superconductor with a high transition temperature, $T_{cr} = 13.8$ K. $Mo_6S_6Br_2$ belongs to the class of Chevrel phases, which are well known for their exceptional high critical field superconductivity (see Chapter 3).[60] This property is closely related to the presence of Mo_6X_8 clusters in a densely packed arrangement. MCE can be varied in several ways, such as (*i*) intercalation of metal atoms between the clusters, (*ii*) substitution of metal atoms in the clusters, or (*iii*) substitution of non-metal atoms. $PbMo_6S_8$ is an example of (*i*). The lead donates 2 electrons to the cluster, which makes 22 electrons per cluster available for M–M bonding, as in the case of $Mo_6S_6Br_2$ which exemplifies type (*iii*). The metal intercalation has several interesting aspects, such as the high ionic mobility, e.g. of Cu, Li, Ni, which facilitates topochemical redox reactions. Another aspect is the coexistence of magnetic ordering and superconductivity, e.g. with $HoMo_6S_8$. Increasing the d electron concentration via (*ii*) to 24 per cluster in the compounds $Mo_2Re_4Se_8$ or $Mo_4Ru_2Se_8$ leads to filled conduction bands and semiconducting properties. At one stage, $PbMo_6S_8$ set a record for the superconducting transition temperature and critical magnetic field ($T_{cr} \approx 15$ K, $H_{c2} \approx 600$ kG) for Chevrel phases, and initiated tremendous interest which led to the synthesis and investigation of more than a hundred phases utilizing procedures (*i*) to (*iii*) or combinations of these, e.g. in phases like $Cu_xMo_6S_6I_2$.

The geometry of the M_6X_8 cluster is specifically influenced by the number of valence electrons for M–M bonding.[14,61] The Mo_6 octahedron becomes elongated along a 3-fold axis when the number of d electrons decreases. The intracluster Mo–Mo distances are a sensitive probe for the electron transfer from the intercalated metal atoms to the cluster. Less obvious is the change of the intercluster Mo–X distances, which tend to be larger with higher values for MCE.

The analysis of chemical bonding in Chevrel phases presented in several papers starts from the isolated, regular Mo_6X_8 cluster and investigates the changes to the system of molecular orbitals when the clusters are distorted and brought to bonding distances in the solid.[49,62,63] Assigning a local coordinate system to the Mo atoms in the regular unit such that the z axes are normal to the faces of the X_8 cube, as discussed earlier, the d_{xy} orbitals are involved in bonding of the 4 X atoms around each Mo atom and the remaining $d_{x^2-y^2}$, $d_{xz,yz}$ and d_{z^2} contribute to 12 Mo–Mo bonding MOs which belong to the representations a_{1g}, t_{1u}, t_{2g}, t_{2u} and e_g (and the antibonding states with the representations e_g, t_{1u}, a_{2g}, t_{1g} and t_{2u}). These metal based states lie above the states which essentially involve bonding of the non-metal atoms. The electronic states of the cluster are split in the presence of any distortions that occur in the real systems. When building a crystal from such clusters, each Mo atom becomes coordinated by an additional X atom from a neighbouring cluster (along its z axis in X^a position). This additional coordination does not influence the position of the HOMO significantly, because the highest occupied e_g (as well as the t_{2u} orbitals) have $d_{x^2-y^2}$ character and therefore only δ-type interactions with the p orbitals of the X atoms on the z axis. In contrast, both LUMOs, e_g and t_{1u}, have $sp_zd_{z^2}$ character and are therefore pushed to higher energies. This donor–acceptor interaction is the main contribution to intercluster bonding and leads to a widening of the HOMO–LUMO gap. From an overlap population analysis the intercluster Mo–X bonding is seen to be as strong as the intracluster bonding, whereas the bond strength for the intercluster Mo–Mo bonds is approximately a tenth of the intracluster bonds. This simplified picture is in agreement with full band structure calculations. The fact that the δ-type intercluster interaction of the cluster e_g state keeps the conduction band narrow and results in a high density of states at the Fermi level seems to be the reason for the high T_{cr} values of some Chevrel phases. The significant t_{2u} character of the bottom of the conduction band explains why the Mo_6 octahedron becomes elongated with fewer electrons in the conduction band. But there are difficulties in accounting for certain properties of Chevrel phases in terms of this e_g band picture.

The large pressure dependence of T_{cr} for $PbMo_6S_8$, as well as the strong electron–phonon coupling with low frequency phonons indicated by tunneling experiments, are incompatible with this picture. The band should not be changed significantly by static or dynamic changes of the intercluster contacts. Band structure calculations for different Chevrel compounds with 20 to 23 d-electrons per cluster revealed the composite nature of the conduction band, which has e_g character ($x^2 - y^2$) throughout most of the Brillouin zone but due to admixing of t_{1u} (xz, yz) becomes e_u at the zone centre (Fig. 4.4). The non-rigid band behaviour is evident from the progressive lowering of the band energy at the zone centre compared with the e_g band energy at the zone boundary, when the number of d electrons is increased. Antibonding π-type interactions between adjacent clusters involving the (bonding) Mo d-like t_{1u} and (antibonding) X p-like t_{1u} orbitals, which repel each other, are responsible for the t_{2u} (e_u) character of the conduction band at Γ. The conduction band level at Γ is therefore low for large intercluster distances as these antibonding states are raised in the crystal, and it is lower for X=S than Se and Te. With 24 electrons in the cluster, all intercluster antibonding states in the conduction band are occupied. Reducing the electron number depletes the band near Γ and allows the clusters to come closer; this in turn increases the band energy near Γ. The pronounced non-rigid band behaviour has interesting consequences. In the band structure of $PbMo_6S_8$ a pocket of the conduction band near Γ is filled. As the energy level at Γ is sensitive to the intercluster separation, close distances d(Mo–S) between clusters will deplete and long distances fill this pocket, giving rise to strong electron–phonon coupling as well as to the observed anisotropy in the critical field. Still more sophisticated band structure calculations seem necessary to understand the bonding of Chevrel phases in detail. In particular, the role of the intercalated A atoms, so far seen as electron donors only, has to be analyzed. Recent Knight shift measurements on $TlMo_6Se_8$ show that the Tl s band lies at the Fermi level.[64]

The Chevrel phases are a fascinating compound class due to their physical properties and their comparatively simple crystal structures, simple in the sense that they represent a primitive packing of quasi-molecular units. An insight to the chemical bonding via band structure calculations is possible, because the calculations are feasible and because the band structure can be made clear through the local bond picture for the molecular unit. These phases are interesting in yet another respect. They represent the borderline between molecular and infinitely extended units in a solid. The M_6 core is completely surrounded by X atoms and intercluster bonding essentially occurs via M–X interaction. M–M bonding between clusters is very weak.

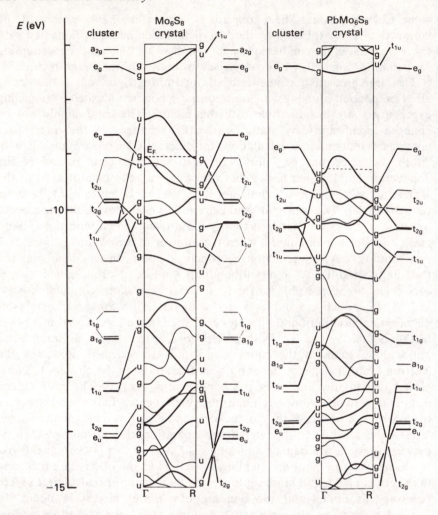

Fig. 4.4. Molecular orbitals for the (distorted) Mo_6S_8 clusters and energy bands along the $\bar{3}$ axis for Mo_6S_8 and $PbMo_6S_8$. Doubly degenerate bands are indicated by strong lines.[62]

What happens if the envelope of X atoms around the M_6 core is opened further by increasing the M/X ratio?

4.5   CONDENSED CLUSTERS

In the preceding section, a brief outline of the interconnection principles for M_6X_8 and M_6X_{12} clusters has been given with different types of linkages via X atoms shared between adjacent clusters (X^{a-a}, X^{i-a}, X^{i-i}). So far, the metal core of the cluster has remained as a discrete unit. The next stage of approach between clusters leads to a new situation in which the metal atoms become shared.

4.5.1 Face-sharing M_6 clusters

The crystal structures of a number of metal-rich compounds between d and p elements have been interpreted in terms of M_6X_8 and M_6X_{12} clusters condensed via sharing M atoms.[16] The different types of condensation involve apices, edges, and faces of the M_6 octahedron. The concept of using these clusters (as well as other cluster types) to build up infinitely extending M–M bonded structures has found recent support in the preparation of compounds which contain quasi-molecular units of a varying number of condensed clusters. To start with the case of face-sharing M_6 octahedra, the stepwise one-dimensional condensation of the M_6X_8-type cluster can be beautifully illustrated by the ternary molybdenum chalcogenides.[27] The condensation finally leads to the infinite chain of composition M_3X_3. Figure 4.5 shows the series of 'oligomeric' cluster anions which have been found and characterized by crystal structure analysis so far, and which lie between the monomeric unit and the polymeric chain. The known compounds are summarized in Table 4.1. The compositions follow the general formula $Mo_{3n+3}X_{3n+5}$, where n denotes the number of Mo_6 octahedra (or rather trigonal antiprisms) in cluster anions which become progressively more metal-rich and electron-rich with length. The condensation of M_6X_8-type clusters via faces of the octahedra calls for the loss of X atoms above those faces which are involved in condensation. A reorientation of the bond directions at Mo atoms now bonded to three Mo atoms instead of one X atom means a significant change of the electronic structure when going from the complete Mo_6X_8 unit to the Mo_6X_6 fragment.

For the description in the oligomeric units it is advantageous to treat them as extended Mo_6X_8 clusters: the Mo_6X_8 cluster is cut by a plane which is normal to one 3-fold axis, leaving two fragments Mo_3X_4. Varying numbers of star-shaped planar Mo_3X_3 units are inserted between these two fragments. All compounds $A_x[Mo_{3n+3}X_{3n+5}]$ crystallize in the same space group, $R\overline{3}$, as the starting member of the family, $A_xMo_6X_8$. The structural principle of the whole family is simply the elongation of the clusters by insertion of Mo_3X_3 fragments and the simultaneous addition of A atoms into the developing channels between the clusters, e.g eight Cs^+ ions in a row with $Cs_8Mo_{30}Se_{32}$. It is clear from what has been said about the structural principle that the Mo atoms in the Mo_3X_4 fragment (terminal Mo atoms) have the same coordination, by five X atoms and three Mo atoms, as in the Chevrel phases, $A_xMo_6X_8$. The intermediate Mo atoms in the chain are coordinated only by X atoms of their own cluster since the intercluster distances $d(Mo–X)$ are larger than 600 pm. But it is interesting to note that

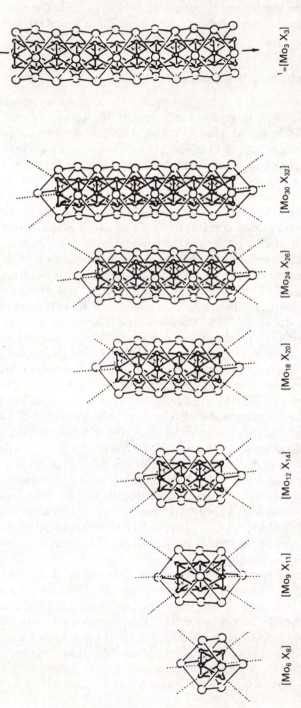

Fig. 4.5. Cluster condensation via opposite faces of the M_6X_8-type cluster. Oligomeric clusters of 2, 3, 5, 7, and 9 octahedral units have been observed as intermediates between the single Mo_6X_8 cluster and the infinite $^1_\infty[Mo_3X_3]$ in ternary molybdenum chalcogenides.[27] The dotted lines indicate the inter-cluster coordination of the terminal Mo atoms.

$[Mo_6X_8]$ $[Mo_9X_{11}]$ $[Mo_{12}X_{14}]$ $[Mo_{18}X_{20}]$ $[Mo_{24}X_{26}]$ $[Mo_{30}X_{32}]$ $^1_\infty[Mo_3X_3]$

Table 4.1 *Ternary molybdenum chalcogenides $A_xMo_{3n+3}X_{3n+5}$:[27] relations between the number of metal-centred electrons (MCE) per cluster or the averaged oxidation state of the Mo and the preferred cluster type. The estimate values are derived from the number of Mo_3X_4 and Mo_3X_3 fragments which are contained in a special cluster*

X/Mo	Compound	Cluster	Experimental MCE (oxidation state Mo)	Estimated MCE (oxidation state Mo)
1.33	$A_xMo_6X_8$ (X = S, Se, Te)	$[Mo_6X_8]^{0\ to\ 4-}$	20–24 (2.67–2.00)	
1.26_7	$In_2Mo_{15}Se_{19}$	$[Mo_6Se_8]^{0\ to\ 2-}$	20–22 (2.67–2.22)	
		$[Mo_9Se_{11}]^{0\ to\ 2-}$	32–32 (2.44–2.33)	33–37 (2.33–1.89)
	$Ba_2Mo_{15}Se_{19}$	$[Mo_6Se_8]^{0\ to\ 4-}$	20–24 (2.67–2.00)	
		$[Mo_9Se_{11}]^{0\ to\ 4-}$	32–36 (2.44–2.00)	33–37 (2.33–1.89)
1.22_2	$Ag_{3.6}Mo_9Se_{11}$	$[Mo_9Se_{11}]^{3.6-}$	35.6 (2.04)	33–37 (2.33–1.89)
	$Tl_4Mo_{18}S_{22}$	$[Mo_6S_8]^{0\ to\ 4-}$	20–24 (2.67–2.00)	
		$[Mo_{12}S_{14}]^{0\ to\ 4-}$	44–48 (2.33–2.00)	46–50 (2.17–1.83)
1.16_7	$Cs_2Mo_{12}Se_{14}$	$[Mo_{12}Se_{14}]^{2-}$	46 (2.17)	46–50 (2.17–1.83)
1.11_1	$Rb_4Mo_{18}Se_{20}$	$[Mo_{18}Se_{20}]^{4-}$	72 (2.00)	72–76 (2.00–1.78)
1.08_3	$Cs_6Mo_{24}Se_{26}$	$[Mo_{24}Se_{26}]^{6-}$	98 (1.92)	98–102 (1.92–1.75)
1.06_6	$Cs_8Mo_{30}Se_{32}$	$[Mo_{30}Se_{32}]^{8-}$	124 (1.87)	124–128 (1.87–1.73)
1.00	$Tl_2Mo_6Se_6$	$^1_\infty[Mo_{6/2}Se_{6/2}]$	13/Mo_3 (1.67)	

each Mo atom has three X atoms arranged around it as in the Mo_6X_8 cluster and one X atom in an X^a position.

The final members of the series, AMo_3X_3, represent a hexagonal close rod-packing of $\frac{1}{\infty}[Mo_3X_3^-]$ chains with A atoms (e.g. Tl) filling the channels between the chains. $TlMo_3Se_3$ is a metal and according to the 1-dimensional character of the structure, the conductivity in the chain direction is 1000 times larger than in the orthogonal directions. At 6 K the compound becomes superconducting and the anisotropy of the critical field is again very large.[65] All compounds containing oligomeric clusters are also metallic at room temperature. At low temperatures, metal–insulator transitions ($Cs_6Mo_{24}S_{26}$ at 112 K) and superconducting transition (e.g. with $Cs_2Mo_{12}Se_{14}$ and $Cs_6Mo_{24}Se_{26}$, $T_{cr} = 4$ and 3 K, respectively) have been observed.

The occurrence of different cluster sizes in the series $A_xMo_{3n+3}X_{3n+5}$ and the metallic properties of the compounds found so far raise important questions. What are the factors determining the cluster size? How many electrons in M–M bonding states correspond to closed-shell configurations? Both questions are intimately related. Inspection of the limiting phases $A_xMo_6X_8$ and AMo_3X_3 allows a pretty clear answer to these questions on the basis of experimental facts. The same answer is reached by molecular orbital calculations for the oligomeric clusters.[66] The isolated M_6X_8 cluster accommodates 24 electrons in M–M bonding states. The possible reduction to 20 electrons per cluster in Chevrel phases is due to the intercluster Mo–X interactions which add antibonding character to the conduction band, as mentioned earlier. We can therefore draw two conclusions.

1. The terminal Mo atoms in the oligomeric clusters are bonded as in $A_xMo_6X_8$, both concerning intra- and intercluster bonding. The fragment Mo_3X_4 will therefore accommodate 12 to 10 electrons in M–M bonding states.

2. In AMo_3X_3, there is no interaction between Mo and X atoms of different chains, and the compounds seem to be strictly stoichiometric. All M–M bonding states of the $\frac{1}{\infty}[Mo_3X_3^-]$ chain must be occupied and obviously 13 electrons per Mo_3X_3-fragment can be accommodated.

The decomposition of the oligomeric clusters into Mo_2X_4 and Mo_2X_3 fragments leads to the values for MCE and corresponding oxidation states for Mo listed in Table 4.1. These values differ from those predicted by the '12n rule'.[67,68] The upper limits of MCE indicate the closed-shell configurations. All compounds with oligomeric clusters characterized so far have less electrons than would correspond to a closed-shell configuration and hence are metallic. In the alkali metal compounds containing the larger clusters, the numbers of electrons transferred from A to the clusters are well defined and in all cases the clusters are 4 electrons short. This value for the electron deficiency seems to be more fixed with the larger clusters than for Mo_6X_6 itself. This observation might be of help in finding the charge distribution in those compounds which contain different kinds of clusters, e.g. $Tl_4Mo_{18}S_{22}$.

One would expect a charge balance close to $(Tl^+)_4(Mo_6S_8)^{2-}(Mo_{12}S_{14})^{2-}$, which corresponds to a slight charge transfer from the M_6S_8 to the $M_{12}S_{14}$ cluster. In fact, the elongation of the Mo_6 unit, which is a good measure for the number of electrons in M–M bonding states, indicates an oxidation state of 2.4 for the Mo atoms in the M_6X_8 cluster, corresponding to a formal charge of 1.6−.

Molecular orbital calculations for the $Mo_9S_{11}^{4-}$ and $Mo_{12}S_{14}^{6-}$ clusters and band structure calculations for the chain $_\infty^1[Mo_3X_3^-]$ lead to a more quantitative understanding of the chemical bonding.[62,63] The use of structural fragmentation to estimate the MCE values in Table 4.1 is rationalized in terms of the fragment molecular orbital approach. The frontier orbitals of the oligomeric clusters are essentially the same as in Mo_6X_8 itself. They are localized on the terminal Mo atoms and are involved in the Mo–X^a-type donor–acceptor interactions between the clusters. The inserted Mo_3X_3 fragments would have 15 metal-centred MOs as isolated units; these interact differently with adjacent Mo_3X_3 fragments. The interaction is weakest with the orbitals oriented in plane and is strongest with the a_1'' fragment orbital which is oriented out of plane and points approximately towards the Mo atoms of adjacent Mo_3X_3 slabs. Whenever the cluster contains an odd number of Mo_3X_3 slabs (the only known example is the Mo_9X_{11} unit) a non-bonding state in the HOMO–LUMO gap will make the electron count ambiguous. For the Mo_9X_{11} cluster the closed-shell electron concentration could be 36 to 38. For the $Mo_{12}X_{14}$ cluster with an even number of slabs it is 50, in agreement with the estimated value in Table 4.1. For the infinitely extending $_\infty^1[Mo_3X_3^-]$ chain the a_2 band originating from the a_1'' fragment orbital corresponds to the 6 equivalent Mo–Mo bonds formed between adjacent Mo_3 units and results in 6 helical bonds along the $_\infty^1[Mo_3]$ columns which make one band (a_2) due to phase locking. The a_2 band shows strong dispersion, as expected. Filling all M–M bonding states up to the non-bonding level leaves the a_2 band half filled, which is the case with 13 electrons per Mo_3X_3 unit. Compounds AMo_3X_3 (A = univalent metal) must be metallic with a very high electron mobility in the chain direction.

The half-filled conduction band in a 1-dimensional metal should lead to a Peierls distortion upon cooling. So far, no structural phase transition has been observed with the chain compounds. The compounds with A = In, Tl stay metallic to the lowest temperatures, whereas with alkali metals a gradual change from metals to semiconductors is observed over a broad temperature range.[70] As the inter-chain coupling is weak in the alkali metal compounds, the smooth changes of the resistivity and magnetic susceptibility could be due to one-dimensional fluctuations of the Peierls order parameter within the chains.

Before leaving the field of oligomeric clusters and chains of face-sharing Mo_6X_8-type clusters, one should mention the interesting solution chemistry with such systems which is just emerging. One approach starts from solid state chemistry. It has been shown that $Li_2Mo_6Se_6$ and $Na_2Mo_6Se_6$ dissolve or form colloidal suspensions with highly polar solvents such as dimethyl sulphoxide or N-methylformamide.[71] By electron microscopy and light scattering it could be demonstrated that some of these solutions contain individual $^1_\infty[Mo_3X_3]$ chains. Together with the possibility of preparing solids with oligometric clusters of defined chain length, a fascinating solution chemistry of inorganic polymers, and perhaps liquid crystals or micelle systems, seems to be opening. Another approach in the field between molecules and extended structures comes from organometallic chemistry and involves the synthesis of condensed cluster compounds in solution. The Chini type carbonyl clusters must be mentioned.[72] Recently, compounds such as $Co_6Se_8(PPh_3)_6$ and $Co_9Se_{11}(PPh_3)_6$ have been prepared and characterized (Co_{12} and Co_{15} species are accessible), and are electron-rich analogues of the Chevrel type clusters Mo_6Se_8 and Mo_9Se_{11}.[73]

4.5.2   Edge-sharing M_6 clusters

We need not leave the element molybdenum when seeking for an alternative way of cluster condensation which involves the edges of the Mo_6 octahedron. A growing class of ternary molybdenum oxides contains chains of *trans*-edge-sharing Mo_6 octahedron with oxygen atoms above all (free) edges as in the Mo_6X_{12} cluster. Five different structure types have been described so far and are summarized in Table 4.2.[26]

Structure 1, $A_xMo_4O_6$, is most suited to the demonstration of the principle (Fig. 4.6). Chains of composition $Mo_2Mo_{4/2}O_2O_{8/2}$ are formed from fragments of the Mo_6O_{12} cluster which omit the O atoms above the edges involved in condensation, similar to the situation with face-sharing Mo_6X_8 clusters. Those Mo and O atoms which are not shared by adjacent clusters in the chain provide the interconnection between the parallel but rotated chains according to $Mo_4O_4^iO_2^{i-a}$. The connecting O atoms exhibit planar coordination by three Mo atoms (as in rutile) and the other O atoms define channels of compressed face-sharing cubes between the chains. These channels are occupied by the cations A. The alkali metal atoms are in the centres of the compressed cubes, Pb and In (disordered) have square pyramidal coordination, and Sn is in the plane of four O atoms. The square planar and nearly square planar coordinations of Sn and In, respectively, together with the fact that these atoms come very close (e.g. $d(In–In) = 286.3$ pm) is indicative for M–M bonding between them in addition to Mo–Mo bonding in the cluster chains. In the alkali metal compounds MCE is 13 per repeat unit Mo_4O_6 (oxidation state 2.75 for Mo). $InMo_4O_6$ is a metal, whereas $NaMo_4O_6$ is a narrow-bandgap semiconductor, or passes through a smooth metal-to-semiconductor transition at low temperatures.

The prototype of compounds with structure 2 is $Sc_{0.75}Zn_{1.25}Mo_4O_7$. The same basic structural units as in 1, namely parallel chains of composition

Table 4.2 *Ternary molybdenum oxides containing chains of edge-sharing Mo_6O_{12} type clusters; the numbers of metal centred electrons, MCE per Mo_4O_6 unit, are derived[2] from (a) the formula assuming O^{2-} ions, (b) from $\sum s_i(Mo-O)$ and (c) from $\sum s_i(Mo-Mo)$*

			MCE		
Structure	Compounds		(a)	(b)	(c)
1	$A_xMo_4O_6$	$A_x = Na$	13.0	12.8	12.7
		In	13.0	13.1	12.3
		$Pb_{0.8}$	13.6	13.3	12.7
		$Sn_{0.9}$	13.8	13.4	13.3
2	$A_{2-x}A'_xMo_4O_7$	$A_{2-x}A'_x = Sc_{0.75}Zn_{1.25}$	$\geqslant 14.7$	14.5	13.6
		$Ti_{0.5}Zn_{1.5}$	$\geqslant 14.5$	14.4	13.4
		$Sc_{0.5}Fe_{1.5}$	$\geqslant 14.4$	14.3	13.5
3	$A_xMo_8O_{11}$	$A_x = Mn_{1.5}$	14.5	14.0	13.4
4	$A_xMo_8O_{10}$	$A_x = Li$	14.5	14.3	13.9
		Zn	15.0	14.7	14.1
5	$A_xMo_{18}O_{32}$	$A_x = Ca_{5.45}$	≈ 14.6	14.6	13.6

Mo_4O_6 are present, and have additional O atoms in X^a position and are interconnected in layers according to $Mo_4O_4^iO_{2/2}^{i-i}O_2^a$. The chains have identical orientations and the interconnecting O atoms enjoy planar coordination by four Mo atoms. The Zn and Sc atoms are in octahedral and tetrahedral voids between the layers. Assuming that these atoms denote 2 and 3 electrons, respectively, the value for MCE in the Mo_4O_6 unit is 14.75 (oxidation state 2.31). The value is not definite, as in all compounds of this type 10 per cent of the Mo atoms are also found in the octahedral voids.

A third type of interconnection of the Mo_4O_6 chains occurs with the structure of **3**, $Mn_{1.5}Mo_8O_{11}$. In **1** and **2**, four O atoms have no bridging function, but in **3** one of these coordinates to an apex Mo atom of a neighbouring chain to form an SF_4-like arrangement of Mo atoms around this O atom, which becomes O^{i-a}. The two O atoms which bridge the chains in **1** and **2** are interconnecting in **3** in the same way. The interconnecting pattern can be formulated as $Mo_4O_3^iO^{i-a}O_{1/2}^{i-i}O^{i-a}$. Anticipating the Mn atoms in the distorted trigonal prisms between the chains to be Mn^{2+} ions, MCE per Mo_4O_6 unit is 14.5 (oxidation state 2.37 for Mo).

In the structure of **4**, parallel Mo_4O_6 chains are linked via O^{i-i} bridges as in **2**, but such layers are interconnected crosswise with respect to the chain directions of adjacent layers. The interconnection is similar to that in $Mo_4O_4^{i-a}O_{2/2}^{i-i}$ and the $i-a$ type bridging O atoms again have the SF_4-type coordination of Mo atoms.

The complicated structure of **5**, $Ca_{5.45}Mo_{18}O_{32}$, shows that M_6X_{12}-type clusters are not the only ones which occur as condensed, infinitely extending

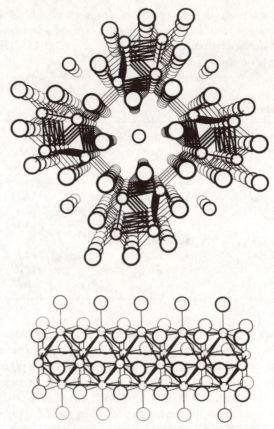

Fig. 4.6. The structure of $NaMo_4O_6$ viewed down the tetragonal c axis and a single chain of trans-edge-sharing Mo_6O_{12}-type clusters extended parallel to the c axis.[74]

units, in **5**, Mo_4O_6 chains are linked via all O atoms to chains of single Mo atoms and to chains of edge-sharing rhomboidal clusters which are the only structural units in $NaMo_2O_4$.

The last structure raises particular problems in establishing the values of MCE for the different units. McCarley has shown in very detailed studies that these values can be derived with satisfying accuracy via the bond length/bond strength formalism using Mo–Mo and Mo–O, as well as A–O, distances.[26, 2] The Mo–Mo distances are analyzed with the Pauling formula, $d = 261.4 - 60\log(s)$, which is checked with those compounds where the counting of electrons on the Mo atoms seems unambiguous; these cases yield $s = 0.9$ to 1.0 per valence electron on Mo. The bond length/bond strength relation $s(Mo–O) = [d(Mo–O)/188.2]^{-6}$ is used to establish the oxidation states of the Mo atoms as $\sum s_i(Mo–O)$ for all surrounding O atoms and $MCE = 6 - \sum s_i(Mo–O)$ can be derived. This bond order summation yields

$Mo_{18}O_{32}^{10.6-} = (Mo_4O_6^{2.6-})_2(Mo_2O_{3.5}^{0.24-})_4(MoO_3^{2.2-})_2$ for compound **5**, which comes very near the charge of the cations $(Ca^{2+})_{5.45}$. The value MCE = 14.6 for the Mo_4O_6 unit in **3** is reasonable.

The electron counts for the different compounds are summarized in Table 4.2. The agreement between the values for MCE derived in different ways is convincing, although MCE derived from the Mo–Mo bonds is rather too small with perhaps the exception of $NaMo_4O_6$. These deviations might be real, indicating the occupation of Mo–Mo antibonding states at high electron counts associated with chain distortions. Structures **2** and **4** are significantly distorted in a rather obvious way. In **2** the octahedra are slightly tilted about the shared edges, resulting in alternatively long and short distances between the apex atoms of the Mo_6 octahedra. In **4** an even more pronounced pairwise coupling of apex atoms from one octahedron to its adjacent cluster is found, and is accompanied by an opening of the corresponding shared edge. Of course, the interconnection pattern and the distribution of cations around the chains determine the kind of distortion in both cases, but obviously the distortion of the chains leads to an electron stabilization at high electron counts. This view is substantiated by the results of extended Hückel calculations, which have been performed for the undistorted Mo_4O_6 chain and the chain which is distorted as in **2**.[66] As the dispersion of all bands is small in directions that are orthogonal to the chain axis, the calculations have been performed for the one-dimensional system of the chain only. For the undistorted chain the decomposition of the calculated density of states curve into crystal orbital overlap population curves (COOP), i.e. into bonding and antibonding contributions of the different kinds of Mo–Mo contacts as a function of MCE, shows that the Mo–Mo bonds along the shared octahedra edges are strongest and the bonds between apex atoms weakest. The electron count MCE = 13 per Mo_4O_6 unit is optimal since all bands up to the gap between the bonding and antibonding states are filled. An increase of MCE to 15, though, would not change the net Mo–Mo bonding drastically. The Mo–Mo bonds along the shared edged would even be slightly strengthened at the expense of weakened bonds in the chain direction. At that high electron count, energy can be gained by the distortion of the chain making one strong bond between apex atoms out of two weak bonds. Bonding between these atoms is mainly based on the interaction of their $d_{x^2-y^2}$ orbitals. The bands corresponding to combinations of orbitals which are symmetric or antisymmetric with respect to the mirror plane through the bases of the octahedra are crossing near the Fermi level in the case of the undistorted chain. Tilting of the octahedra removes the mirror plane, the band crossing is avoided and a gap is opened which leads to a lower energy for the occupied band in the case of the distorted chain.

The oxomolybdates discussed so far contained infinite chains of edge sharing Mo_6O_{12}-type clusters. Quite recently, the first compound with oligomeric units was synthesized.[75] $In_{11}Mo_{40}O_{62}$ contains clusters which

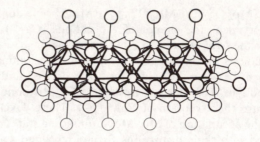

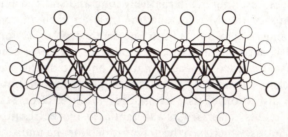

Fig. 4.7. Oligomeric clusters in the structure of $In_{11}Mo_{40}O_{62}$ composed of 4 and 5 Mo_6 octahedra which are condensed via opposite edges and surrounded by O atoms as in the Mo_6X_{12}-type cluster. All Mo atoms are coordinated by O atoms of neighbouring clusters in X^a position.[75]

are formed from four and five Mo_6 units, respectively. As Fig. 4.7 shows, all free edges of the clusters are bridged by O atoms and most of these O atoms interconnect neighbouring clusters according to $Mo_{18}O_{14}^i O_{4/2}^{i-i} O_{24/2}^{i-a}$ and $Mo_{22}O_{18}^i O_{4/2}^{i-i} O_{28/2}^{i-a}$. The i–a type interaction leads to a coordination of the apex Mo atoms as in $A_xMo_4O_6$ (**1**). In fact, slabs of the $InMo_4O_6$ structure which extend over a length of four and five Mo_6 octahedra, respectively, are stacked in an alternating sequence. Due to a relative shift of the slabs, the Mo atoms at the heads and tails of the clusters are coordinated by O atoms in X^i and X^a positions exactly as they are in an isolated $Mo_6X_{12}^i X_6^a$ cluster. As the Mo_4O_6 chain is cut into pieces of 4 and 5 octahedra, respectively, the same happens to the chain of In atoms, which is cut into pieces of 5 and 6 atoms that occupy the channels between the $(Mo_6)_4$ and $(Mo_6)_5$ units. The characteristic coordination of the In atoms by both O and In atoms shows that $In_{11}Mo_{40}O_{62}$ is metal-rich in two respects. Besides extended clusters with strong Mo–Mo bonds, approximately linear In clusters with strong In–In bonds exist in this compound. These bonds are evident from the fact that the central atom in the In_5 cluster lies exactly in a plane with four O atoms, and two In atoms at the shortest known distance of 262 pm are situated above and below the plane. The bond-order sum $\sum s_i(\text{In–O})$ for the central In atom is exactly 1.0. It is therefore an In^+ ion which uses its

2 valence electrons in bonds to 2 In^+ neighbours. The terminal In atoms show approximately tetrahedral coordination by three O and one In atom, corresponding to In^{2+} which again is M–M bonded to the neighbouring In atom at a distance of 266 pm. In the compound $In_{11}Mo_{40}O_{62}$ we therefore find In_5^{7+} and In_6^{8+} cations whose charges are counterbalanced by the $Mo_{18}O_{28}^{7-}$ and $Mo_{22}O_{34}^{8-}$ anions, respectively. The In–In bonds found in these polycations make In–In bonding in the compound $InMo_4O$ very plausible. The metallic conductor $InMo_4O_6$ (possibly also $SnMo_4O_6$) offers a fascinating system of two distinct pathways for conductivity via Mo–Mo and In–In bonding.

The formal description of $In_{11}Mo_{40}O_{62}$ as $In_{11}^{15+}(Mo_{40}O_{62})^{15-}$, derived from In–O bonding, is supported by the results of the summation over all Mo–O bonds which leads to another interesting finding. The sums $\sum s_i(\text{Mo–O})$ are 48.4 and 59.2 for the $Mo_{18}O_{28}^{7-}$ and the $Mo_{22}O_{34}^{8-}$ cluster, respectively. The difference, 10.80, which corresponds to the bond order sum for a $Mo_4O_6^-$ fragment, is nearly the same as that calculated from the structure of $InMo_4O_6$ itself (4×6—13.1; see Table 4.2).

Obviously a fragmentation of the extended edge-sharing M_6X_{12}-type clusters is possible, corresponding to the procedure with extended face-sharing M_6X_8-type clusters, and yields an estimate for optimal MCE values. For this purpose the M_6X_{12} cluster (closed shell configuration with MCE = 16) is cut into M_4X_7 and M_2X_5 fragments, and M_4X_6 fragments are inserted between them (compare Figs. 4.6 and 4.7). Now, the electron count for the M_4X_6 fragments presents a problem as it ranges from MCE = 13 (undistorted) to 15 (distorted chain). The distortion of the $Mo_{18}O_{28}^{7-}$ cluster exhibits the same kind of pairwise bonding between apex atoms as is found in the infinite chain compounds **4**, justifying a high count of MCE $\approx$ 14.5 for the inserted Mo_4O_6 fragments in this cluster, whereas the additional Mo_4O_6 fragment in the $Mo_{22}O_{34}^{8-}$ ion shows no or only weak bonding between the apex atoms, consistent with MCE = 13. The electron counts for the $Mo_{18}O_{28}^{7-}$ and $Mo_{22}O_{34}^{8-}$ ions estimated via such a fragmentation, 59.5 and 72.5, respectively, are nearly identical to those derived directly from the formula $In_{11}Mo_{40}O_{62} = In_5^{7+}(Mo_{18}O_{28})^{7-}In_6^{8+}(Mo_{22}O_{34})^{8-}$. Obviously, in this compound the electron counts for both the In atoms and the clusters of $n = 4$ and 5 condensed octahedra are ideal. Without changing the distortion pattern, slabs of general compositions

$$(In_{n+1})^{(n+3)+}(Mo_{4n+2}O_{6n+4})^{(n+3)-}$$

would become increasingly electron deficient in the Mo clusters with $n \geqslant 6$.

Our reasoning raises more questions than it answers for the moment. Why is the structure of $In_{11}Mo_{40}O_{62}$ composed of alternating slabs with $n = 4$ and 5 more stable than the 'pure' compounds with only 4 ($In_5Mo_{18}O_{28}$) and 5 ($In_6Mo_{22}O_{34}$) condensed octahedra? Is it possible to increase n by partial substitution of In, for example by Sn? Which parameters actually determine the cluster size: the number of electrons donated by the A atoms, the size of the A atoms, the Mo/O ratio? All aspects seem to be influential; larger clusters than $(Mo_6)_4$ and $(Mo_6)_5$ are observed with Ba and other In oxomolybdates, and 'incommensurate' superstructure reflections in samples of '$InMo_4O_6$' indicate the occurrence of giant clusters of condensed Mo_6 octahedra. The gap between molecular clusters and solid state structures becomes smaller.

4.5.3   Apex-sharing M_6 clusters

Finally, the condensation of M_6 octahedra via apex M atoms has to be discussed. If opposite vertices are involved, a chain of composition $M_{2/2}M_4X_{8/2} = M_5X_4$ with the M_6X_8 cluster, or a chain of composition $M_{2/2}M_4X_{8/2}X_4 = M_5X_8$ with the M_6X_{12} cluster results. Both types of infinite chains are found in actual compounds. Figure 4.8 shows the M_5X_4 chains occurring in a number of compounds which have the Ti_5Te_4-type structure.[76] The range of MCE for this structure is remarkable. It occurs with MCE = 12 (Ti_5Te_4), 13 (Ta_5As_4), 17 (Nb_5Te_4) and up to 18 (Mo_5As_4).

The octahedra are always compressed in the chain direction, e.g. by approximately 17 per cent in the structure of Ti_5Te_4. The 8 octahedral edges involving the apex atoms are 295 pm long, whereas the 4 edges in the octahedron base are 322 pm, slightly shorter than the interchain distances $d(Ti–Ti) = 343$ pm. The elongation of the Ti–Ti distances in the octahedron

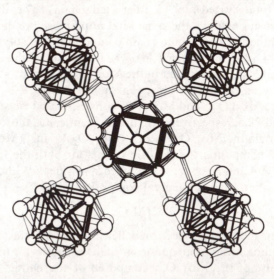

Fig. 4.8. The structure of Ti_5Te_4 viewed down the tetragonal c axis.[76] The structure consists of parallel chains of *trans*-corner sharing, compressed M_6 octahedra which are surrounded by chalcogen atoms as in the M_6X_8-type cluster.

bases can be related to a packing of chains that is comparable to the packing of M_6X_8 clusters in the Chevrel phases. All Te atoms coordinate Ti atoms of a neighbouring chain in X^a positions at distances $d(\text{Ti–Te}) = 277$ pm, which are even slightly shorter than the intrachain distances 277, 282, and 295 pm, respectively.

Strong donor–acceptor interactions between the chains are present and the arguments used for the Chevrel compounds may again explain the cluster distortions as being due to interchain interactions. An alternative explanation of the distortion is based on the assumption that a *bcc* type environment is preferred by the shared M atoms which are coordinated by eight M atoms.[77] In spite of the distortion of the M_6 octahedra, the band structure of the Ti_5Te_4-type compounds can be easily related to the molecular orbital diagram of the M_6X_8 cluster, as shown by Andersen, because the condensation via apex atoms keeps the M_6X_8 cluster complete (in contrast to the condensation via faces or edges). The equivalent 12 M–M bonding orbitals, formed from linear combinations of the four d orbitals of each M atom which are directed approximately along the octahedra edges, are used essentially unchanged for bonding in the M_5X_4 chain. To make this statement more explicit: interactions between the orbitals of the 4 atoms in the octahedron bases lead to 4 bonding and 4 antibonding states (2 centre bonds). The shared atoms use their orbitals to bond to the 4 bases atoms, but also—via the back-loops of these orbitals—to bond to the bases atoms of the adjacent unit in the chain, resulting in 4 bonding, 4 non-bonding and 4 antibonding combinations for these 3 centre bonds. The M_5X_4 chain with regular M_6 octahedra should accommodate 16 electrons per formula unit in bonding states. This simple pattern is, of course, somewhat obscured in the full band structure, but still recognizable.[78] The determination of the optimal MCE for the M_5X_4 chain is ambiguous, as one would expect from the experimentally observed range. The ambiguity comes from the changes of the energies of all cluster states (including the non-bonding ones) due to dipersion, but even more from the interchain interactions which lengthen the bonds in the octahedron bases and thus shift the corresponding bonding states to higher energies. Taking the 'heat of fusion' associated to the condensation of M_6X_8 clusters into a chain as a measure of stability, the M_5X_4-type compounds appear probable at MCE values between 8 and 18, which is an even larger range than is experimentally found.[79, 80] One-dimensional band structure calculations for the (distorted) M_5X_4 chains lead to small gaps (or regions of low density of state) for 12/13 and 17/18 electrons per formula unit, in good agreement with the observed phase stabilities.

So far, oligomeric clusters of M_6X_8 units condensed via the octahedron apices are not known and there is no straightforward way to search for them as the appropriate electron count cannot be predicted. One has to await the fortunate experimentalist.

It has to be mentioned that chains of M_6X_{12}-type clusters condensed via

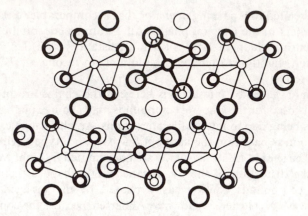

Fig. 4.9. Structure of the low-temperature modification of TiO projected along the short c axis of the monoclinic unit cell.[81] Parallel chains of *trans*-corner sharing M_6X_{12}-type clusters result from the special vacancy ordering. The Ti and O atoms lie in $z = 0$ and 0.5.

apex atoms are also known. In Fig. 4.9 we look down such chains in the structure of the low temperature form of TiO. The O atoms are shared in different ways between neighbouring chains as X^{i-i} and X^{i-a} type atoms which gives the Ti atoms in the octahedron bases the characteristic $X^a + 4 X^i$ coordination. The structure of TiO is closely related to that of NbO, where all cluster atoms are involved in condensation according to the formula $Nb_{6/2}O_{12/4}$. Clearly, both structures are defect rocksalt structures, with 1/6 and 1/4 of the atoms of both sublattices missing in an ordered way in TiO and NbO, respectively. The TiO structure is intermediate between the rocksalt and NbO structures.

It was not until recently that the specific vacancy ordering in TiO and NbO was taken into account as an essential feature of chemical bonding for band structure calculations,[50, 82] nicely confirming the earlier interpretation of the structures in terms of condensed clusters.[7, 16] The energy levels of the isolated M_6X_{12} cluster are preserved in the density of states curves decomposed into different projections according to the orbital symmetry characteristics for TiO and NbO. Energetically, the NbO structure is favoured at an electron count of 3 per M atoms. The TiO structure becomes more stable near 2 electrons per M atom and loses against the rocksalt structure when a d^1 system is approached. Qualitatively speaking, at low values of MCE the clusters become unstable and M–M bonding is then assisted by additional M–X bonding involving interstitial atoms. A delicate balance between these two kinds of bonding is reached with TiO, which crystallizes with empty clusters at low temperatures but with filled clusters in a random rocksalt structure at high temperatures. The broad range of homogeneity of TiO, in contrast to NbO, is another indication of this balance.

4.6 VARIATIONS

The above discussion of condensed cluster systems has been greatly simplified in so far as it was restricted for the sake of clarity to only linear systems formed from M_6X_8- and M_6X_{12}-type clusters. In discussing more complicated stuctures of metal-rich compounds between d metals and p elements, the range of applicability of the condensed cluster concept becomes obvious but alternative viewpoints which help to describe and interpret these structures can also be introduced. The chain of condensed M_6X_8 clusters in the Ti_5Te_4-type compounds is a good starting candidate to explain general features such as coordination, the degree of condensation and the development of network structures.

The characteristic packing of the Ti_5Te_4 chains has been described as giving the M atoms of the octahedron bases a $X^a + 4X^i$ type coordination, as in Chevrel phases $A_xM_6X_8$. As in these phases, A atoms can also be 'intercalated' in the M_5X_4 structure and these atoms act as electron donors. The compound $Nb_5Si_4Cu_4$ is a good illustration.[83] The chain has the very low value MCE = 9. The stability of the chain is derived from the 4 electrons donated by the Cu atoms between the chains (anticipating Cu^+), making them isoelectronic with the chains in, for example, Nb_5Sb_4. Compared to the situation with the Chevrel phases, our knowledge of the intercalation chemistry with M_5X_4 compounds is extremely fragmentary. It has been pointed out, though, that the structures of V_3As_2 ($= V_5As_4 \cdot V$) and Nb_7P_4 ($= Nb_5P_4 \cdot Nb_2$) are derived by simply inserting metal atoms into the channels which exist between the chains in the Ti_5Te_4-type structure.

Arguing in terms of the condensed cluster concept, V_3As_2 is a filled V_5As_4, which does not exist as such (Fig. 4.10(b)).[84] One can take a second point of view and examine the coordination of the non-metal in V_3As_2. Obviously the structure consists of a network of trigonal prisms, AsV_6, which are seen projected partially down their quasi-trigonal axes and partially rotated by 90° in Fig. 4.10(a). The orientation of the prisms influences the M–M distances in the chains of (distorted) condensed octahedra. They are 268/297 and 279/318 pm, respectively, for the two kinds of chains in V_3As_2. Clearly, M–M bonding is subordinated to M–X bonding. On the other hand the rather unusual coordination of the X atoms in the unfilled Ti_5Te_4 structure itself might be taken as evidence for the importance of M–M bonding or the M_6X_8-type cluster configuration in this structure, if it is not interpreted in terms of necessary 'void' space for the non-bonding X–X interactions.[13]

The (capped) trigonal prismatic coordination occurs with metal-rich chalcogenides, pnictides, etc. whose structures have been extensively analysed in terms of such interconnected polyhedra.[6,9] The analysis of all homonuclear contacts between the metal atoms (which are in the majority) seems to be of great importance for an understanding of these structures and becomes more important with increasing M/X ratio.

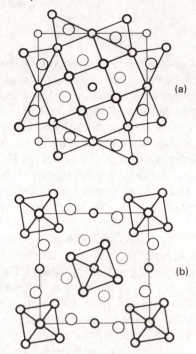

Fig. 4.10. Structure of V_3As_2 projected down the tetragonal c axis. The atoms (small circles V) lie in $z = 0$ and 0.5. (a) The framework of trigonal prisms V_6As is outlined. (b) The relationship with the Ti_5Te_4 structure $(V_5As_4 \cdot V)$ is enhanced by outlining the chains of trans-corner-sharing (compressed) M_6X_8 type clusters.[84]

The structures of the metal-rich sulphides $Nb_{14}S_5$, $Nb_{21}S_8$, and Ti_8S_3 are shown in Fig. 4.11. They are rather complicated, but the lines drawn in Fig. 4.11 reveal a comparatively simple pattern. In the centre of the unit cell of $Nb_{21}S_8$ the single M_5X_4-type chain is recognized, and this chain is the essential feature in all the structures. Two, three, and four such chains are condensed with S atoms arranged above all free octahedron faces. These units may be isolated as in $Nb_{21}S_8$, or further condensed via apices as in $Nb_{14}S_5$ and Ti_8S_3. The units of two chains condensed as in $Nb_{14}S_5$ are the only structural element of Nb_2Se, and the twin-chains condensed in a parallel mode in Ti_8S_3 occur as the main unit in the structure of Ti_2S (Ta_2P). One can imagine other units formed by condensation of M_5X_4 chains and other ways of arranging these units in a crystal which also provide appropriate coordination for the non-metal atoms. These structures are likely to be the tip of an iceberg.

Obviously, M–M bonding is extremely important in $Nb_{14}S_5$, $Nb_{21}S_8$, and Ti_8S_3, since parts of their structures are purely M–M bonded. These parts look like cuttings from a *bcc* metal.[13] So, we reach a third viewpoint

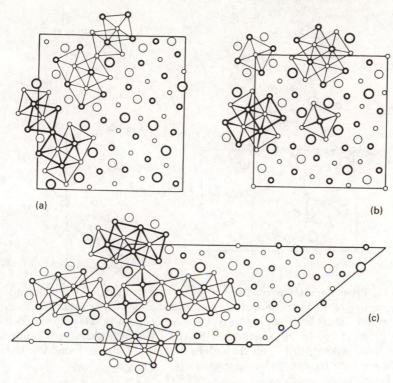

Fig. 4.11. The crystal structures of metal-rich sulphides of Nb and Ti projected down the short axes of the unit cells. The atoms (M atoms represented by small spheres) lie in heights 0 and 0.5. (a) $Nb_{14}S_5$ (orthorhombic, down [010]),[85] (b) $Nb_{21}S_8$ (tetragonal, down [001]),[86] and (c) Ti_8S_3 (monoclinic, down [010]).[87]

for analysing structures of metal-rich compounds. Instead of building up these structures from quasi-molecular units, i.e. clusters, the lattice of the metallic elements are cut into pieces. Non-metal atoms then substitute for missing metal atom neighbours. As the cube, which is the preferred coordination polyhedron around the M atom, is nicely compatible with the trigonal prism, which is the preferred coordination polyhedron of the X atom, many metal-rich borides, silicides, pnictides, and chalcogenides have structures of combined cubes and prisms.

Both procedures, starting from a cluster or starting from the infinite structure of the metallic element, of course, are not contradictory. Figure 4.12 illustrates the two viewpoints once more. The structures of Gd_2Cl_3, Sc_7Cl_{10}, and ZrBr are projected along chains of edge-sharing M_6 octahedra which are surrounded by X atoms above all free faces, as in the isolated M_6X_8 cluster. In the structure of Sc_7Cl_{10}, additional single Sc atoms occupy octahedral voids between Cl atoms and the role they play must be seen in

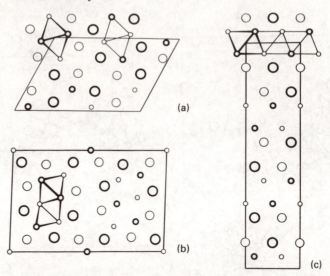

Fig. 4.12. Projections of the structures of some metal-rich halides which illustrate an increasing degree of condensation. Chains of edge-sharing Gd_6 octahedra surrounded by Cl atoms similar as in the M_6X_8 type cluster occur in (a) Gd_2Cl_3 (monoclinic, down [010]).[88] (b) In the structure of Sc_7Cl_{10} two such chains are condensed (monoclinic, down [010]).[89] (c) An infinite number of chains forms the layered structure of ZrBr (rhombohedral, along [11.0]).[90]

the context of the extremely low value for MCE in the cluster unit. Obviously these 'intercalated' Sc atoms donate electrons to the cluster chains. Details of the structures will not be discussed here. Figure 4.12 just demonstrates the stepwise increase in the degree of condensation, starting from the single chain of octahedra and ending with layers of edge-sharing octahedra formally resulting from a condensation of an infinite number of chains. It is a question of taste whether one discusses the structure of ZrBr in an alternative way as a sliced metal (zirconium) which has been enclosed between layers of non-metal atoms. The van der Waals bonds between the Br–Zr–Zr–Br sheets make the graphite-like compound ZrBr a two-dimensional metal.

The brittle metallic compound Hf_2S has a closely related structure to ZrBr.[91] In the sulphide, twin-layers of close-packed Hf atoms are stacked in such a way as to yield trigonal prismatic voids which are occupied by the S atoms. The metal atoms, which are purely M–M bonded in the twin-layers, have M–M distances (3×306 and 6×337 pm) comparable to those in elemental (hexagonal) hafnium (313 and 319 pm). Hf_2S leads back to Ba_2N, and again we may take different viewpoints. What has been discussed in terms of condensed filled clusters with Ba_2N has changed to condensed empty clusters of the M_6X_8 type with Hf_2S. We need to link both views, but before moving to the large group of compounds containing

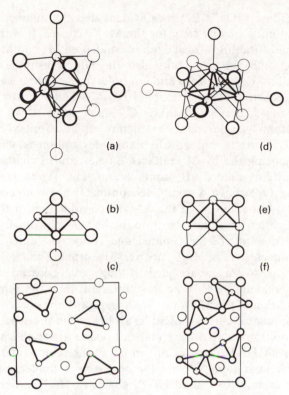

Fig. 4.13. Fragment clusters related to the $M_6X_8^iX_6^a$ arrangement. (a) $Mo_4I_{11}^{2-}$ cluster ('arachno' cluster),[92] (b) same cluster without X^a atoms; (c) condensed arachno clusters in the structure of Hf_3P_2 (orthorhombic, down [010]);[93] (d) $Mo_5Cl_{13}^{2-}$ cluster ('nido' cluster);[94] (e) same cluster without X^a atoms; (f) condensed nido clusters in the structure of Rh_5Ge_3 (orthorhombic, down [001]).[95]

filled metal clusters, two other aspects of cluster condensation must be mentioned briefly.

In Section 4.5 the cutting of infinite chains of condensed clusters into finite oligomeric clusters has been discussed. As an alternative, the cutting could be done parallel to the chain direction leading to infinite chains formed from incomplete M_6M_8- or M_6X_{12}-type clusters. Such incomplete clusters are well known as isolated complex ions.

Figure 4.13 shows the ions $Mo_4I_{11}^{2-}$ and $Mo_5Cl_{13}^{2-}$, which correspond to arachno- and nido-structures, respectively, of the $Mo_6X_8^iX_6^a$ cluster. Apart from the experimental fact that these ions have been found with MCE = 15 and 19, respectively, there is some ambiguity concerning the optimal electron count for them, because M–M interactions in the nido-cluster comprise 7 bonding and 4 non-bonding orbitals, and in the arachno-cluster 6 bonding and 2 non-bonding MOs.[92] But it is evident that the number of bonding states per M atom is smaller than for the M_6X_8 cluster. It is therefore not surprising to find structures which contain condensed M_5 and M_4 units with compounds of both the early and also the late transition metals with multivalent anions. Few electrons and few holes in the d band lead to similar structural consequences. The structure of Hf_3P_2—similar cluster chains, but differently arranged, are found in Cr_3C_2—may serve as an example of condensed arachno type clusters. The structure can be described as a network of trigonal Hf_6P prisms, but an alternative description is shown in Fig. 4.13(c) by enhancing the M–M contacts. Single parallel chains formed by condensing butterfly-shaped Hf_4 units occur. The P atoms of the X^{i-i} interconnecting type make a complete non-metal environment, as in the isolated M_4X_7 cluster, and even the X^a-type coordination of the unshared M atoms by P atoms is found. In spite of the very convincing agreement between the geometries of the isolated and condensed units, one should mention that the clusters in Hf_3P_2 are very distorted. The Hf–Hf distances range from 306 to 344 pm and are partially longer than between neighbouring chains. M–M bonding seems to be less important than the achievement of special coordination of the M atoms. With respect to the observed distances the model of condensed clusters might be arbitrary. On the other hand, close structural relationships with other metal-rich phases are established, e.g. with Hf_2P. The 'partial' Hf_5P_4 chain occurs in Hf_3P_2. The compound Hf_5P_4 with the complete chain is unknown, but the next step of condensation is again realized in the units of two fused Hf_5P_4 chains in Hf_2P, which is isotypic with Ta_2P.

Finally, a structure containing M_5 cluster fragments is shown in Fig. 4.13(f). In Rh_5Ge_3 chains of the partial cluster M_5X_8 are further condensed via apex atoms. It is interesting to see an additional X atom take up the position above the base of the nido cluster, which means—when reduced to the situation of the isolated cluster—that the 4 non-bonding states in the nido-cluster are used for bonding this X atom which, in turn, approximately takes the position of the missing M atom.

These structures of metal-rich compounds composed of condensed clusters are similar to SiO_2 and silicate structures in the sense that characteristic building units are condensed to form 0-, 1-, 2-, or 3-dimensionally extending structures. But in contrast to the silicates, which—with very few exceptions—are formed from apex-sharing SiO_4 tetrahedra, the field of condensed cluster structures is more complicated for a number of reasons. A wide variety of different clusters can be used. They can be interconnected in several different ways, e.g. via apices, edges, or faces, as we have seen. Due to intercluster interactions the distances in the clusters can vary greatly, and, last but not least, different clusters may occur in one compound. The recently determined structure of $Zr_{14}P_9$ (Fig. 4.14) affords a nice example of the intergrowth of different structural units, as well as a good candidate to illustrate once more

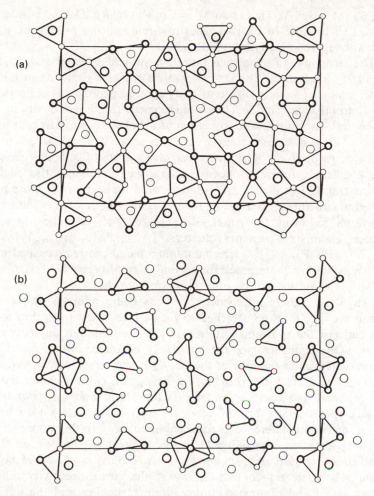

Fig. 4.14. Alternate description of the structure of $Zr_{14}P_9$.[96] The orthorhombic cell is projected down [001]; all atoms (Zr small circles) are in $z = 0$ and 0.5. (a) Framework of trigonal Zr_6P prisms, and (b) cluster structures as in Ti_5Te_4, Hf_3P_2, and θ-Ni_2Si.

the two viewpoints: 'network of trigonal prisms around the X atoms' and 'chains of condensed closo- and arachno- clusters of the M_6X_8 type'. Both seem to be important for an understanding of the structure.

The condensed cluster systems discussed so far are all based on the M_6X_8- and M_6X_{12}-type clusters or parts thereof. Other types, e.g clusters with tetrahedral M_4 cores, are also found as condensed systems. The characteristic structural units in $NdCo_4B_4$ are parallel chains of edge-sharing Co_4 tetrahedra. As described in Chapter 3, borides of the general composition

MM'_4B_4 (M = Ln, Y, Th, U and M' = Co, Rh, Ir, Ru, Os) have been much investigated because of their unusual low temperature properties, e.g. the re-entrant superconductivity in $ErRh_4B_4$.[23]

The structures of metal-rich tantalum sulphides, Ta_2S and Ta_6S, are particularly interesting. The characteristic structural elements in both are parallel chains of slightly distorted body-centred pentagonal antiprisms or, in an alternative description, chains of interpenetrating icosahedra in which each central Ta atom is also an apical atom of the next icosahedron. The distances between the central atoms are very short, 279 and 264 pm in Ta_2S and Ta_6S, respectively. The structures of Ta_2S and Ta_6S link seemingly entirely different fields of chemistry. On the one hand, the chains of interpenetrating icosahedra are 1-dimensional realizations of close-packed tetrahedral structures and therefore related to the vast number of intermetallic compounds following the Frank–Kasper concept.[98] On the other hand, complex chemistry presents clusters, $[Au_{13}(PMe_2Ph)_{10}Cl_2]^{3+}$ and $[Au_{13}Ag_{12}(Ph_3P)_{12}Cl_6]^{3+}$, which contain a metal-centred icosahedron and an extended unit of 3 interpenetrating icosahedra, respectively, thus representing the first steps in the formation of the infinite chain.

In Ta_2S the rods of interpenetrating icosahedra adopt a simple square packing with the S atoms attached to the surface of the rods, thus forming voids between the rods and resulting in low coordination numbers for the S atoms. One could imagine the possibility of 'intercalating' metal atoms into the voids, as in the case of the M_5X_4-based structures, V_3As_2, etc. In the structure of Ta_6S the rods have a hexagonal close-packing, the voids have vanished and the S atom has its more common (capped) trigonal prismatic coordination. The difference in sulphur coordination is combined with an interesting difference in chemical bonding.[77] In the more metal-rich Ta_6S the p–dπ interaction will be large due to the larger proportion of metal d band states per sulphur atom and the broad d band due to M–M bonding. As the relative metal content increases, the electronegativity difference between metal and non-metal atoms becomes smaller and the non-metal becomes 'metallic' by contributing to the conduction band. This 'metallization' of the sulphur atoms can express itself in a fascinating way.[77] In the structures of Zr_9S_2, α- and β-V_3S, the metal and sulphur atoms together form polyhedra of coordination number cn = 14 around central metal atoms. Units consisting of 22 atoms from 2 such interpenetrating Frank–Kasper polyhedra with cn = 14 are characteristic of all three structures. These compounds are interesting borderline cases to intermetallic compounds.

The structures above have been discussed in a rather qualitative way. The discussion corresponds to the state of the art in this broad research area, where numerous compounds have been characterized by their crystal structures and some physical properties are known but where a quantitative treatise of chemical bonding is missing, partly due to the fairly complicated structures. Perhaps the concept of condensed clusters, within its indicated limits, provides some basis for a quantitative approach and may stimulate further chemical work aimed, for example, at the synthesis of (ternary) compounds which contain fractions of the known, infinitely extending units of condensed clusters.

4.7 CLUSTERS WITH INTERSTITIAL ATOMS

4.7.1 Transition metal clusters

At the beginning of this text, weakly metal–metal bonded clusters of the electron-poor alkali metals were discussed. Stable configurations are found in the gas phase as well as in the solid alkali metal suboxides when the cavities in the clusters are filled with non-metal atoms. The cluster stability is due to strong metal–non-metal bonding assisted by some metal–metal bonding. What seems to be rather unique at first sight with clusters of the alkali metals also occurs widely with transition metals.

The first example of the introduction of a non-metal atom into the centre of a preformed metal cluster was found in the reaction of Nb_6I_{11} with hydrogen to yield HNb_6I_{11}. The uptake of hydrogen is accompanied by a change from Curie–Weiss paramagnetism to nonmagnetic behaviour at low temperatures.[99] The hydrogen is strongly bonded and cannot be released by heating the compound *in vacuo*, because of decomposition into gaseous NbI_4 and a residue of Nb/NbH_x. If the decomposition reaction is suppressed by enclosing the sample in an ampoule of niobium which is only permeable to hydrogen, desorption is possible and the compound Nb_6I_{11} with empty clusters is again formed.

The uptake of hydrogen by Nb_6I_{11} is closely related to the electron deficiency in the cluster which, as mentioned already, has only 19 instead of 24 electrons to fill the M–M bonding states. Formally the H atom is adding an electron to the cluster, although detailed calculations show considerable charge transfer to occur from the metal atoms to the interstitial atom.[100] The apparent contradiction is easily solved.[78] The interactions of the H 1s orbital and the M–M bonding orbitals of the cluster are essentially non-bonding except for the a_{1g} orbital which is greatly lowered in energy to a bonding hydrogen-like state; the corresponding antibonding state lies above the Fermi level. The number of bonding orbitals therefore does not change for the cluster, but one more electron enters a bonding state just below the Fermi level. It is not the occupation of this bonding state that stabilizes the cluster, but rather the low energy of the hydrogen-like bonding state. The iodine ligand orbitals do not participate significantly in the bonding inside the octahedron, which behaves rather like a small piece of niobium metal. In fact, the cluster bond model corresponds to the generally accepted description of the bonding of hydrogen in metallic hydrides, e.g. PdH_x and NiH_x. The 'impurity band' of metal–hydrogen bonding states below the conduction band originates from low-lying, filled band states of the metal, and the added electrons enter band holes at the Fermi level.[101]

In recent years a growing number of metal clusters with interstitial atoms have become known, both in the field of organometallic chemistry[102,103] and solid state chemistry. In the following section, a brief outline of developments in the latter field will be given, starting with discrete clusters.

HNb_6I_{11} and $HCsNb_6I_{11}$, which also contain Nb_6I_8 clusters,[104] are the only halides with interstitial non-metal atoms formed by a Group V metal. No such compounds with filled clusters could be prepared with a metal of Group VI (and beyond). Earlier results on the hydrogen absorption by Nb_6Cl_{14} and Mo_6Cl_{12} had to be revised, and it is now known that halide and chalcogenide clusters are always empty with these metals. The situation changes with the Group IV metal, zirconium.

The valence electron concentration of zirconium is, of course, too low to allow the formation of (isolated) Zr_6X_8 type clusters, even with halides. But in spite of the small number of electrons in M–M bonding states (MCE $\leqslant$ 12 per cluster), several different phases containing the Zr_6X_{12} type cluster have been isolated from what was thought to be binary metal-halide reactions. Recent work, however, has shown that the cluster centres in all of these phases are occupied by a non-metal atom (preferably N and C) whose presence seems essential for the stability of the compounds.[105,106] A variety of second period elements (including B and even Be) could be incorporated. Obviously clusters with 14 electrons are preferred, counting all valence electrons of the interstitial atom together with the cluster. Some phases are isostructural with compounds of a Group V metal except for the occupied cluster centre, e.g. $Zr_6Cl_{14}C$ (Nb_6Cl_{14}), $Zr_6Cl_{15}N$ (Ta_6Cl_{15}), or $KZr_6Cl_{15}C$ ($CsNb_6Cl_{15}$). Others, like $Zr_6I_{12}C$, $Zr_6Cl_{13}B$, $K_2Zr_6Cl_{15}B$, or $Ba_4Zr_6Cl_{16}Be$, exhibit new structures.

Extended Hückel calculations performed for the $Zr_6I_{16}C^{4-}$ cluster show that the bonding of the C atom can be discussed along the lines explained already for hydrogen.[106] The carbon 2s (a_{1g}) and 2p (t_{1u}) orbitals strongly interact with cluster orbitals of the same symmetries derived from the d_{z^2} and $d_{xz,yz}$ states of the metal atoms. The new states have essentially C–Zr σ and π bonding character, respectively. Again the antibonding combinations are raised above the Fermi level, keeping the total number of bonding states the same as for the empty cluster, M–M bonding states near the Fermi level accommodate the additional 4 electrons in the carbide cluster. The stabilization of the cluster is, of course, not due to filling these weakly bonding states, but due to the formation of strong polar bonds between the central atom and the surrounding ones. Considerable charge transfer from the rather electropositive zirconium to the carbon atom provides the main quantitative difference between the MO diagram for the zirconium cluster and the known interaction diagrams for organometallic clusters containing C atoms in octahedral units of platinum metals.[107,108]

So far, only interstitial atoms in isolated clusters have been discussed, but structures with condensed clusters may also serve as hosts for interstitial atoms. Reversible hydrogen absorption has been demonstrated with some of the metal-rich phases mentioned earlier, e.g. $Nb_{21}S_8$, Ta_6S, V_3S, etc.[109] The hydrogenation products of the zirconium halides, ZrCl and ZrBr, are especially interesting. Both host structures consist of close-packed layer slabs X–Zr–Zr–X with a stacking sequence ABCA; the structure was described earlier as being derived from edge-condensed M_6X_8-type clusters. Within the same space groups, $R\bar{3}m$, the structures of ZrCl and ZrBr differ only in the stacking sequence of such composite slabs. With hydrogen the phases $ZrXH_{0.5}$ and ZrXH are formed.[110] The structure of $ZrBrH_{0.5}$ has monoclinic symmetry, but the layer sequence ABCA of the host structure is retained with minor distortions. It is assumed that the hydrogen atoms occupy half of the tetrahedral interstices between the metal layers in an ordered manner (zig-zag chains) which causes the distortions. In contrast, the heavy atom layering in ZrBrH has changed from ABCA into ABAB, which corresponds to a two-dimensional edge condensation of M_6X_{12}-type clusters. A neutron diffraction study of ZrBrD reveals a pair-wise occupation of all octahedral or, better, trigonal antiprismatic voids.[111] The D atoms have non-bonding distances $d(D–D) = 220.4$ pm and are nearly in the centre of opposite triangular faces (Fig. 4.15). This kind of pair-wise occupation of the octahedral voids has also been demonstrated for the layer compound $TbBrD_2$ (see below). It is not a special electron concentration, as in isolated clusters, which moves the X atoms from positions above faces to edges of the metal octahedra, but rather the electrostatic repulsion between the hydridic hydrogen inside the octahedra and the halogen atoms outside.

Atoms other than hydrogen may occupy interstices in condensed cluster structures. The compound Ta_2C_2C is nicely related to the preceding example. Carbon atoms take the central 'octahedral' positions in S–Ta–Ta–S slabs with the layer sequence ABAB, and one or three such slabs form the repeat unit in the 1s and 3s modifications, respectively.[112] Carbon atoms are located in the octahedral voids of the Hf_2S-like structure of Ti_2SC and similarly in chains of face-sharing M_6X_8 clusters in, for example, Mo_5Si_3C or V_5P_3C, which have a filled Mn_5Si_3 structure.[113] These few examples again look like the tip of an iceberg.

4.7.2 Group III and rare earth compounds

The stabilizing effect of the interstitial atoms in these structures is essentially the same as explained before for compounds with discrete metal clusters, namely the formation of strong polar bonds between the metal and the interstitial non-metal atoms. It is admitted that the formulation of a compound $Zr_6I_{14}C$ as $(Zr^{4+})_6(I^-)_{14}C^{4-}(e^-)_6$ is exaggerated. According to the population analysis, the carbon atom has a charge of only $1.8-$.[106] But

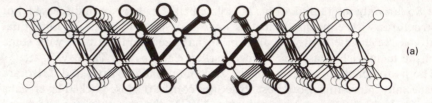

(a)

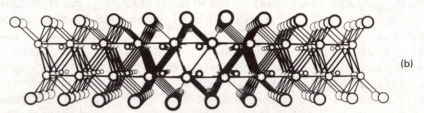

(b)

Fig. 4.15. (a) One slab of close-packed Br–Zr–Zr–Br layers in the structure of ZrBr.[114] The positions of atoms are ABCA related to the M_6X_8 cluster. (b) A corresponding slab in the structure of ZrBrD.[108] The positions of the heavy atoms are ABAB, related to the M_6X_{12} cluster and two D atoms occupy each trigonal antiprismatic void. They lie approximately in the centres of Zr_3 triangles. The experimentally determined partial occupation of tetrahedral voids is omitted from the drawing.

such descriptions become more realistic for the electron-poor electropositive metals of Group III, including the lanthanides, and serve as a valuable guideline for understanding the structure and properties of the cluster compounds of these Group III metals. A variety of different metal-rich halides MX_n ($n < 2$) of these Group III metals have been prepared, many of them only in low yields. They all contain octahedral M_6 units, which are isolated for $n = 1.71$, condensed via edges into chains for $n = 1.60$, 1.50, and 1.25, into twin-chains for $n = 1.43$ and 1.17, and into layers for $n = 1.00$.[16, 20] Free faces or edges of the octahedra are capped by the X atoms, as they are in the M_6X_8- and M_6X_{12}-type cluster. Most of these phases are not— as originally thought—binary compounds, but contain interstitial atoms which stabilize the structure as already discussed for zirconium compounds. There is a good criterion, beyond chemical analysis, for judging the necessity of impurity atoms for compound stability: usually the yield of a reaction increases with the experience of the experimentalist; it should not decrease!

The complete range from purely M–M bonded cluster structures, via structures with strong heteropolar bonding between interstitial atoms and weakly M–M bonded surrounding metal atoms, to purely salt-like structures with no electrons left for M–M bonding, can be found with halide compounds of the rare earth metals.

Discrete M_6X_{12} clusters only occur with compounds $M_7X_{12}X'$, for example, $Sc_7Cl_{12}N$[115] and the briefly mentioned isotypic Ln iodides,[16] which are formed in the presence of carbon. In $NSc_6Cl_{12} \cdot Sc$, single Sc atoms enter octahedral voids in the $Zr_6I_{12}C$-type structure. If we assume full charge transfer from the single metal atom to the cluster—as with the Chevrel phases—we obtain a 14-electron $Sc_6Cl_{12}N$ unit analogous to many zirconium halide clusters. According to $(Sc^{3+})_6(Cl^-)_{12}Sc^{3+}N^{3-}(e^-)_6$, the Sc_6 octahedron is mainly held together by strong ionic Sc–N bonds assisted by weak Sc–Sc bonds.

The extreme situations, namely structures with empty metal clusters on the one hand and salt-like compounds with only heteroatom bonds on the other hand, are illustrated by the compound pair Gd_2Cl_3 and Gd_2Cl_3N. Gd_2Cl_3 forms black needles with a metallic lustre but semiconducting properties ($E_g = 0.85$ eV). The structure (Fig. 4.12) contains parallel chains of *trans*-edge-sharing Gd_6 octahedra which can be derived by condensation of M_6X_8-type clusters. The shared edges are short ($d(Gd-Gd) = 337$ pm) compared to elemental gadolinium (357/363 pm), the octahedron base is elongated (390 pm) in the chain direction, and the Gd–Gd distances to the apex atoms are intermediate in lengths (373 and 378 pm). The octahedra are empty, and M–M bonding according to $(Gd^{3+})_2(Cl^-)_3(e^-)_3$, with a rather low level of MCE = 1.5 per Gd atom, stabilizes the cluster structure. Gadolinium acts as a d metal in Gd_2Cl_3 according to its photoelectron spectra, with the narrow 4f band which is typical for a $4f^7$ configuration, as in the Gd^{3+} ion. The 4f band lies 10 eV below the Fermi level.

Several properties of Gd_2Cl_3, for example, mechanical behaviour, optical reflectivity, static and dynamic magnetic properties, are quasi-one-dimensional in character. Band-structure calculations have been performed for an isolated cluster chain and lead to a chemical interpretation of the observed (and calculated) band gap in Gd_2Cl_3.[116] Three d bands are split off to low energies from the block of Gd d bands. In a rather simplified view, the lowest lying band corresponds to a σ bond between the atoms belonging to the shared edge of the Gd_6 octahedron; it involves the d_{z^2} orbitals of these atoms. The other two bands represent π bonds between the same atoms, one in the plane of the octahedron base (d_{xy}) and the other in the plane perpendicular to the chain direction (d_{xz}). The mixing of (especially) the first mentioned π bonding band with the $d_{x^2-y^2}$ band leads to $x^2 - y^2$ character near the zone centre and results in bonding interactions between the Gd atoms along the chain direction. The observed gap opens to avoid crossing of these bands. In terms of the above band structure, the original interpretation of the structure of Gd_2Cl_3 with 'ladders' of Gd_2 pairs seems quite realistic.

The structure of Gd_2Cl_3N can be formally related to that of Gd_2Cl_3. If two N atoms are inserted into each Gd_6 octahedron in the chains of Gd_2Cl_3

($=Gd_2Gd_{4/2}Cl_6$), electrons will be transferred to the interstitial atoms leading to $(Gd^{3+})_2(Cl^-)_3N^{3-}$. Due to electrostatic repulsion between the N^{3-} ions, the Gd_6 octahedron is drastically deformed resulting in two edge-sharing N-centred Gd_4 tetrahedra. In fact, the structure of Gd_2Cl_3N is made up of parallel chains of *trans*-edge-shared tetrahedra, $^1_\infty[Gd_{4/2}N]$. Of course, no electrons for M–M bonding are left and, as one could expect, Gd_2Cl_3N is a colourless ionic compound. The salt-like bonding also manifests itself structurally. The arrangement of anions is such that each Gd atom is surrounded by a (distorted) tricapped trigonal prism of 2 N and 7 Cl atoms, which is the characteristic coordination of Gd^{3+}, e.g. in the ionic compound $GdCl_3$. Essentially the same coordination is found for the colourless Gd_3Cl_6N, which contains Gd_6N_2 units (two Gd_4N tetrahedra) as fragments of the $^1_\infty[Gd_{4/2}N]$ chain.

The band structure calculation for a single chain of Gd_2Cl_3N confirms the picture of ionic bonding.[116] A large gap of 4.1 eV separates the valence band of occupied anion states from the conduction band of empty Gd d states, clearly showing that no M–M bonds exist. There is, however, a substantial metal d orbital contribution, especially to the bottom of the valence band, which gives one a feeling for the simplification of the formulation as $(Gd^{3+})_2(Cl^-)_3N^{3-}$. A Mulliken population analysis attributes the charges $(Gd^{2+})_2(Cl^{0.7-})_3N^{1.8-}$. Both direct and indirect interactions between Gd d orbitals are made responsible for the short Gd–Gd distances. But following the discussion with alkali metal suboxides, these short distances are easily explained as being due to the attraction between the highly charged anions and cations and the constriction of the coordination polyhedron around the nitride ions. The shared edges of the Gd_4N tetrahedra are 335 pm long and the other edges range from 383 to 397 pm. In spite of the nearly identical Gd–Gd distances in Gd^2Cl_3 and Gd_2Cl_3N, their shortness alone does not indicate M–M bonding and bond length/bond strength calculations, which only focus on the M–M bond lengths, must lead to wrong results. The metal–non-metal bonds are a better indicator for M–M bonding. In the case of Gd_2Cl_3N the sum of all Gd–Cl and Gd–N bond orders for the Gd atom is 3.1, clearly showing that, in spite of the short distances, no Gd–Gd bonding is present. By contrast, the bond orders for the Gd–Cl bonds in Gd_2Cl_3 sum up to 1.6 and 1.7, respectively, for the two independent Gd atoms, and therefore valency is left for M–M bonding.

The earlier mentioned 'insertion' of N atoms into the Gd_2Cl_3 structure to yield Gd_2Cl_3N does not, of course, occur as a topochemical reaction. Such a reversible reaction is, however possible with hydrogen and the layered halides $LnXH_n$ (X = Cl, Br, I). When reacting, for example, $GdCl_3$ or Gd_2Cl_3 and Gd in sealed tantalum ampoules, small amounts of crystals are formed which are isotypic with ZrCl or ZrBr, depending on the reaction conditions. It transpires that these 'monohalides' of Gd and all other rare earth metals are always stablized by hydrogen, which is introduced into the system via the metals (including the tantalum) or by handling the components in a glove box. If hydrogen is added to the system on purpose, the 'monohalides', which are in fact LnXH, are easily prepared as single-phase products. When these compounds, e.g. GdClH, are heated in a sealed tantalum container *in vacuo*, using the tantalum as a membrane for hydrogen, as with Nb_6I_{11}/HNb_6I_{11}, a partial but incomplete loss of hydrogen is possible. A range of homogeneity, $GdClH_n$ ($0.7 \leqslant n \leqslant 1.0$), exists and a further decrease of the hydrogen content at temperatures above 830°C results in the formation of Gd and $GdCl_3$.[117]

It is interesting that GdClH has the same electron count as the binary compound ZrCl if the formalism adopted for interstitial atoms in discrete clusters is used. But, of course, the hydrogen atom is rather anionic and therefore localizes electrons, and the number of electrons delocalized in M–M bonding states is significantly higher in ZrCl ($= Zr^{4+}Cl^-(e^-)_3$) than in GdClH ($= Gd^{3+}Cl^-H^-e^-$). Yet, GdClH is still a 2-dimensional metal. Further addition of hydrogen to GdClH at 400°C under 1 atm H_2 leads to $GdClH_2$, which is an insulating transparent compound as all the electrons now are withdrawn from M–M bonding states and localized according to $Gd^{3+}Cl^-(H^-)_2$. Hydrogen desorption at higher temperatures recovers the metallic GdClH. According to neutron diffraction studies on $TbClD_{0.8}$ the hydrogen atoms in the phases $LnXH_n$ ($n \leqslant 1.0$) occupy tetrahedral interstices,[118] in contrast to the situation found with ZrBrH. The formation of $LnXH_2$ from LnXH is accompanied by two significant changes to the structure: the c lattice constant increases by 7 per cent, mainly due to larger interlayer Ln–Ln distances, and the sequence of layers within the X–Ln–Ln–X slab changes from ABCA to ABAB, as in ZrBr/ZrBrH. Neutron diffraction experiments on $TbBrD_2$ show that each 'tetrahedral' interstice is occupied by one D atom and each 'octahedral' site by two with a non-bonding distance $d(D–D) = 298$ pm.[119] The metallic hydride halides offer interesting possibilities for different kinds of intercalation chemistry involving the voids between metal atom layers on the one hand and the voids between the non-metal atom layers on the other.

Atoms other than hydrogen can be incorporated between the metal atom layers in the X–Ln–Ln–X slabs, though not in a reversible reaction. Compounds like Gd_2Br_2C and $Gd_2Br_2C_2$ belong to a growing class of carbide halides of the Group III metals and lanthanides, with single C atoms or molecular C_2 units in octahedral interstices. The compounds with known crystal structures are summarized in Table 4.3. Similiar phases with scandium, Sc_2Cl_2C and $Sc_7Cl_{10}C_2$, are also known.[115]

From a purely structural point of view all the compounds in Table 4.3 can be formally described as condensed cluster systems, based on the M_6X_{12}-type clusters. Discrete units formed by condensing two M_6X_{12}-type clusters via an edge of each octahedron are found in the structures of $Gd_{10}Cl_{18}C_4$, $Gd_{10}Cl_{17}C_4$, and $Gd_{10}I_{16}C_4$ (Fig. 4.16). The centres of the octahedra are

Table 4.3. *Carbide halides of gadolinium;*[18] *the number of electrons donated (per C_2 unit) from the framework of heavy atoms is based on the assumption of* M^{3+} *and* X^- *ions*

Compound	Framework	Carbon species	$d(C–C)$ (pm)
$Gd_2Br_2C_2$	$(Gd_2Br_2)^{4+}(e^-)_4$	C_2	127
$Gd_2Cl_2C_2$	$(Gd_2Cl_2)^{4+}(e^-)_4$	C_2	133
$Gd_{10}Cl_{18}C_4$	$(Gd_5Cl_9)^{6+}(e^-)_6$	C_2	147
$Gd_{12}I_{17}C_6$	$(Gd_4I_{5.67})^{6.33+}(e^-)_{6.33}$	C_2	145
$Gd_{10}Cl_{17}C_4$	$(Gd_5Cl_{8.5})^{6.5+}(e^-)_{6.5}$	C_2	147
$Gd_{10}I_{16}C_4$	$(Gd_5I_8)^{7+}(e^-)_7$	C_2	143
Gd_2Br_2C	$(Gd_4Br_4)^{8+}(e^-)_8$	C	–
$Gd_6Cl_5C_3$	$(Gd_4Cl_{3.33})^{8.67+}(e^-)_{8.67}$	C	–
$Gd_6I_7C_2$	$(Gd_6I_7)^{11+}(e^-)_{11}$	C	–
Gd_3Cl_3C	$(Gd_6Cl_6)^{12+}(e^-)_{12}$	C	–
Gd_4I_5C	$(Gd_8I_{10})^{14+}(e^-)_{14}$	C	–

occupied by C_2 groups in a parallel orientation, each pointing towards the octahedron apex atoms. A similar bi-octahedral cluster is observed in the $Ru_{10}C_2(CO)_{24}^{2-}$ anion, where single C atoms occupy the octahedral interstices.[120] In $Gd_{10}Cl_{18}C_4$, the units are discrete and the compound has the same composition as the cluster. In $Gd_{10}Cl_{17}C_4$ they are connected via $X^{i–i}$ bridges, these X atoms having the same planar coordination by 4 metal atoms as in the cluster-chain compound $Sc_{0.75}Zn_{1.25}Mo_4O_7$. This type of interconnection, via SF_4-like coordinated I atoms in $Gd_{10}I_{16}C_4$, has already been described for a molybdate, $Mn_{1.4}Mo_8O_{11}$. The structural relationship between these molybdates (with empty clusters) and lanthanide compounds (with filled clusters) is really striking. Thus, the interconnection between the $(M_6)_4$ and $(M_6)_5$ units in $In_{11}Mo_{40}O_{62}$ is exactly the same as shown in Fig. 4.16 for the $(M_6)_2$ units in $Gd_{10}I_{16}C_4$.

Different kinds of chains of edge-sharing Gd_6 octahedra are found in Gd_4I_5C, $Gd_6I_7C_2$, and $Gd_{12}I_{17}C_6$. The chains are related to those in $NaMo_4O_6$, but with filled octahedra. These chains are connected according to $Gd_4I_2^iI_2^{i-a}X_{2/2}^{i-i}$. The twin-chains of octahedra in the structure of $Gd_6I_7C_2$, similar to those in Sc_7Cl_{10} (Fig. 4.12) but based on the M_6X_{12} type arrangement of the I atoms, are not yet paralleled in the chemistry of molybdates. Finally, the structure of $Gd_{12}I_{17}C_6$ contains zig-zag chains of edge-sharing octahedra filled with C_2 groups.

Layered structures are found with Gd_2X_2C and $Gd_2X_2C_2$ which are related to 1s Ta_2Sc_2C. The X–Gd–Gd–X slabs exhibit a layer sequence ABAB and the octahedral interstices are filled with C atoms or C_2 groups.[121] As all the C_2 units are aligned parallel, oriented towards apex atoms as in

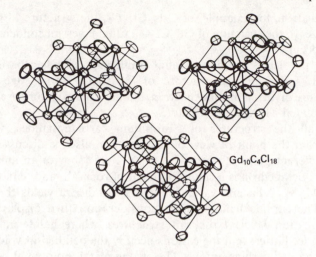

Gd$_{10}$C$_4$Cl$_{18}$

Gd$_{10}$C$_4$Cl$_{17}$

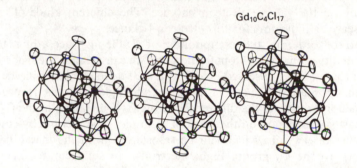

Gd$_{10}$C$_4$I$_{16}$

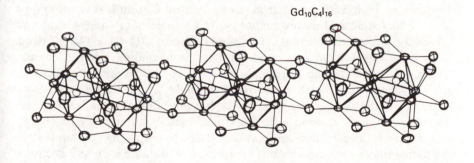

Fig. 4.16. Gd$_{10}$X$_{18}$(C$_2$)$_2$ units and their interconnection in Gd$_{10}$Cl$_{18}$C$_4$, Gd$_{10}$Cl$_{17}$C$_4$, and Gd$_{10}$I$_{16}$C$_4$.[18]

the bi-octahedron, for example, of $Gd_{10}Cl_{18}C_4$, the structure of $Gd_2X_2C_2$ is monoclinic. In the structure of $Gd_6Cl_5C_3$, tilted layers of condensed Gd_6C octahedra occur.

Last but not least, a 3-dimensional framework of edge-sharing Gd_6C octahedra is found in the structure of Gd_3Cl_3C. Three edges of each octahedron are involved in condensation, but all other Cl atoms are arranged as in the isolated M_6X_{12} type cluster.

Clearly, all the structures of gadolinium carbide halides, which we discussed from the point of view of clusters, can also be discussed from a distinctly different point of view. The M_6X_{12} cluster with an anion in the centre of the octahedron is a small fragment of rocksalt, and 'condensation' of such $X'M_6X_{12}$ 'clusters' via edges of the octahedra yields the rocksalt structure. The carbide halides under consideration then display different variants of cation-deficient rocksalt structures, where halide and carbide anions form *fcc* lattices and the cations occupy the octahedral voids around the highly charged carbide anions. The strong metal–non-metal bonding in these compounds may or may not be assisted by some metal–metal bonding, depending on the valence electron balance. The different kinds of carbon species serve as indicators for this electron balance.

Instead of ordering the compounds of Table 4.3 according to the dimensionality of the metal atom framework, we can arrange them according to the carbon species observed. We find a well-defined borderline between phases that contain molecular C_2 groups and single C atoms, respectively. Both this borderline and the differences in the C–C distances in the C_2 units can be easily understood on the basis of a simple ionic picture, by counting the number of electrons which can be transferred from the metal–halogen framework to the C_2 group. Eight electrons are sufficient to fill all the antibonding states of the 'molecule' C_2 and bond rupture is expected. Accordingly, with 8 and more electrons only single C atoms (formally C^{4-} ions) occur. Transferring 6 electrons to the neutral C_2 unit leaves one empty antibonding state and corresponds to a C–C single bond, and 4 electrons allows for a C=C double bond. The bond distances $d(C-C) = 145(\pm 2)$ and $d(C=C) = 130(\pm 3)$ pm are found for the corresponding electron counts. The single bond distance seems short when compared with distances in organic compounds. Such a comparison might be misleading due to a difference in the electronic ground states, but one could also understand the shortening of the C–C bond as arising from a back-bonding effect, i.e. charge transfer from filled carbon antibonding states to empty metal states. It is interesting that, within experimental error, the C–C distance for 6 to 7 electrons made available to each C_2 group is constant. Obviously, any additional electrons beyond 6 enter M–M bonding states until the critical value 8 is reached. Thus we have a third way to order the compounds in Table 4.3.

Compounds with 4, 6, and 8 electrons transferred to 2 carbon atoms are normal valence compounds which can be formulated as, for example,

$(Gd^{3+})_2(Br^-)_2C_2^{4-}$, $(Gd^{3+})_{10}(Cl^-)_{18}(C_2^{6-})_2$, or $(Gd^{3+})_2(Br^-)_2C^{4-}$. They all fulfill the necessary condition to be semiconductors due to filled (anionic) valence bands. In fact, the dark red $Gd_{10}Cl_{18}C_4$ is semiconducting. The black conducting crystals of Gd_2Br_2C are obviously doped, but the corresponding dark red scandium compound Sc_2Cl_2C is a semiconductor.

Counting electrons unfortunately does not tell us the width of the band gap. $Gd_2Br_2C_2$ and $Gd_2Cl_2C_2$ are golden coloured metallic compounds in which the vanishing band gap is due to the previously discussed back-bonding effect. Extended Hückel calculations on $Gd_2Cl_2C_2$ indicate strong mixing of the $C_2\pi^*$ levels (60 per cent) and the Gd d orbitals (40 per cent), and the Fermi level falls in the midst of the $C_2\pi^*$ band.[122] From COOP curves two facts are evident, namely that filling of the Gd–C bonding and C–C antibonding bands occurs as a compromise, and that any significant Gd–Gd bonding is missing in $Gd_2Cl_2C_2$ (as it is in $Gd_{10}Cl_{18}C_4$ and Gd_2Br_2C).

The compounds $Gd_{10}Cl_{17}C_4$ and $Gd_{10}I_{16}C_4$ belong to a second category which is characterized by the presence of weak M–M bonding at very low concentrations of metal-centred electrons, MCE = 1 and 2, respectively, per 10 Gd atoms. The localization of these electrons in discrete clusters—a similar situation to Nb_6I_{11}—makes the compounds semiconducting.

In a third category, infinitely extended M–M bonding occurs in one (e.g. Gd_4I_5C), two ($Gd_6Cl_5C_3$), or three dimensions (Gd_3Cl_3C). All these compounds are 'true' metals, with partially filled conduction bands which have mainly metal d character.

Recently theoretical work on the chemical bonding in $Gd_{10}Cl_{18}C_4$ clearly demonstrates the borderline character of lanthanide halide carbides between cluster compounds and simple ionic solids.[123] The density of states calculated for the $Gd_{10}Cl_{18}C_4$ unit is shown in Fig. 4.17, which also indicates the levels for the two weakly interacting C_2 groups, neglecting any interaction with Gd and Cl atoms. Apart from the small splitting, the levels of these decoupled C_2 groups are the same as for the free C_2 molecule (the σ_s and σ_p^* levels being omitted from the figure). The covalent interaction, which is mainly with the Gd orbitals, lowers the energies of the carbon electronic states considerably. Six electrons are transferred to each C_2 unit, the HOMO being the π_p^* state, in accordance with the electron counting scheme used in Table 4.3. Of course, the purely ionic picture is oversimplified. Figure 4.17 shows strong covalent mixing of all C_2-like states with orbitals of the gadolinium; the HOMO has about 30 per cent Gd d contribution. But as the π_p^* character is spread over more or less the entire d bands, no direct M–M bonding results. According to a Mulliken population analysis, the chemical bonding is described by the formula $(Gd^{1.3+})_{10}(Cl^{0.45-})_{18}(C_2^{2.4-})_2$.

M–M bonding is introduced by occupying the state which is the LUMO in $Gd_{10}Cl_{18}C_4$. This rather narrow band, which is split off from the block of Gd 5d bands, has 90 per cent Gd d character and is occupied by

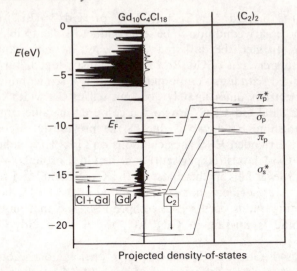

Fig. 4.17. Projected densities of states for the $Gd_{10}Cl_{10}C_4$ cluster.[104] The right panel shows energy levels for the $(C_2)_2$ systems alone (no interactions with the Gd and Cl atom) σ_s and σ_p^* levels omitted.

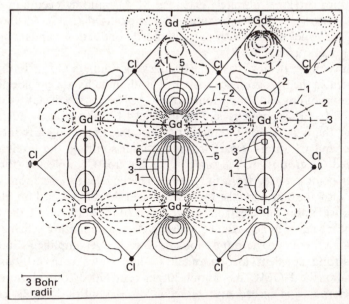

Fig. 4.18. Lowest unoccupied molecular orbital (LUMO) of $Gd_{10}Cl_{18}C_4$ plotted in the bases of the bi-octahedron. Occupation of the LUMO in $Gd_{10}Cl_{17}C_4$ (or $Gd_{10}I_{16}C_4$) results in weak M–M bonding, especially between the two atoms which belong to the shared edge of the octahedra. The contours refer to units of 0.2 [Bohr radius]$^{-3/2}$.

one electron in $Gd_{10}Cl_{17}C_4$. The orbital is plotted in Fig. 4.18, which shows that its occupation leads to M–M bonding in the bases of the bi-octahedron, especially along the shared edge in the centre. The apex atoms contribute very little to this orbital. The result of the calculation agrees very well with experiments. The main difference in Gd–Gd distances is found with the shared octahedron edge, which is 9 pm shorter in $Gd_{10}Cl_{17}C_4$ than in $Gd_{10}Cl_{18}C_4$.

These gadolinium carbide halides take us back to the starting point of the discussion: the structures and bonding in these solids seem more complex but, they are really not too far away from bonding in 'normal' and 'hypermetalated' gas phase molecules such as Li_4C and Li_6C.

ACKNOWLEDGEMENT

Thanks are due to B. Krauter, Hj. Mattausch, and C. Schwarz for help with the figures, and I. Remon for typing.

4.8 REFERENCES

1. Kjekshus, A. and Rakke, T., *Structure and Bonding (Berlin)*, **19**, 45–83 (1974).
2. McCarley, R. E., *Polyhedron*, **5**, 51 (1986).
3. Parthé, E., *Cristallochimie des Structures Tétraédriques*, Gordon and Breach, Paris (1972).
4. Parthé, E., *Acta Cryst.*, **29**, 2808–15 (1973).
5. Nesper, R. and v. Schnering, H. G., *TMPM Tschermarks Min. Petr. Mitt.*, **32**, 195–208 (1983).
6. Rundquist, S., *Arkiv Kemi*, **20**, 67–113 (1962).
7. Schäfer, H. and Schnering, H. G. *Angew. Chem.*, **20**, 833–49 (1964).
8. Schubert, K., *Kristallstrukturen zweikomponentiger Phasen*, Springer-Verlag, Berlin, Göttingen, Heidelberg, pp. 432 (1964).
9. Hulliger, F., *Structure and Bonding (Berlin)*, **4**, 83–229 (1968).
10. Flahaut, J., Transition metal chalcogenides, *MTP. Int. Rev. Sci., Inorg. Chem.*, Ser. 1 (ed. Roberts, L. E. J.), Butterworths, London, pp. 189–241 (1972).
11. Jellinek, F., Sulphides, selenides and tellurides of transition elements. *MTP Int. Rev. Sci., Inorg. Chem.*, Ser. 1 (ed. D. W. A. Sharp), Vol. 5, Butterworths, London, pp. 339–96 (1972).
12. Taylor, M. J., *Metal-to-metal bonded states of main group elements*, Acad. Press, London, New York, San Francisco, pp. 211 (1975).
13. Franzen, H. F., *Progr. Solid State Chem.*, **12**, 1–39 (1978).
14. Yvon, K., *Curr. Top. Mater. Sci.*, **3**, 53–129 (1979).
15. Simon, A., *Structure and Bonding (Berlin)*, **36**, 81–127 (1979).
16. Simon, A., *Angew. Chem.* [Intern. Ed. Engl.], **20**, 1–22 (1981).
17. Simon, A., *Naturwiss.*, **71**, 171–80 (1984).
18. Simon, A., *J. Solid State Chem.*, **57**, 2–16 (1985).

19. Corbett, J. D., *Adv. Chem. Ser., Solid State Chem.*, **186**, 329–47 (1980).
20. Corbett, J. D., *Acc. Chem. Res.*, **14**, 239–46 (1981).
21. Corbett, J. D., *Chem. Rev.*, **85**, 383–97 (1985).
22. v. Schnering, H. G., *Angew. Chem.* [Intern. Ed. Engl.], **20**, 33–51 (1981).
23. Fischer, Ø. and Maple, M. B. (eds), *Superconductivity in ternary compounds I, II: Topics in Current Physics*, 32, Springer-Verlag, Berlin, Heidelberg, New York, 283 pp. (1982).
24. McCarley, R. E., *Phil. Trans. R. Soc. Lond.* A, **308**, 141–57 (1982).
25. McCarley, R. E., *ACS Symp. Ser.*, **211**, 273–90 (1983).
26. McCarley, R. E., Lii, K.-H., Edwards, P. A., and Brough, L. F., *J. Solid State Chem.*, **57**, 17–24 (1985).
27. Chevrel, R., Gougeon, P., Potel, M., and Sergent, M., *J. Solid State Chem.*, **57**, 25–33 (1985).
28. Martin, T. P., *J. Chem. Phys.*, **83**, 78–84 (1985).
29. Martin, T. P., The structures of elemental and molecular clusters. In *Festkörperprobleme XXIV, Adv. Solid State Physics*, (ed. P. Groose), Vieweg & Sohn, Wiesbaden, pp. 1–24 (1984).
30. Herrmann, A., Schumacher, E., and Wöste, L., *J. Chem. Phys.*, **68**, 2327–36 (1978).
31. Würthwein, E.-U. and v. Rague-Schleyer, P., *J. Am. Chem. Soc.*, **106**, 6973–78 (1984).
32. v. Rague-Schleyer, P., Are CLi_6, NLi_5, OLi_4, etc., hypervalent? In *New horizons of quantum chemistry* (eds Löwdin, P.-O. and Pullman, B.), Reidel Publ. Comp., pp. 95–109 (1983).
33. Martin, T. P., *J. Chem. Phys.*, **81**, 4426–32 (1984).
34. Corbett, J. D., *J. Solid State Chem*, **37**, 335–51 (1981).
35. Pauling, L., *The nature of the chemical bond*, 3rd. edn, 8.print, Cornell Univ. Press, Ithaca, N. Y., pp. 644 (1973).
36. Brown, I. D., The bond-valence method: An empirical approach to chemical structure and bonding. In *Structure and Bonding in Crystals*, Vol. II (eds O'Keeffe, M. and Navrotsky, A.), Acad. Press, New York, London, pp. 1–30 (1981).
37. Metsch, G., Bauhofer, W., and Simon, A. *Z. Naturforsch.*, **40a**, 303–12 (1985).
38. Vohra, Y. K., Weir, S. T., Brister, K. E., and Ruoff, A. L., *Phys. Rev. Lett.*, **55**, 977–9 (1985).
39. Satpathy, S., Christensen, N. E., and Jepsen, O., *Phys. Rev.* B, **32**, 6793–88 (1985).
40. Ebbinghaus, G. and Simon, A. *Chem. Phys.*, **43**, 117–33 (1980).
41. Gmelin, E., Simon, A., Brämer, W., and Villar, R., *J. Chem. Phys.*, **76**, 6256–61 (1982).
42. Steeb, S. and Renner J., *J. Less Common Met.*, **10**, 246–9 (1966).
43. Burt, M. G. and Heine, V., *J. Phys.* C, **11**, 961–68 (1978).
44. Brice, J.-F., Motte, J.-P., and Aubry, J., *Rev. Chim. Minér.*, **12**, 105–12 (1975).
45. Holmberg, B., *Acta Chem. Scand.*, **16**, 1245–50 (1962).
46. Ott, H. and Seyfarth, H., *Z. Kristallogr.*, **67**, 430–33 (1928).
47. Cotton, F. A. and Haas, T. E., *Inorg. Chem.*, **3**, 10–17 (1964).
48. Bursten, B. E., Cotton, F. A., and Stanley, G. G., *Israel J. Chem.*, **19**, 132–42 (1980).
49. Andersen, O. K., Klose, W., and Nohl, H., *Phys. Rev.* B, **17**, 1209–37 (1978).
50. Andersen, O. K. and Satpathy, S., Calculation of the electronic band structure

for the 3d-monoxides and the vacancy compound niobium oxide (Nb_3O_3). In *Basic properties of binary oxides* (eds Dominguez Rodriquez, A., Castaing, J., and Marquez, R.), Serv. Publ. Univ. Sevilla, pp. 21–42 (1983).

51. Kettle, S. F. A., *Theor. Chim. Acta*, **3**, 211–12 (1965).
52. Edwards, P. P. and Sienko, M. J., *Int. Rev. Phys. Chem.*, **3**, 83–137 (1983).
53. Finley, J. J., Nohl, H., Vogel, E. E., Imoto, H., Camley, R. E., Zevin, V., Andersen, O. K., and Simon, A., *Phys. Rev. Lett.*, **46**, 1472–5 (1981).
54. Finley, J. J., Camley, R. E., Vogel, E. E., Zevin, V., and Gmelin, E., *Phys. Rev. B*, **24**, 1323–31 (1981).
55. Imoto, H. and Simon, A., *Inorg. Chem.*, **21**, 308–19 (1982).
56. Gütlich, P., *Structure and Bonding (Berlin)*, **44**, 83, 195 (1981).
57. Leduc, L., Perrin, A., and Sergent, B., *Acta Cryst. C*, **39**, 1503–06 (1983).
58. Perrin, C., Potel, M., and Sergent, M., *Acta Cryst. C*, **39**, 415–18 (1983).
59. Perrin, C., Chevrel, R., and Sergent, M., *Mat. Res. Bull.*, **14**, 1505–15 (1979).
60. Chevrel, R. and Sergent M., Chemistry and structure of ternary molybdenum chalcogenides. In *Superconductivity in ternary compounds I—Topics in current physics* (eds Fischer, Ø and Maple, M. B.), Springer-Verlag, Berlin, Heidelberg, New York, pp. 25–86 (1982).
61. Corbett, J. D., *J. Solid State Chem.*, **39**, 56–74 (1981).
62. Nohl, H., Klose, W., and Andersen, O. K., Band structures of $M_xMo_6X_8$ and $M_2Mo_6X_6$ cluster compounds. In *Superconductivity in ternary compounds I—Topics in current physics* (eds Fischer, Ø. and Maple, M. B.), Springer-Verlag, Berlin, Heidelberg, New York, pp. 166–221 (1982).
63. Hughbanks, T. and Hoffmann, R., *J. Am. Chem. Soc.*, **105**, 1150–62 (1983).
64. Bell, R. F., Di Salvo, F. J., Dunn, R. G., and Tarascon, J. M., *Phys. Rev. B*, **32**, 1461–3 (1985).
65. Armici, J. C., Decroux, M., Fischer, Ø., Potel, M., Chevrel, R., and Sergent, M., *Solid State Comm.*, **33**, 607–11 (1980).
66. Hughbanks, T. and Hoffmann, R., *J. Am. Chem. Soc.*, **105**, 3528–37 (1983).
67. Hönle, W., v. Schnering, H. G., Lipka, A., and Yvon, K., *J. Less Common Met.*, **71**, 135–45 (1980).
68. Yvon, K., Structures and bonding of ternary superconductors. In *Superconductivity in ternary compounds I—Topics in current physics* (eds Fischer Ø. and Maple, M. B.), Springer-Verlag, Berlin, Heidelberg, New York, pp. 87–111 (1982).
69. Hughbanks, T. and Hoffmann, R., *Inorg. Chem.*, **21**, 3578–80 (1982).
70. Tarascon, J. M., Di Salvo, F. J., and Waszczak, J. V., *Solid State Comm.*, **52**, 227–31 (1984).
71. Tarascon, J. M., Di Salvo, F. J., Chen, C. H., Carroll, P. J., Walsh, M., and Rupp, L., *J. Solid State Chem.*, **58**, 290–300 (1985).
72. Chini, P., *J. Organomet. Chem.*, **200**, 37–61 (1980).
73. Fenske, D., Ohmer, J., and Hachgenei, J., *Angew. Chem.*, **97**, 993–5 (1985).
74. Torardi, C. C. and McCarley, R. E., *J. Am. Chem. Soc.*, **101**, 3963–4 (1979).
75. Mattausch, Hj., Simon, A., and Peters, E.-M., *Inorg. Chem.*, **25**, 3428 (1986).
76. Grønvold, F., Kjekshus, A., and Raaum, R., *Acta Cryst.*, **14**, 903–34 (1961).
77. Chen, H.-Y. and Franzen, H. F., *Natl. Bur. Stand. Spec. Publ.*, **364**, 65–62 (1972).
78. Simon, A., *Ann. Chim. France*, 1982, 539–51 (1982).
79. Kumar, V. and Heine, V., *Inorg. Chem.*, **23**, 1498–99 (1984).
80. Kumar, V. and Heine, V., *J. Phys. F*, **14**, 365–79 (1984).

81. Watanabe, D., Castles, J. R., Jostsons, A., and Malin, A. S., *Acta Cryst.*, **23**, 307–13 (1967).
82. Burdett, J. K. and Hughbanks, T., *J. Am. Chem. Soc.*, **106**, 3191–13 (1984).
83. Ganglberger, E., *Monatsh. Chem.*, **99**, 549–56 (1968).
84. Berger, R., *Acta. Chem. Scand.* A, **31**, 287–91 (1977).
85. Chen, H.-Y., Tuenge, R. T., and Franzen, H. F., *Inorg. Chem.*, **12**, 552–5 (1973).
86. Franzen, H. F., Beineke, T. A., and Conard, B. R., *Acta Cryst.* B, **24**, 412–16 (1968).
87. Owens, J. P. and Franzen, H. F., *Acta Cryst.* B, **30**, 427–30 (1964).
88. Simon, A., Holzer, N., and Mattausch, Hj., *Z. Anorg. Allg. Chem.*, **456**, 207–16 (1979).
89. Poeppelmeier, K. R. and Corbett, J. D., *Inorg. Chem.*, **16**, 1107–11 (1977).
90. Adolphson, D. G. and Corbett, J. D., *Inorg. Chem.*, **15**, 1820–3 (1976).
91. Franzen, F. and Graham, J. Z. *Kristallogr.*, **123**, 133–38 (1966).
92. Stensvad, S., Helland, B. J., Babich, M. W., Jacobson, R. A., and McCarley, R. E., *J. Am. Chem. Soc.*, **100**, 6257–8 (1978).
93. Lundström, T., *Acta Chem. Scand.*, **22**, 2191 (1968).
94. Jödden, K., v. Schnering, H. G., and Schäfer, H., *Angew. Chem.*, **87**, 594–5 (1975).
95. Geller, S., *Acta Cryst.*, **8**, 15–21 (1955).
96. Ergenius, L.-E., Noläng, B. I., and Lundström, T., *Acta Chem. Scand.* A, **35**, 693–9 (1981).
97. Pearson, W. B., *The crystal structures and physics of metals and alloys*, Wiley-Interscience, New York, London, Sydney, Toronto, pp. 806 (1972).
98. Teo, K. and Keating, K., *J. Am. Chem. Soc.*, **106**, 2224–6 (1984).
99. Simon, A., *Z. Anorg. Allg. Chem.*, **355**, 311–22 (1967).
100. Fritsche, H. G., Dübler, F., and Müller, H., *Zn. Anorg. Allg. Chem.*, **513**, 46–56 (1984).
101. Switendick, A. C., The change in electronic properties on hydrogen alloying and hydride formation. In *Hydrogen in metals I* (eds Alefeld, G., and Völkl, J.), Springer-Verlag, New York, pp. 101–29 (1978).
102. Chini, P., Longoni, G., and Albano, V. G., *Adv. Organomet. Chem.*, **14**, 285–344 (1976).
103. Albano, V. G. and Martinengo, S., *Machr. Chem. Techn. Lab.*, **28**, 654–64 (1980).
104. Imoto, H. and Corbett, D. J., *Inorg. Chem.*, **19**, 1241–45 (1980).
105. Ziebarth, R. P. and Corbett, J. D., *J. Am. Chem. Soc.*, **107**, 4571–73 (1985).
106. Smith, J. D. and Corbett, J. D., *J. Am. Chem. Soc.*, **107**, 5704–11 (1985).
107. Lauher, J. W., *J. Am. Chem. Soc.*, **100**, 5305–15 (1978).
108. Wijeyesekera, S. D. and Hoffmann, R., *Organometallics*, **3**, 949–61 (1984).
109. Rundquist, S., Tellgren, R., and Andersson, Y., *J. Less Common Met.*, **101**, 145–68 (1984).
110. Struss, A. W. and Corbett, J. D., *Inorg. Chem.*, **16**, 360–2 (1977).
111. Wijeyesekera, S. D. and Corbett, J. D., *Solid State Comm.*, **54**, 657–60 (1985).
112. Nowotny, H., Boller, H., and Beckmann, O., *J. Solid State Chem.*, **2**, 462–71 (1970).
113. Nowotny, H., *Berg-Hüttenmänn. Monatsh.*, **110**, 171–82 (1966)).
114. Daake, R. L. and Corbett, J. D., *Inorg. Chem.*, **16**, 2029–33 (1977).
115. Hwu, S.-J., Corbett, J. D., and Poeppelmeier, K. R., *J. Solid State Chem.*, **57**, 43–58 (1985).

116. Bullett, D. W., *Inorg. Chem.*, **24**, 3319–23 (1985).
117. Mattausch, Hj., Schramm, W., Eger, R., and Simon, A., *Z. Anorg. Allg. Chem.*, **530**, 43–59 (1985).
118. Ueno, F., Ziebeck, K., Mattausch, Hj., and Simon, A., *Rev. Chim. Minér.*, **21**, 804–8 (1984).
119. Mattausch, Hj., Simon, A., and Ziebeck, K., *J. Less Common Met.*, **113**, 149–55 (1985).
120. Hayward, S.-M.T. and Shapley, J. R., *J. Am. Chem. Soc.*, **104**, 7349–51 (1982).
121. Schwanitz-Schüller, U. and Simon, A., *Z. Naturforsch.*, **40b**, 710–16 (1985).
122. Miller, G. J., Burdett, J. K., Schwarz, C., and Simon, A., *Inorg. Chem.*, **25**, 4437 (1986).
123. Satpathy, S. and Andersen, O. K., *Inorg. Chem.*, **24**, 2604–8 (1985).

5 Introduction to heterogeneous catalysts

Arthur W. Sleight

5.1 INTRODUCTION

An aspect of inorganic solids that has frequently been neglected by solid state chemists is the catalytic property. The reason for this neglect can be primarily attributed to the great difficulty of achieving a detailed understanding of surface structure and reactivity. Certainly, it has been true that reliable mechanistic information on heterogeneously catalysed reactions is rarely achieved. Even on an empirical level, good correlations between catalytic properties and the chemistry of the catalyst itself have been largely lacking. Fortunately, advanced techniques and instrumentation for microstructure and surface studies are now allowing rapid strides to be made in the area of heterogeneous catalysis. Indeed, many obvious and potentially highly revealing experiments within our grasp are waiting to be done.

Surface properties of solids are of paramount importance for heterogeneous catalysis. One would like to know the active site density on each crystallographic surface of a solid, and one would also like to know if these active sites are related to dangling bonds due to coordinatively unsaturated surface sites or if they are related to special defects. In a few cases, we are beginning to obtain answers to these questions, and the rate of progress is increasing.

Bulk properties of solids can also frequently be of considerable importance for heterogeneous catalysis. In the case of oxidation or reduction reactions, the bulk catalyst may serve as a reservoir of electrons for these processes. Thus, the electronic properties of the solid become critical. Diffusion of oxygen or hydrogen through the bulk solid may also be important. Zeolites (Section 5.4) represent a unique class of microporous materials used as heterogeneous catalysts in which it is not meaningful to distinguish between bulk and surface properties.

This chapter is divided into sections on metal, oxide, zeolite, and sulphide catalysts. In each case, the solid state chemistry of that class of catalyst is related to catalytic properties. There are other inorganic materials of interest as heterogeneous catalysts such as, for example, halides as halogenation catalysts. However, the overwhelming majority of catalysts fall within the four categories to be discussed, since other types of materials are generally not stable under the conditions normally employed for catalysis.

5.2 METAL CATALYSTS

5.2.1 Bulk metal

With silver metal we have both a good example of the use of a bulk metal catalyst and an example of a metal used as a selective oxidation catalyst. Silver is one of the most commonly used catalysts today for the conversion of methanol to formaldehyde. It is used in low surface area with no support at a reaction temperature of about 600°C. The reactant stream is a mixture of methanol and oxygen in which the oxygen-to-methanol ratio is kept low enough to avoid an explosive mixture. The reaction products are primarily CH_2O, H_2, and H_2O.

The mechanism of the reaction of methanol to formaldehyde over silver has been elucidated in some detail.[1] Firstly, we know that essentially every exposed silver atom is a potential active site. However, methanol will not dissociatively adsorb on a silver surface unless oxygen has been preadsorbed (Fig. 5.1). The first chemically significant interaction of methanol with a silver

$$1/2\ O_2 \Rightarrow O(s)$$

$$2\ CH_3OH + 2\ O(s) \Rightarrow 2\ CH_3O(s) + 2\ OH(s)$$

$$2\ OH(s) \Rightarrow H_2O + O(s)$$

$$2\ CH_3O(s) \Rightarrow 2\ CH_2O + 2H(s)$$

$$2H(s) \Rightarrow H_2$$

$$2\ CH_3OH + 1/2\ O_2 \Rightarrow 2\ CH_2O + H_2 + H_2O$$

Fig. 5.1. Elementary steps of methanol oxidation over a silver catalyst. Surface species indicated by (s).

surface involves the abstraction of the hydroxyl hydrogen by a surface oxygen. A surface hydroxyl group is formed, and two such groups combine to form water which readily desorbs. The slow step in the overall reaction is the breaking of a carbon–hydrogen bond. This hydrogen abstraction is carried out by a surface silver atom and is followed by rapid desorption of formaldehyde. The abstracted hydrogen combines with another such hydrogen and desorbs as H_2. Thus, from the basic mechanism one expects a stoichiometric reaction yielding a H_2O/H_2 ratio of one. Somewhat lower ratios may be observed in practice due to some formaldehyde decomposition.

The selectivity of the silver catalyst may be improved if it is alloyed with gold,[2] but the expense of such a modification has so far precluded commercialization of this improved catalyst. A supported silver catalyst could offer an advantage in cost reduction over a bulk silver catalyst. However, supports have not been employed commercially because they result in unacceptable selectivity losses.

Most metals that are active oxidation catalysts tend to promote fully oxidized products rather than to selectively produce partially oxidized products. Silver is an exception, as are some of its near neighbours in the periodic table, e.g. copper and palladium.

Different crystallographic faces of the same metal can show significantly different catalytic activity.[3] Generally, the more open, higher energy faces (e.g. FCC (100)) show higher activity than the more dense, lower energy faces (e.g. FCC (111)). This is presumably due to a higher activity of metal atoms as their coordination number decreases.

5.2.2 Supported metals

There are several reasons for dispersing metal catalysts on supports. The most obvious is to utilize an expensive metal more effectively. However, the support can substantially modify the catalytic properties of the metal.

Metals are normally supported using wet-impregnation methods. A salt of the metal is contacted with a high surface area support which is then dried, calcined, and reduced. In the incipient wetness impregnation method, the amount of liquid used is just that which is required to fill the pore structure of the support. This method tends to yield metal particle sizes which are restricted by the pore structure of the support. Even higher dispersions may be obtained by ion exchange techniques. Many supports have surface hydroxyl groups which can act as exchange sites for metal, and if the pH of the solution is properly controlled, very high metal dispersions can be obtained. Such high metal dispersions are easier to produce than to maintain during calcination and under catalytic conditions. Metal particles containing just a few atoms tend to be highly mobile on support surfaces and thus readily agglomerate. In supported metal catalysts, the actual metal particle size in working catalysts ranges from about 5 Å to about 100 microns.

The catalytic properties of a metal may depend on its particle size. In the case of reactions involving carbon–carbon bond cleavage such as hydrogenolysis, small particles tend to be more active than large particles, even after normalization for metal surface area. The reason for this dependence on particle size could be due to a change in the electronic properties on going from bulk metal to small metal clusters. However, it is generally believed that the increased activity of very small crystallites is instead due to a greater proportion of edge sites, which are more active.

Methods for determining metal surface area for supported metal catalysts have become routine and generally reliable. For many metals, either hydrogen or CO chemisorption measurements can be used. The support is assumed not to interfere, although effects such as hydrogen spillover can cause misleading results in some cases. As indicated above, there is no assurance that the number of available metal atoms is the same as the number of active sites. For the CO/H_2 reaction on supported Ni, Co, and Ru, it has been estimated that only about 10 per cent or less of the available metal sites are active sites for catalysis.[4] Although all of the sites can chemisorb both CO and H_2, and can also dissociate H_2, most of the sites that adsorb CO do not dissociate it.

5.2.3 Metal–support interactions

Numerous different kinds of metal–support interactions are possible. The effects are well understood in some cases, but certainly not in all cases. An interaction between a metal and a support may be sufficiently strong so that the oxidation state of the metal is affected. For example, when iron is highly dispersed on alumina, it is very difficult to reduce iron to the zero valent state. With increasing metal loading, the metal–support interaction becomes less important and reduction of most of the iron becomes easier. On the other hand when iron is supported on SiO_2, reduction to the metal is relatively easier due to the much weaker metal–support interaction. This difference in behaviour of iron on the Al_2O_3 vs. the SiO_2 surface is presumably related to the ease of iron cations substituting into the Al_2O_3 lattice as opposed to the SiO_2 lattice.

Another type of metal–support interaction involves an actual change in composition of the metal particles. For example, in Pt/Al_2O_3 reduced at high temperatures, some aluminium is reported to have alloyed with the Pt.[5] Metals may also scavenge impurities in the support. This effect can be especially pronounced when carbon supports are used since the impurity content of common carbon supports is high. On the other hand, a carbon support might trap an impurity in the reactant stream which might otherwise poison the catalytic activity of the metal particles.

The morphology of metal particles may depend on the support which is used. For example, a particular metal will wet some support surfaces but not others.

The support may play a very direct role in a catalytic reaction. The petroleum reforming reaction is carried out using a catalyst with metal supported on alumina, and much of the catalysis takes place on the surface of the alumina. When catalysis is occurring at sites on both the metal and the support surface, various mechanisms are possible. In the classic hydrogen spillover case, hydrogen adsorbs and dissociates on metal particles, and then spills over onto the support surface where it reacts with adsorbed organic molecules. Another possibility is that a reaction intermediate may desorb from one site (e.g. metal) and readsorb at another site (e.g. support) where it reacts further. Still another possible way of using active sites on both the metal and the support surface is to have activation at the metal–support interface. High metal dispersions are necessary for this mode of activation to be pronounced, and this may well account for CO activation over some catalysts.

The shape, size, and pattern of metal particles may be strongly dominated by the pore structure of the support. The pore structure of the support may also strongly affect the diffusion rates of reactants or products. This effect is most pronounced in the case of zeolite supports, but it has also been reported for amorphous materials.[6]

The electronic properties of a metal are influenced by a support when the metal particles are small. Since metals are normally good electron donors relative to other materials, there will usually be some flow of electrons from the metal to the support. The metal thus takes on a small positive charge. This could be accentuated by electron-withdrawing surface states (e.g. Lewis acid sites) created by dangling bonds at the surface.

Recently, much attention has been focused on a strong metal–support interaction which occurs when a metal is supported on a reduced metal oxide.[7] The most pronounced manifestation of this interaction is a dramatic loss of the H_2 and CO chemisorption capacity of the metal. As a consequence, such chemisorption measurements can no longer be used to assess metal surface area. The effect of the strong metal–support interaction on catalysis depends on the reaction being considered. For example, the effect can be very pronounced for reactions involving carbon–carbon bond breaking, but is generally rather small for hydrogenation reactions.

In the case of a strong metal–support interaction, one expects electron flow from the support to the metal so that the metal takes on a negative charge.[8] There are indications that some metals wet the surfaces of reduced metal oxides whereas they do not wet the fully oxidized surface. There is also evidence that in some cases the metal particles may become partially buried in the support material. However, at the present time a good universal explanation for why the capacity for chemisorption of H_2 and CO has so dramatically decreased in this situation is lacking.

Despite the significance of the metal–support interaction, there are situations where the catalytic reactions on supported catalysts are essentially identical to those on single-crystal metal surfaces. This is well documented by the work of Goodman,[9] who studies the reaction of CO and H_2 to form CH_4 over (100) and (111) surfaces of Ni. The reaction rates, normalized to exposed Ni, were the same as those observed for Ni supported on alumina. Having established the relevance of single-crystal studies, definitive studies of the poisoning effects of sulphur, phosphorous, and carbon were accomplished utilizing well-defined single-crystal surfaces.[9]

5.2.4 Multimetallic catalysts

Another area that has recently received much attention is multimetallic catalysts in which more than one metal is dispersed on a given support.[10]

This interest was prompted largely by the discovery at Chevron that the Pt/Al_2O_3 reforming catalyst could be significantly improved by additions of rhenium.[11] The nature of the interaction between Pt and Re is controversial.

5.3 OXIDE CATALYSTS

Oxides are commonly used catalysts for a wide variety of chemical reactions (Table 5.1). Oxide catalysts are generally used without supports, but there are exceptions. Molybdenum oxide catalysts are frequently supported on alumina, and vanadium oxide catalysts are sometimes supported on titania. Commercially used ammoxidation catalysts generally contain about 50 per cent silica by weight, but this silica might be regarded more as a binder providing strength than as a support providing dispersion.

Elegant studies of oxide catalysts by Stone and Cimino[12] have utilized a solid solution approach to the investigation of the catalytic properties of transition metal cations. Various amounts of transition metal ions (e.g. Ni^{2+}, Co^{2+} and Mn^{2+}) were substituted for Mg^{2+} in MgO, and the resulting solid solutions were evaluated for N_2O decomposition activity. In all cases it was found that transition metal ions isolated from one another were more active than when concentrated (Fig. 5.2). Furthermore, it was found that Ni^{2+} in MgO was considerably more active for N_2O decomposition than Ni^{2+} in ZnO, suggesting that Ni^{2+} ions exposed from octahedral sites are more active than those exposed from tetrahedral sites.

Similarly, the $Cr_xAl_{2-x}O_3$ (hex) system was evaluated for H_2–D_2 exchange. Here it was found that concentrated Cr^{3+} ions are more active than dilute ones. A study by Marcilly and Delmon[14] showed that Cr ions in Al_2O_3 (hex) are more active than those in Al_2O_3 (cub) for isobutane dehydrogenation.

Table 5.1 *Oxide catalysts*

Reaction type	Catalyst
Dehydration of alcohols	Al_2O_3, Al/Si/O
Dehydrogenation	Cr_2O_3, $(Cr,Al)_2O_3$, Mo/Al/O, Zn/Cr/O
Cracking of hydrocarbons	Al_2O_3, Si/AlO, Si/Mg/O
Reforming	Mo/Al/O, $(Cr,Al)_2O_3$
Methanol synthesis	Cu/Zn/O
Sulphuric acid synthesis	K/V/O
Selective oxidation	Fe/Mo/O, Bi/Mo/O, Cu/O, Fe/Sb/O, Zn/Fe/O, U/Sb/O, V/P/O
NO reduction	Perovskites
Olefin disproportionation	Mo/O, W/O, Re/O

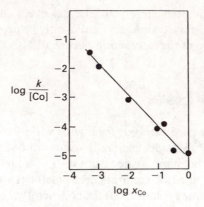

Fig. 5.2. Specific activity, $k/[Co]$, for N_2O oxidation vs. log of cobalt concentration in $Co_xMg_{1-x}O$ catalysts.

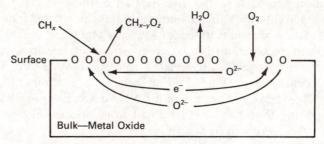

Fig. 5.3. Generalized schematic of selective hydrocarbon oxidation over oxide catalysts such as molybdates.

Many of the reactions shown in Table 5.1 are mainly dependent on surface properties of the catalysts. Surface properties are always important, but bulk properties are particularly important for the selective oxidation and ammox-idation catalysts. In these cases, the catalyst can be regarded as a regenerable reagent and as a reservoir for both electrons and oxygen (Fig. 5.3). Hydrocarbon oxidation and catalyst reduction occur at one type of site, and reoxidation of the catalyst may occur at another site. It is essential that electrons and oxygen anions flow readily from one site to another. Thus, electron and oxygen mobility must be high. In the case of selective olefin oxidation, these processes are generally sufficiently facile that the overall rate of reaction is dependent on the activation of the olefin molecule. If it were otherwise, these activated molecules might undergo undesired reactions to a much greater extent.

The most widely used selective oxidation catalysts are molybdates and vanadates. The most selective catalysts for the oxidation of methanol to formaldehyde are simple compounds such as $Fe_2(MoO_4)_3$ and MoO_3, and some of the best catalysts for the oxidation of propylene to acrolein or the ammoxidation of propylene to acrylonitrile are based on bismuth molybdates. However, modern day acrylonitrile catalysts have become very complex and contain transition metal molybdates in addition to bismuth molybdates. These multiphase catalysts are much more active than any of the components, but the origin of this dramatic synergism has not been definitely established.

Much is known about the mechanism of olefin oxidation over molybdate catalysts (Fig. 5.4). The rate limiting step is the dissociative adsorption of

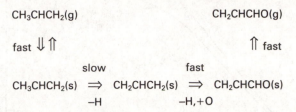

Fig. 5.4. Propylene oxidation over a molybdate catalyst. Gas (g) and surface (s) species indicated.

the olefin to form a surface allyl species.[15] The hydrogen is abstracted from the olefin by a basic surface oxygen, and a hydroxyl group is formed. The hydroxyl groups readily combine, and water desorbs. The allyl species also readily reacts further forming acrolein in the case of propylene oxidation. Tagged oxygen studies show that the oxygen in the product acrolein and water derives from the lattice and not from adsorbed surface oxygen or gas phase oxygen.[16] Tagged carbon studies show that oxygen adds with equal probability to either end of the propylene molecule.[17] Thus, a symmetric surface intermediate is inferred.

Further elucidation of the mechanism is very difficult. Since the allyl species quickly reacts once formed, one cannot perform chemisorption measurements with propylene to determine the number of sites where the olefin activation occurs. Thus for olefin reactions of molybdates, we have no good estimate of the number of active sites. However, studies with the scheelite structure[18] showed that the olefin oxidation activity, and presumably the number of active sites, correlated very well with the number of point defects (Fig. 5.5). On the other hand, electron microscopy studies of Gai indicate that extended defects (e.g. crystallographic shear planes) are essentially absent in optimum bismuth molybdate catalysts.[19] Thus shear planes are not characteristic of good selective oxidation catalysts,[20,21] since they presumably reduce catalytic activity by reducing the point defect concentration.

Another consequence of the mechanism for olefin oxidation is that the lifetime and concentration of the surface intermediates is so low that normal spectroscopic methods cannot be used to observe these intermediates.

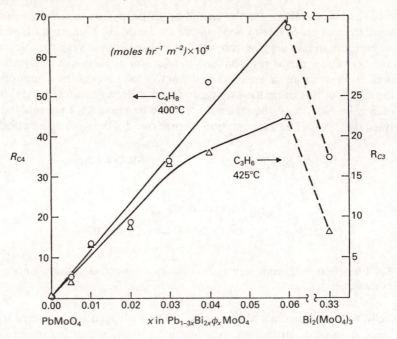

Fig. 5.5. Specific activity for propylene and 1-butene oxidation vs. x for $Pb_{1-3x}Bi_{2x}MoO_4$.

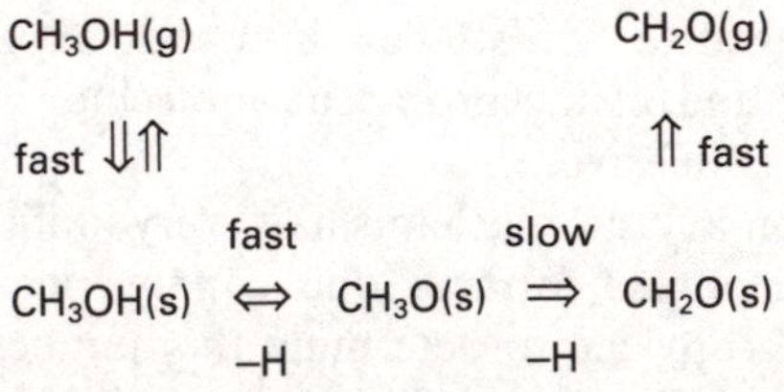

$$CH_3OH(g) \qquad\qquad CH_2O(g)$$

$$\text{fast} \Downarrow\Uparrow \qquad\qquad\qquad \Uparrow \text{ fast}$$

$$\qquad\qquad \text{fast} \qquad\quad \text{slow}$$

$$CH_3OH(s) \Leftrightarrow CH_3O(s) \Rightarrow CH_2O(s)$$

$$\quad -H \qquad\qquad -H$$

Fig. 5.6. Methanol oxidation over a molybdate catalyst. Gas (g) and surface (s) species indicated.

Fortunately, this situation does not exist for all selective oxidation reactions over molybdate catalysts.

In the case of methanol oxidation over molybdate catalysts, it is again the breaking of carbon–hydrogen bonds of the methyl group that is rate limiting (Fig. 5.6),[22] but in this case the initial dissociative chemisorption involves the breaking of oxygen–hydrogen bonds. At room temperature this reaction proceeds rapidly and the methoxy groups bound to the surface are relatively stable. As in the propylene oxidation reaction, the surface hydroxyl groups formed readily combine to liberate water. Chemisorption measurements with methanol can be used to determine the number of active sites, and the present evidence is that most molybdenum cations of most crystallographic surfaces are active in the cases of MoO_3 and ferric molybdate. Special surface defects are not required to activate methanol. The very stable (010) surface of MoO_3 is, however, found to be inactive.[23] Molybdenum trioxide has a layered structure with very easy cleavage in the (010) direction. The layers are held together only by van der Waals forces, and there are no dangling bonds or unsaturated coordinations on the (010) surface. We may, therefore, conclude that although no special defects are required for methanol chemisorption, it is necessary to have a relatively high energy surface where coordinations are unsatisfied.

Since surface methoxy species are numerous and are stable at room temperature, spectroscopic methods can be used to study them. Infrared studies show that methanol initially adsorbs in an undissociated state involving hydrogen bonding with the molybdate surface.[24] However, the molecules readily dissociate and water desorbs. If this surface is exposed to high concentrations of water, the reaction can be reversed (Fig. 5.6).

Commercial processes for the conversion of methanol to formaldehyde utilize either a silver catalyst or a molybdate catalyst. The reaction mechanism in these two cases is nearly identical. The major difference is that the hydrogen from the methoxy group is abstracted by silver metal in the one case and by a surface oxygen in the case of the molybdate catalyst. Thus, all hydrogen removed from methanol forms water in the case of the oxide catalyst whereas both hydrogen and water are produced with a silver catalyst. The commercial oxide catalyst for methanol oxidation to formaldehyde contains ferric molybdate and molybdenum trioxide. Ferric molybdate is somewhat more active than molybdenum trioxide because, unlike molybdenum trioxide, all of its crystallographic faces are active. A more important consideration in terms of activity is that high surface area is easier to obtain and maintain for ferric molybdate than for molybdenum trioxide. The selectivity to formaldehyde and the actual reaction mechanism are essentially identical for ferric molybdate and molybdenum trioxide.[24] Nonetheless, ferric molybdate is not a viable commercial catalyst unless additional molybdenum trioxide is present. This is because molybdenum oxide slowly volatilizes from the ferric molybdate surface leaving behind an iron-rich surface which is a very non-selective oxidation catalyst. In the presence of excess MoO_3, this depletion and loss of selectivity does not occur.

Another excellent example of the catalytic importance of crystallite morphology is the V/P/O catalyst used for the oxidation of linear four-carbon hydrocarbons to maleic anhydride. Selective catalysis to maleic anhydride is assumed to take place on certain surfaces of $(VO)_2P_2O_7$. In good catalysts, this phase has a plate-like morphology despite the fact that it does not have a layer structure. This habit results from the precursor, $(VO)_2H_4P_2O_9$, which has a layered structure and which dehydrates topotactically to $(VO)_2P_2O_7$ resulting in the desired morphology.[25] Thus, the best $(VO)_2P_2O_7$ catalysts are prepared through layered percursors such as $(VO)_2H_4P_2O_9$.

5.4 ZEOLITES

Zeolites are crystalline materials with open framework structures. Their pores are so open that various organic molecules can diffuse in and out of the channel structures. Strictly speaking, zeolites are aluminosilicates, but the term is also applied to other compositions with the same or related structures, such as the SiO_2 and $AlPO_4$ molecular sieves. The zeolite pores of molecular dimensions can be used for shape-selective catalysis since some molecules or intermediates are too large for the pores of the zeolite structure. Another important property of some zeolites is their exceptional acidity.

The way in which the AlO_4 and SiO_4 tetrahedra are linked together by corners to form frameworks is illustrated in Fig. 5.7. There are many different ways in which the tetrahedra may be linked together. About 35 zeolites occur naturally, and about 25 of these have been synthesized. Another hundred or so new zeolites have been synthesized over the last 30 years.

Since the framework composition of zeolite is MO_2, it will be neutral for an all-silica zeolite or for a strictly stoichiometric $AlPO_4$ zeolite. Most zeolites have a negatively charged network, i.e. $(Si_{1-x}Al_xO_2)^{x-}$, which is balanced by interstitial cations or protons, e.g. $NaSiAlO_4$

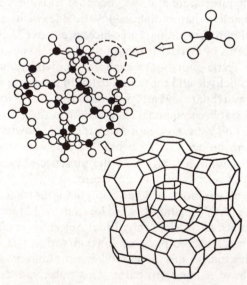

Fig. 5.7. Structure of zeolite X or Y showing tetrahedra linkage after Haynes.[26]

Zeolites are basically metastable phases, but for kinetic reasons they may be very stable, even up to temperatures of 800°C. They are generally synthesized from aqueous, alkaline gel precursors which are slowly crystallized at 100–400°C. The gel itself is of course highly disordered, but it already contains some zeolite structural features such as corner-shared tetrahedra linked to form rings. Crystallization is an ordering of the tetrahedral linkages and involves expelling considerable water from the network. Generally, the structure that first forms from the gel later converts to another framework structure from which even more water has been expelled. Thus, different phases may result as a function of synthesis time. Each phase is successively less open and more stable than its predecessor. Given enough time, the phase or phases formed will be thermodynamically stable, but then they may not be zeolites.

Zeolites may also be synthesized utilizing large bulky cations such as quaternary ammonium hydroxides as templates. This approach led to the first laboratory synthesis of certain naturally occurring zeolites, and also led to the synthesis of some new and technically important zeolites such as ZSM5. Another approach has been to use quaternary ammonium polymers as organic cations. Stacking faults in some zeolites cause much of the pore space to be unavailable for catalysis, but by using long-chain polymer cations these faults can be prevented so that all the pore space becomes available.[27]

The tunnels which exist in zeolite structures can be imaged by electron microscopy. This can be useful for direct structure determination and more importantly for investigation of defects and intergrowths which may have significant impact on catalytic properties.[28,29]

The number of acid sites within zeolite pores is proportional to the aluminium content in the framework providing all interstitial cations have been eliminated.[30] These are Brønsted acid sites when the proton concentration equals the aluminium concentration, i.e. $H_xSi_{1-x}Al_xO_2$. However, the Brønsted acid sites become converted to Lewis acid sites when the proton concentration drops below the aluminium concentration, as occurs during severe dehydration. Either Brønsted or Lewis acid sites may be utilized for catalysis. The acid strength of some sites in zeolites is higher than that of sulphuric acid.

In view of the importance of aluminium in determining the nature of acid sites in zeolites, the actual distribution of aluminium and silicon in the framework is of considerable interest. One would expect aluminium cations to avoid each other.[31] Thus, Al–O–Al linkages tend not to occur, and zeolites with an Al/Si ratio greater than 1 tend not to form. When the Al/Si ratio approaches 1, long-range order of Si and Al occurs to avoid the Al–O–Al linkage. Although structures with long-range order do not exist at lower Al/Si ratios, Si NMR studies have confirmed the absence of Al–O–Al linkages in such cases.[32] The variation in the Al/Si ratio from the centre to the exterior of a zeolite crystallite depends on synthesis conditions, but can increase, decrease, or remain nearly constant.[33]

Zeolites possess a special type of catalytic selectivity due to their regular pores of molecular dimensions. This shape selectivity can be exhibited in at least three different ways.

1. If a mixture of different gases is passed over a zeolite catalyst, only the smaller molecules will enter the pores and react. Thus, the zeolite selectivity catalyses the reaction of certain molecules in a mixture. For example, a metal loaded zeolite can selectively oxidize CO in a mixture containing both CO and n-butane.[34]

2. In passing a gas over a zeolite catalyst, some products of reaction may be too large to leave the zeolite pores. Only the smaller molecules leave, and the larger molecules stay behind and may continue to react. For example in the synthesis of xylenes, zeolites may be used to selectively produce the *para* isomer. The *ortho* and *meta* isomers are more bulky and tend to remain in the zeolite until they are converted to the *para* form.

3. There are also cases where the reactant and products can easily diffuse in and out of the zeolite pores, but where the transition state to produce a certain product involves an intermediate which is too large for any cavity within the pore structure of the zeolite. Thus, a particular reaction is effectively blocked, and shape selectivity results.

Many different cations can be exchanged into zeolites. This is done to adjust the acidity, to increase stability, and also to adjust the shape selective properties. Transition metal cations may be exchanged into zeolites to add additional catalytic functions. For example, cations which are ion exchanged into the zeolites and then reduced to the metal can be utilized for hydrogenation reactions.

Essentially all cracking of petroleum now utilizes a zeolite based catalyst which is several orders of magnitude more active than the previously used amorphous aluminosilicates. It is primarily the acidity which is important for this application; thus, a large pore zeolite which occurs naturally, faujasite (Fig. 5.7), is used. Due to coking, the lifetime of this catalyst is only seconds, but the catalyst is continually renewed by passing it into a regeneration zone where the coke is burned off and then returning it to the reaction zone.

A synthetic zeolite not known in nature is currently having a great impact on catalysis. This is the ZSM5 zeolite discovered at Mobil.[35] It has channels with diameters of about 5 Å, and is currently being used in a process to convert methanol directly to gasoline, and for xylene isomerization.

5.5   SULPHIDE CATALYSTS

Sulphide catalysts find wide application only for the hydrodesulphurization reaction. However, this reaction is of great importance since natural gas, petroleum, and coal all contain sulphur which would ideally be removed

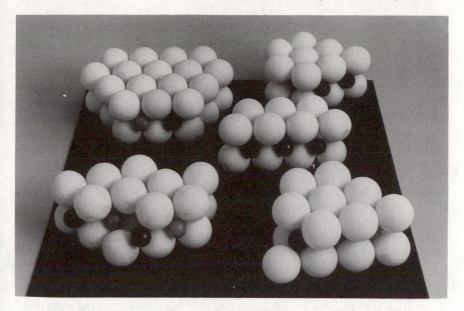

Fig. 5.8. Schematic of $Co/MoS_2/Al_2O_3$ catalyst after Topsoe *et al.*[36]

before chemical processing or fuel utilization. This reaction can be schematically represented as

$$-CH-S- + 2H_2 \rightarrow -CH_3 + H_2S.$$

Other reactions, such as hydrogenations or hydrodenitrogenations, normally occur simultaneously.

The usual catalyst for hydrodesulphurization is commonly referred to as a cobalt molybdate catalyst. In fact, the catalyst consists of highly dispersed Co/Mo/S on a Al_2O_3 (cub) support in which some of the cobalt finds its way into the Al_2O_3 (cub) phase. The sulphide phase itself consists of small slabs of the MoS_2 structure which may be only one layer thick.[36] Low-spin Co^{2+} is located at the surfaces of the MoS_2 slabs, probably at the edges (Fig. 5.8). One could then describe this catalyst as Co^{2+} supported on MoS_2, which in turn is supported on Al_2O_3 (cub). It should be noted that this allows for a straightforward solution to charge balance on surface termination of the MoS_2 phase. One may regard the situation formally as $Mo_{1-x}^{4+}Co_{2x}^{2+}S_2$ even though Co^{2+} predominantly occurs only at surface sites. The point is that there are extra cations available so the edge surfaces may be terminated without the necessity of cation vacancies. This is especially important if the catalytically active sites are surface cations which might exist only at slab edges. Such surface cation sites are almost certainly required both for H_2 dissociation and for abstraction of sulphur from the organic reactant.

Another role of Co^{2+} ions could be to keep the MoS_2 slabs small and thus present a higher surface area. Once Co^{2+} ions have situated themselves at edge sites, the slab will not tend to continue growing because local charge neutrality has been achieved.

5.6 REFERENCES

1. Barteau, M. A. and Madix, R. J., Chapter 4 (p. 95) in *The chemical physics of solid surfaces and heterogeneous catalysis* (eds King, D. A. and Woodruff, D. P.), Elsevier Pub. Co. (1982).
2. Rao, V. N. M., *Proceedings of the 8th North American Meeting of the Catalysis Soc.*, Paper D-23 (1983); and Adams, M. A., Roselaar, L. C., and Webster, D. E., *ibid.*, Paper D-24.
3. Ertl, G., *Catal. Rev. Sci. Eng.*, **21**, 201 (1980).
4. Biloen, P., Helle, J. N., van der Berg, F. G. A., and Sachtler, W. M. H., *J. Catal.*, **81**, 450 (1983).
5. den Otter, G. J. and Dautzenberg, F. M., *J. Catal.*, **53**, 116 (1978).
6. Inui, T., Sezume, T., Miyaji, K., and Takegami, J. C. S., *Chem. Comm.*, **874** (1979).
7. Tauster, S. J., Fung, S. C., and Garten, R. J., *J. Am. Chem. Soc.*, **100**, 170 (1978).
8. Sleight, A. W., *Science*, **208**, 895 (1980).
9. Goodman, D. W., *Acc. Chem. Res.*, **17**, 194 (1984).
10. Ponec, V. in *Advances in catalysis*, Vol. 32 (eds Eley, D. D., Pines, H., and Waisz, P. B.), Academic Press, New York (1983).
11. Jacobson, R. L., Kluksdahl, H. E., McCoy, C. S., and Davis, R. W., *Proc. Am. Pet. Inst. Sec. 3*, **49**, 504 (1969).
12. Stone, F. S., *J. Solid State Chem.*, **17**, 271 (1975).
13. Stone, F. S., *Chem. Phys. Aspects Catal. Oxid.*, **459** (1980).
14. Marcilly, Ch. and Delmon, B., *J. Catal.*, **24**, 336 (1972).
15. Adams, C. R. and Jennings, T. J., *J. Catal.*, **3**, 549 (1969); **2**, 63 (1962).
16. Keulks, G. W., *J. Catal.*, **19**, 232 (1970).
17. Sachtler, W. H. and deBoer, N. H., *Proc. Int. Congr. Catal.*, 3rd 1964, p. 252 (1965).
18. Sleight, A. W. in *Advanced materials in catalysis* (eds Burton, J. J. and Garten, R. L.), Academic Press, New York (1977).
19. Gai, P. L., *J. Solid State Chem.*, **49**, 25 (1983).
20. Sleight, A. W., *Farad. Disc. Chem. Soc.*, **72**, 305 (1981).
21. Gai, P. L., Boyes, E. D., and Bart, J. C. J., *Phil. Mag. A*, **45**, 351 (1982).
22. Machiels, C. J. and Sleight, A. W., *J. Catal.*, **76**, 238 (1982).
23. Firment, L. E. and Ferretti, A., *Surface Science*, **129**, 155 (1983); Ohuchi, F., Firment, L. E., Chowdhry, U., and Ferretti, A., *Proc. Am. Vacuum Soc. Meet.*, Boston (1983).
24. Machiels, C. J., Chowdhry, U., Staley, R. H., Ohuchi, F., and Sleight, A. W. in *Catalytic conversions for synthesis gas and alcohols to chemicals* (ed. Herman, R. G.), Plenum Pub. Corp., New York, p. 413 (1984).
25. Torardi, C. C. and Calabrese, J. C., *Inorg. Chem.*, **23**, 1308 (1984).
26. Haynes, H. W., Jr., *Catal. Rev.*, **17**, 276 (1978).

27. Rollman, L. D., *Adv. Chem.*, **173**, 387 (1979).

28. Thomas, J. M., *Proceedings 8th International Congress on Catalysis*, Berlin, I31 (1984).

29. Bursill, L. A., Lodge, E. A., and Thomas, J. M., *Nature*, **286**, 111 (1980); Millward, G. R. and Thomas, J. M., *J. Chem. Soc. Chem. Commun.*, **77** (1984); Terasaki, O., Thomas, J. M., and Ramdas, S., *ibid.*, **216** (1984).

30. Hagg, W. O., Lago, R. M., and Weisz, P. B., *Nature*, **309**, 589 (1984).

31. Pauling, L., *The nature of the chemical bond*, 3rd edn, Cornell University Press, New York, p. 550 (1960).

32. Vega, A. J., *ACS Symposium Series*, 218, 217 (1983).

33. Lyman, C. E., Betteridge, P. W., and E. F. Moran, *ACS Symposium Series*, **218**, 199.

34. Chen, N. Y. and Weisz, P. B., *Chem. Eng. Progr., Symp. Ser.*, **63**, 86 (1967).

35. Meisel, S. L., McCullough, J. P., Lechthaler, C. H., and Weisz, P. B., *Chem. Technol.*, **6**, 86 (1976).

36. Topsoe, H., Clausen, B. S., Candia, R., Wivel, C., and Morup, S., *Bull. Soc. Chim. Belg.*, **90**, 1189 (1981).

6 Intercalation reactions of layered compounds

Allan J. Jacobson

6.1 INTRODUCTION

Intercalation reactions are reactions of solids in which a guest molecule or ion is inserted into a solid lattice without major rearrangement of the solid structure. Unlike normal solid state processes, intercalation reactions do not involve extensive bond breaking and consequently require special chemical and structural features in order to occur. In particular, intercalation requires that the host lattice has a strong covalent network of atoms which remains unchanged on reaction and that there are vacant sites in the structure which are interconnected and of suitable size to permit diffusion of the guest species into the solid. Reactions may occur near ambient temperature but, in general, the reaction temperature needs to be sufficiently high to ensure reasonable diffusivities of the guest molecules or ions, but not so high that rearrangements of the strong bonds in the structure occur. The requirement that part of the solid structure remains unchanged on reaction implies that intercalation compounds will often be metastable phases. The general reaction can be written schematically, following the notation used by Schöllhorn in ref. 3, as

$$x\mathrm{G} + \square_x[\mathrm{H}] \rightleftarrows \mathrm{G}_x[\mathrm{H}] \qquad (6.1)$$

where G, $\square$, H refer to the guest species, vacant lattice sites, and the host lattice, respectively. As indicated in eqn (6.1), intercalation reactions are usually reversible.

Despite the structural and chemical restrictions on intercalation reactions, a wide range of different types of compounds satisfy the requirements for at least some guest species. Examples include chain structures, layer lattices, and three dimensionally connected frameworks containing either tunnels or intersecting channels. The focus of this chapter is on layered structures, which have the special feature of being able to accommodate very large guest molecules by free adjustment of the interlayer separation. Discussion of reactions of the other different structure types can be found in the general references.[1-21]

A survey of different types of layer structure, classified by the charge on the layers, is given in Table 6.1. All of the compounds are characterized

Table 6.1 *Layered structures for intercalation reactions*

Neutral layers		Refs.
Graphite		11
$Ni(CN)_2$		3, 8
MX_2	(M = Ti, Zr, Hf, V, Nb, Ta, Mo, W; X = S, Se, Te)	1–9
MPX_3	(M = Mg, V, Mn, Fe, Co, Ni, Zn, Cd, In; X = S, Se)	1, 3, 11
MoO_3, V_2O_5		1, 3
$MOXO_4$	(M = V, Nb, Ta, Mo; X = P, As)	12
MOX	(M = Ti, V, Cr, Fe; X = Cl, Br)	1, 3
Negatively charged layers		
$CaSi_2$		13
AMX_2	(A = Group IA; M = Ti, V, Cr, Mn, Fe, Co, Ni; X = O, S)	3
Clays and layered silicates		10, 14, 15
Titanates	e.g. $K_2Ti_4O_9$	16
Niobates	e.g. $K[Ca_2Na_{n-3}Nb_nO_{3n+1}]$ $3 \leqslant n \leqslant 7$	17
$M(HPO_4)_2$	M = Ti, Zr, Hf, Ce, Sn	1, 18
Positively charged layers		
Hydrotalcites	$LiAl_2(OH)_6OH \cdot 2H_2O$, $Zn_2Cr(OH)_6Cl \cdot 2H_2O$	19

by strong intralayer bonding and weak interlayer interactions. In the compounds with neutral layers, the interlayer bonding is primarily of the van der Waals type and the interlayer space is a connected network of empty lattice sites. In the charged layer systems, the layers are held together by weak electrostatic forces and the interlayer sites are partially or completely filled by ions or by a combination of ions and solvent molecules. The compounds in Table 6.1 may be further classified by the type of chemistry that occurs on intercalation. In ion exchange or molecule exchange reactions, the framework layer charge remains unchanged. Such reactions are observed for clays, sheet silicates, ternary alkali metal titanates and niobates, acid phosphates and hydrotalcites. In many reactions of transition metal compounds, however, the layer charge does change. The solid lattice is either oxidized or reduced and overall charge balance is maintained by intercalation or deintercalation of ions into or from the interlayer space. The transition metal compounds (Table 6.1) which show this kind of redox behavior are the main focus of this chapter. Examples of some unusual types of transition metal reactions specific to certain classes of compound are also described.

A redox intercalation reaction of a layered compound is illustrated schematically in Fig. 6.1 for the specific case of the reaction of lithium with titanium disulphide. Lithium metal reduces the layers of titanium disulphide,

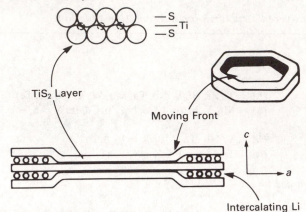

Fig. 6.1. Schematic representation of the process of lithium intercalation into layered titanium disulphide (adapted from ref. 22).

Fig. 6.2. Optical micrograph of a partially intercalated TiS_2 crystal (from ref. 22).

and lithium cations are inserted into the vacant interlayer sites to compensate for the negative layer charge caused by the electron transfer. The reaction starts at the crystal edges and the reaction front moves through the crystal until the final product is obtained. The final product of the reaction has a structure that is expanded in a direction perpendicular to the layers. The intermediate stage, which is shown schematically in Fig. 6.1, is shown for a real system in Fig. 6.2 in an optical micrograph of a partially reacted TiS_2 crystal.[22]

The redox intercalation process can be written in a more general form by an extension of eqn (6.1).

$$xG^+ + xe^- + \square_x[H] \rightleftharpoons G_x[H]. \tag{6.2}$$

Compounds formed by intercalation via eqn (6.2) can subsequently further react by ion exchange and solvent intercalation. The reaction has been written for cation intercalation, but anion insertion is also possible. For example, the intercalation of oxygen into non-stoichiometric mixed-valence copper oxides has been discussed by Michel and Raveau.[23] The general class of reactions defined by eqn (6.2) have been called reversible topotactic redox reactions[7] and are currently of interest for several reasons. Intercalation reactions provide new routes for the synthesis of solids with kinetic rather than thermodynamic stability, they permit controlled systematic changes in the physical properties, particularly the electronic and magnetic properties of the host lattices, and they are also of practical use in various kinds of electrochemical devices such as sensors, electrochromic displays, and batteries. More detailed discussion of the uses and potential applications of redox reactions can be found in refs. 2, 3, and 21.

The chemistry of the dichalcogenides is described in some detail in the first part of this chapter in order to illustrate the characteristic features of redox intercalation reactions. Subsequently, reactions of a small number of other classes of transition metal host lattices, including the metal phosphorus trisulphides, the metal oxyhalides and metal oxides, are described. These solids show, in addition to redox chemistry, some other unusual kinds of reactivity and serve to illustrate the diversity of interlamellar reaction pathways.

6.2 TRANSITION METAL DICHALCOGENIDES

6.2.1 Structures

The transition metal dichalcogenides of Groups IV, V, and VI have layered structures in which the transition metal (M) occupies either octahedral (trigonal antiprismatic) or trigonal prismatic sites between two layers of chalcogen atoms (X). Each chalcogen layer is close-packed as shown in Fig. 6.3. The X–M–X sandwiches are then stacked in a direction perpendicular to the chalcogen layers to complete the structure. Bonding within a layer is strong whereas the interactions between layers, which arise from van der Waals forces, are weak. Different combinations of octahedral and trigonal prismatic coordination for M and different stacking sequences of the layers give rise to a variety of polytypes. The different structures can be conveniently represented by projection in the $(11\bar{2}0)$ plane of a hexagonal cell. Each layer of atoms is labelled as A $(0, 0)$, B $(2/3, 1/3)$, C $(1/3, 2/3)$ with respect to its

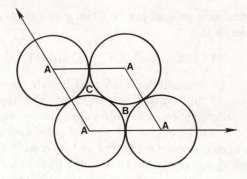

Fig. 6.3. Chalcogen layer indicating the stacking sequence notation.

translation in the xy plane from the origin at A (Fig. 6.3). Uppercase letters denote chalcogen layers, lower case represent the metal sites and empty sites. In this notation, the sequence AbC represents an octahedral site whereas trigonal prismatic coordination is indicated by AbA. Several commonly occurring stacking arrangements are shown in Fig. 6.4. The simplest example is represented by the notation AbC/AbC and designated 1T in the nomenclature of Brown and Beernsten.[24] All of the Group IV dichalcogenides adopt

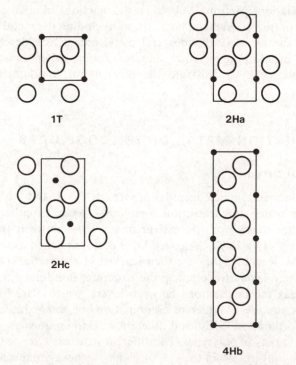

Fig. 6.4. The $(11\bar{2}0)$ projections of the common transition metal dichalcogenide structures.

this structure, in which all of the transition metal atoms have octahedral coordination. The unit cell is trigonal and contains one X–M–X layer. The same structure is adopted by CdI_2 and it is closely related to the structure of NiAs which is obtained by filling all of the interlayer octahedral sites. The most commonly occurring structures with metals in trigonal prismatic sites are designated are 2Ha and 2Hc (Fig. 6.4). These structures, which contain two layers per unit cell, are adopted by the sulphides and selenides of Groups V and VI, respectively, and are often referred to as 2H and 2H–MoS_2. The third possibility for a two-layer trigonal prismatic arrangement, 2Hb, is found for $TaSe_2$ and $NbSe_2$. Two trigonal prismatic three-layer sequences are possible but only the 3R sequence BcB/CaC/AbA has been found; the other sequence BcB/CbC/AbA is unknown.

A summary of examples of different stacking sequences and different transition metal coordinations is given in Table 6.2. In general, the Group IV metals and vanadium are octahedrally coordinated, whereas trigonal prismatic sites are occupied by the Group VI elements. Niobium and tantalum are found in both coordinations and the number of possible structures for the dichalcogenides of these elements is greatly increased by the possibility of alternating octahedral layers, AbC, with trigonal prismatic layers, AbA. An example of a mixed layer sequence, 4Hb, which is observed for both TaS_2 and $TaSe_2$, is shown in Fig. 6.4.

In an intercalation reaction, guest species are inserted into the empty coordination sites between the layers. For small guest ions such as alkali metal cations, octahedral, tetrahedral, or trigonal prismatic sites between the chalcogenide layers can be occupied. Intercalation often produces a change in the layer stacking sequence.

6.2.2 Alkali metal intercalation

6.2.2.1 Synthesis

Intercalation reactions of the dichalcogenides with alkali metals are redox reactions in which the host lattice is reduced by electron transfer from the alkali metal. Electrochemical synthesis is therefore generally applicable. The dichalcogenide is used as the cathode of an electrochemical cell with an alkali metal anode and a non-aqueous solution of the alkali metal salt as the electrolyte. For example, lithium and sodium intercalation reactions have been studied using cells of the type $Li/LiClO_4$–dioxolane/MX_2 and Na/NaI–propylene carbonate/MX_2. The reactions proceed spontaneously to form the intercalation compound if the cell is short-circuited or, alternatively, a reverse potential can be applied to control the composition of the final product. Apart from their application in synthesis, such electrochemical cells can be used to obtain detailed thermodynamic information and to establish phase relations by measuring, at equilibrium, the dependence of the cell voltage (partial molar free energy) on composition.

Table 6.2 *Structures of layered dichalcogenides*

Designation	Sequence	Space group	Examples	M Coordination
1T	AbC|	P$\bar{3}$m1	MX$_2$ (M = Ti, Zr, Hf, V; V = S, Se, Te)	TAP
2Ha	BaB|CaC	P6$_3$/mmc	MX$_2$ (M = Ta, Nb; X = S, Se)	TP
2Hb	BaB|CbC	P$\bar{6}$m2	TaSe$_2$, NbSe$_2$	TP
2Hc	BcB|CbC	P6$_3$/mmc	MX$_2$ (M = Mo, W; X = S, Se), MoTe$_2$	TP
3R	BcB|CaC|AbA	R$\bar{3}$m	MX$_2$ (M = Ta, Nb, Mo; X = S, Se), WS$_2$	TP
4Hb	aB|CaC|BaC|BaB|Ca	P6$_3$/mmc	TaS$_2$, TaSe$_2$	TP, TAP

T = trigonal, H = hexagonal, R = rhombohedral, TAP = trigonal antiprismatic (octahedral), TP = trigonal prismatic

A variety of chemical reagents may also be used to form alkali metal intercalation compounds. The first compounds of the dichalcogenides were prepared by Rüdorff who reacted the host lattices with liquid ammonia solutions of the alkali metals.[25] The compounds formed by this procedure contain co-intercalated ammonia molecules which are difficult to remove completely and the high alkali metal thermodynamic activity can lead to destructive side reactions. However, the method is applicable to all of the alkali metals and has been used successfully for the synthesis of a variety of compounds including the low stability A_xMoS_2 phases (A = K, Rb, Cs).[2,3,5]

Other chemical reagents which give lower alkali metal activities, for example, sodium and potassium naphthalide and benzophenone solutions in tetrahydrofuran, have been used.[26] A hexane solution of n-BuLi has an effective potential of about 1.0 V relative to Li/Li^+ and is a mild reagent for lithium intercalation.[27] The reaction is irreversible and octane is the major organic product though some butene and butane are also formed.

Some non-stoichiometric phases can also be synthesized at high temperatures. For example, the compounds A_xMX_2 (M = V, Nb, Ta, X = S, Se) can be prepared by direct reaction of the elements at 800°C in sealed tubes or by reaction of the dichalcogenides with the alkali metals.[28] Alternatively, mixed oxides or ternary oxide phases are converted to sulphides by reaction at high temperature in H_2S or CS_2.[29] The high temperature intercalation compounds can be used for synthesis of new metastable layered phases by chemical or electrochemical oxidative removal of the guest species. Examples of this approach include the synthesis of VS_2, $CrSe_2$ and $Cr_{0.5}Ti_{0.5}S_2$.[30–32] None of these phases can be obtained by the usual high temperature methods.

6.2.2.2 Phase relations

The phase relations for the alkali metal–TiS_2 and ZrS_2 systems (Fig. 6.5) illustrate most of the general features of alkali metal dichalcogenide systems. Further details and discussion of other dichalcogenides can be found in refs. 2, 5, and 6.

The simplest system is Li_xTiS_2 which is a single phase over the entire composition range, as indicated by the smooth variation of the c axis lattice parameter with composition[33] and by the equilibrium electrochemical data[34] (Fig. 6.6). The intercalation reaction leads to occupation of the octahedral interlayer sites and the final product, $LiTiS_2$, is isostructural with the high temperature phases $LiVS_2$ and $LiCrS_2$.[29] The intercalation compounds formed by the other alkali metals are more complex and show two phase behaviour up to some low value of x, implying that a minimum energy of reaction is required to open up the van der Waals gap. Staging can provide a mechanism to minimize this energy and is observed for all of the alkali metal compounds at low x values except for lithium. Staging results in the formation of compounds with alternating sequences of filled and empty

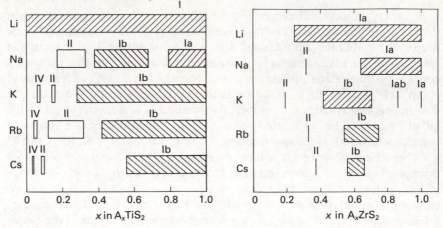

Fig. 6.5. Phase relations for the alkali metal intercalation compounds of TiS_2 and ZrS_2 (data from ref. 2.3(ii)).

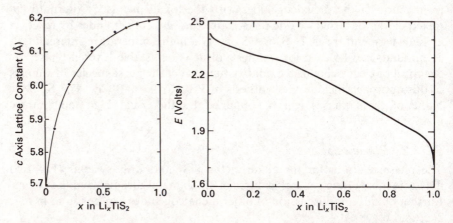

Fig. 6.6. Variation in the c axis lattice parameter (left—from ref. 33) and the cell voltage (right—adapted from ref. 33) for the single phase intercalation compound $Li_x TiS_2$.

van der Waals gaps, as shown in Fig. 6.7, with higher stages correspond to decreasing alkali concentrations. Problems with this simple picture in understanding transformations between odd and even numbered stages are overcome if the stages are assumed to exist within a crystal in the form of local domains (Fig. 6.7).[2,3] In either case, the existence of staging indicates that at low concentrations the energy of a system with a homogeneous distribution of cations throughout the interlayer space is higher than that for a phase with the cations concentrated in a fraction of the interlayer sites.

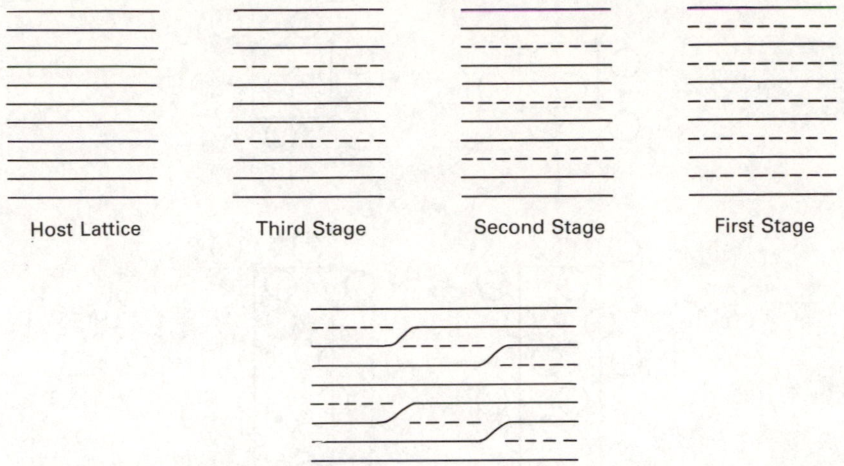

Fig. 6.7. Schematic representation of staging in intercalation compounds. The lower part of the figure represents the domain model proposed by Herold (ref. 2.3). The continuous line denotes an MX_2 layer, the dashed line a layer of alkali metal ions.

The origin of this effect has been discussed in detail for graphite systems.[35] In the TiS_2 systems, second stage compounds are formed for all alkali metals except lithium, and fourth stage phases occur with K, Rb and Cs.

The structures of the first stage compounds illustrate how the dichalcogenide layers are restacked to optimize the guest coordination and the interlayer interactions (Fig. 6.8). Structure I is observed for $LiTiS_2$ and has the same hexagonal sulphur stacking as TiS_2. Structure type Ia, in contrast, involves a restacking such that the chalcogen layers become cubic close-packed. The alkali metals in Ia are in octahedral interlayer sites. Structure Ib is also restacked, but in this case the rearrangement results in trigonal prismatic coordination for the alkali cations.

The alkali metal coordination depends on its ionic radius and also on the particular composition and the ionicity of the specific host lattice. Chalcogen–chalcogen repulsions destabilize trigonal prismatic coordination and are larger at higher layer charges and for the more ionic transition metal–chalcogen bonds. Consequently, the alkali cation coordination generally changes from trigonal to octahedral as the layer charge increases. For similar reasons, the Ib structure with trigonal prismatic alkali cation sites is observed for Na_xTiS_2, but not for Na_xZrS_2 which is more ionic.

Little is known about the distribution of cations in the interlayer space in the non-stoichiometric phases, but studies on Na_xTiS_2 show that ordered arrangements of alkali cations and vacancies can occur.[36] It is likely that other systems, if examined closely, would show similar effects.

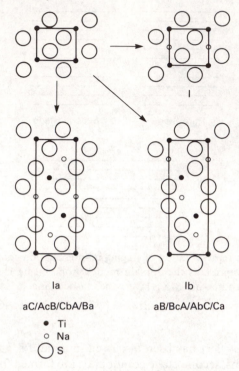

Ia   Ib

aC/AcB/CbA/Ba   aB/BcA/AbC/Ca

● Ti
○ Na
○ S

Fig. 6.8. Structures of the alkali metal intercalation compounds of TiS_2 and ZrS_2.

6.2.3 Hydrated alkali metal intercalation compounds

The alkali metal intercalation compounds of the dichalcogenides readily form hydrated phases on exposure to water.[37-39]

$$AMX_2 + H_2O \longrightarrow A_x(H_2O)_y MX_2 + (1 - x)A^+ + (1 - y)OH^-$$
$$+ (1 - y)/2H_2. \tag{6.3}$$

The value of x in eqn (6.3) varies with host lattice and depends on the reducing power of the intercalated host relative to water. For example, approximate x values of 0.5, 0.33, and 0.1 are found for hydration of the sodium intercalation compounds of TiS_2, TaS_2, and MoS_2, respectively. Hydrated phases may be prepared by several other procedures, including electrochemical reduction in the corresponding aqueous electrolyte and chemical reduction using aqueous reducing agents such as sulphide, borohydride and dithionate. Aqueous sodium dithionite, which is a strong reducing agent in alkaline solution, is particularly convenient. The formation of hydrated intercalation compounds has also been observed in aqueous alkali hydroxide solutions without addition of a reducing agent. It seems likely that the effective reducing agent in these reactions is sulphide or selenide produced by partial hydrolysis of the host lattice.[40]

The water content (y) in eqn (6.3) depends on the ambient water pressure and on the hydration energy of the interlayer cation. The effects of these two variables are illustrated by the data for TaS_2 compounds shown in Fig. 6.9. In both examples, two extremes of behaviour are observed and are illustrated schematically in Fig. 6.10. At low water partial pressures or for cations with low hydration energies (small charge-to-radius ratio) a monolayer of water molecules is present between the chalcogen layers. At higher pressures or at higher cation charge-to-radius ratio a bilayer structure is observed. In the $Na_{1/3}(H_2O)_y TaS_2$ example shown in Fig. 6.9, the α and δ phases, respectively, are the monolayer and bilayer phases of the type shown schematically in Fig. 6.10.[41,42] The β phase is also a monolayer hydrate, but additional water molecules are accommodated by displacing the sodium cations into sites between the sulphur and water layers.

The remaining hydrated phase (γ) in the $Na_{0.33}(H_2O)_y TaS_2$ system is characterized by a random stacking arrangement of monolayer and bilayer structures. Stacking disorder of this type is unusual for the dichalcogenides but has often been observed in hydrated clay minerals.[43] The hydration behaviour of the intercalation compounds of the dichalcogenides is simpler than that observed in layered oxide systems. In clays and transition metal oxides, hydrogen bonding between water molecules and the layers is as important an interaction as cation hydration. As a consequence, and in contrast to the dichalcogenides, several oxide systems with neutral layers form stable hydrates, e.g. $VOPO_4 \cdot 2H_2O$, $MoO_3 \cdot 2H_2O$ and kaolinite.

In the hydrated alkali metal intercalation compounds, both the alkali cations and the water molecules have high mobilities. Consequently, these phases are useful starting materials for the synthesis of new compounds by ion or solvent exchange. Compounds with large interlayer separations can readily be formed, particularly, when the expanded lattice of the hydrated phase is used as a starting material. Intercalation compounds containing large cations or solvent molecules can also be prepared by electrochemical reduction, although kinetic problems are frequently encountered.[44] Some examples of compounds containing other solvents or complex cations are given in Table 6.3.

In solvents with high dielectric constants, e.g. *N*-methyl formamide, the application of weak shear forces to solvated alkali metal systems results in complete dispersion of the solid intercalation compound and the formation of homogeneous dispersions or colloidal solutions of negatively charged dichalcogenide layers.[39] Such colloids follow Beer's law at low concentrations and show streaming birefringence due to the particle anisotropy. Similar behaviour is also observed for $H_{0.25}TaS_2$, which shows reversible swelling in water as a function of pH and gives quite stable colloids when shear forces are applied in the presence of a surfactant.[45]

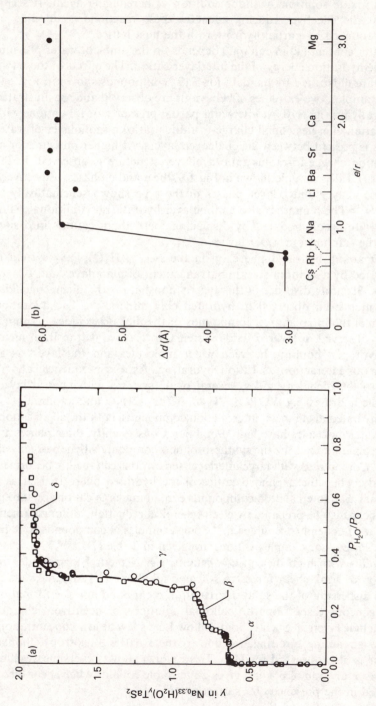

Fig. 6.9. Hydration behaviour of TaS$_2$ intercalation compounds as a function of (a) water pressure for Na$_{0.33}$(H$_2$O)$_y$TaS$_2$ (from ref. 41) and (b) charge-to-radius ratio of the guest cation (from ref. 39).

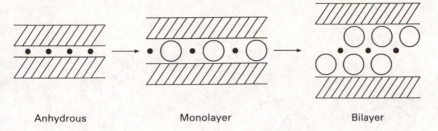

Anhydrous   Monolayer   Bilayer

Fig. 6.10. Formation of a monolayer and bilayer hydrate by progressive hydration of an alkali metal intercalation compound.

Table 6.3 *Intercalation compounds of TaS_2 with different solvents and complex cations (from refs. 39, 44)*

Cation	Solvent	Interlayer spacing (Å)
Na^+	H_2O	11.87
Na^+	Ethanol	10.04
Na^+	Formamide	16.35
Na^+	M-Methyl formamide	18.91
Na^+	Dimethyl sulphoxide	18.63
$Pt(NH_3)_4^{2+}$	H_2O	9.18
$Fe(2,2'\text{-bipyridine})^{2+}$	H_2O	14.92
$(C_5H_5)_2Co^+$	H_2O	11.45
$C_7H_7^+$	H_2O	13.15

Colloidal dispersions can be reflocculated by addition of electrolyte solutions of other cations. Ion exchange occurs at the same time and the overall process is a good way to circumvent problems with ion exchange kinetics. For example, an intercalation compound of the large cluster cation $[Fe_6S_8(P(C_2H_5)_3)_6]^{2+}$ was synthesized by flocculation of a colloidal solution of $Na_{1/3}TaS_2$ in NMF/H_2O.[46] Even though the reflocculation reaction is very rapid, the products may be well ordered. In the above example, the product was sufficiently crystalline to enable the orientation of the Fe_6S_8 core to be determined by X-ray diffraction from an oriented film (Fig. 6.11).

6.2.4 Intercalation of organic molecules

Intercalation compounds of the dichalcogenides with organic molecules can be formed under strictly anhydrous conditions by direct thermal reaction at temperatures up to about 200°C. Early studies of transition metal dichalcogenide intercalation largely focused on this type of reaction after the initial

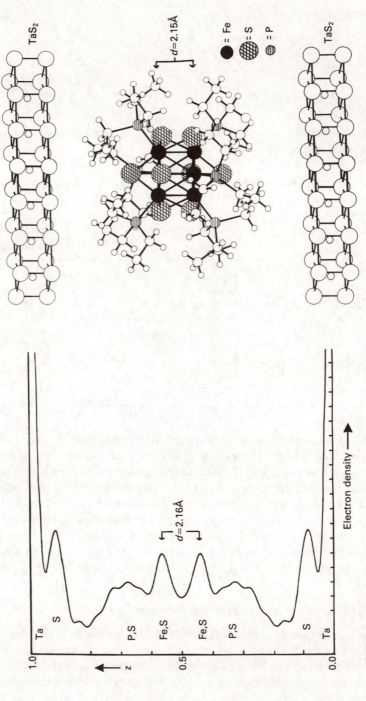

Fig. 6.11. The intercalation compound formed by $[Fe_6S_8(P(C_2H_5)_3)_6]^{2+}$ with TaS_2; right—schematic representation, left—the measured electron density projected perpendicular to the layers (from ref. 46).

Table 6.4   *Types of organic molecules that form intercalation compounds with dichalcogenides*

Amines	RNH_2, R_2NH, R_3N, $H_2N(CH_2)_nNH_2$
Phosphines	R_3P
Amides	$RCONH_2$, $CO(NH_2)_2$
Amine oxides	Pyridine N-oxide
Phosphine oxides	R_3PO
N-Heterocycles	Pyridine
Isocyanides	RNC

discovery of amide intercalation in TiS_2[47] and the subsequent observation of the increase in the superconducting transition temperature of $2H$–TaS_2 on intercalation of primary organic amines.[48] Since the early work, a large number of compounds with several different general classes of organic molecules have been synthesized (see Table 6.4). Several studies of physical properties of organic intercalation compounds, such as optical spectra and magnetic susceptibilities, have established charge transfer from the organic guest to the host layers. It is now generally believed that the origin of the charge transfer is through redox chemistry involving the guest molecules themselves and that the resulting intercalation compounds are chemically similar to the solvated systems discussed in the previous section. Intercalation of ammonia and pyridine have been examined in most detail and illustrate guest molecule redox chemistry.

6.2.4.1   Ammonia

Ammonia may be intercalated into TaS_2 or the other dichalcogenides by direct reaction at ambient temperature. A convenient procedure for carrying out this reaction is to distill liquid ammonia from sodium onto the dichalcogenide and then allow the reaction mixture to warm up to room temperature. Onset of the reaction is marked by an increase in the total sample volume, which if the dichalcogenide is highly stoichiometric occurs very rapidly. The reaction proceeds to a limiting composition NH_3MX_2 which is a first stage compound. This composition, however, is not very stable and ammonia is readily removed under vacuum at room temperature to form a second stage of composition $(NH_3)_{0.5}MX_2$. Further removal of ammonia requires higher temperatures.

The ammonia molecules occupy trigonal prismatic sites between the dichalcogen layers and consequently the layer stacking sequence changes during intercalation. For example, in the 1T structures (TiS_2, ZrS_2) the layers translate to give rhombohedral cells analogous to the 1b-type structure of $KTiS_2$ (Fig. 6.8). NMR measurements have shown that the ammonia molecules located in the trigonal prismatic interlayer sites are oriented with their three fold axes parallel to the dichalcogenide layers, indicating only weak lone pair interactions with the layers.[49] Careful study of the reaction stoichiometry, prompted by this observation, led to the conclusion that ammonia oxidation was involved and that the overall mechanism of reaction could be summarized by[50]

$$(1 + x/3)NH_3 + TaS_2 \longrightarrow (NH_4)_x(NH_3)_{1-x}TaS_2 + x/6N_2. \quad (6.4)$$

The reaction product has x in the range 0.1 to 0.3 and contains ammonium ions solvated by neutral ammonia molecules. A thermal analysis of the second stage NH_3 intercalation compound of TiS_2 shows two distinct steps corresponding to loss of NH_3 and NH_4^+, confirming the reaction mechanism.[51] The ammonia orientation in this model is a natural consequence of ion dipole interactions between cations and solvating molecules.

6.2.4.2   Pyridine

While ammonia can be taken as a model system for primary organic amines, pyridine represents the simplest N heterocyclic Lewis base. The general chemistry is closely analogous to that shown by ammonia. Direct reaction of the dichalcogenide, e.g. $2H–TaS_2$, with pyridine at elevated temperature (200°C) leads to the formation of a first stage intercalation compound with limiting composition $(pyridine)_{0.5}TaS_2$. As in the ammonia system, structural measurements have shown that pyridine is oriented with the long C–N axis and hence the nitrogen lone pair parallel to the layers. Chemical studies established that intercalation involves redox chemistry of the guest molecules leading to the formation of bipyridines.[52] The oxidation of pyridine by TaS_2 is analogous to the oxidation of ammonia.

$$2\ pyridine \longrightarrow bipyridine + 2H^+ + 2e$$

$$pyridine + H^+ \longrightarrow pyridinium$$

$$TaS_2 + xe \longrightarrow TaS_2^{x-}$$

The product of the reaction is $(py)_{0.5-2x}(pyH)_x(bipy)_{x/2}TaS_2$ and the intercalation compound contains solvated cationic species inserted between negatively charged layers.

While the reaction mechanisms for most organic molecule intercalation reactions are not known in as much detail as those for pyridine and ammonia, it appears likely that other neutral molecules react by similar or more complex redox processes involving the guest molecules themselves.

6.2.5   Organometallic intercalation compounds

A second class of molecular intercalation compounds which can be prepared by direct reaction with the dichalcogenides comprises those formed with certain organometallic molecules. The first examples of this reaction were

Table 6.5 *Organometallic intercalation compounds of* ZrS_2

Organometallic guest	Stoichiometry (x in G_xMS_2)	Layer spacing (Å)
Cobaltocene	0.27	5.35
Chromocene	0.25	5.61
bis(η-Benzene) chromium	0.16	5.90
bis(η-Benzene) molybdenum	0.16	5.81
bis(η-Toluene) molybdenum	0.13	5.80
bis(η-Mesitylene) molybdenum	0.08	5.78

observed with chromocene and cobaltocene and a variety of host lattices.[53] The stoichiometries and layer separations for intercalation compounds with these molecules, and for some additional examples, are given in Table 6.5 for the host lattice, ZrS_2.[53,54] All of the organometallic molecules known to form intercalation compounds are good reducing agents and form stable cations. Compounds may be prepared by direct reaction with a hydrocarbon solution of the organometallic molecule or alternatively by electrochemical or ion exchange methods. Electrochemical intercalation of $Co(\eta\text{-}C_5H_5)_2^+$ by TaS_2 gives a compound with the same stoichiometry as that obtained for the compound prepared by direct reaction with neutral cobaltocene and confirms the redox nature of the process.[44] The magnetic susceptibilities of $[Co(\eta\text{-}C_5H_5)_2]_{1/4}TaS_2$ and $[Cr(\eta\text{-}C_5H_5)_2]_{1/4}TaS_2$ are also consistent with complete electron transfer to the TaS_2 layers. The electron transfer mechanism accounts for the correlation which exists between the reducing power of the organometallic (measured in a relative way by its first ionization potential) and its ability to intercalate specific host lattices. For example, $Co(\eta\text{-}C_5H_5)_2$ with an ionization potential (IP) of 5.56 eV will intercalate both TaS_2 and FeOCl, but ferrocene, which is less reducing (IP is 6.88 eV), will only intercalate the more oxidizing host lattice, FeOCl. It should be noted, however, that several systems which meet the IP criterion and form stable cations, for example $(\eta\text{-}C_5H_5)_2MoH_2$, do not form intercalation compounds, either because reaction kinetics are very slow or because side reactions occur at the crystal edges which prevent opening the host lattice interlayers.

6.2.6 Summary

The intercalation reactions of the dichalcogenides are summarized in the reaction scheme below, adapted from ref. 7.

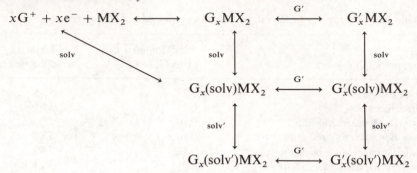

This description emphasizes the fact that all of the reactions are topotactic redox reactions.[7,9] Even when the intercalated species are apparently neutral molecules, guest molecule reactions, such as ammonia decomposition or pyridine coupling, serve to provide the necessary host lattice–guest molecule redox chemistry. A large charge transfer is not necessary for stability, as shown by the formation of the solvated systems $Na_{1/3}(H_2O)_{2/3}TaSe_2$ and $(NH_4)_{0.1}(NH_3)_{0.9}TaS_2$. Further, the electron transfer reaction need not necessarily be complete. Measurement of the lithium Knight shift in Li_xTiS_2 indicates a reduction of the extent of electron transfer at high x values, suggesting that the alkali metal lattice may be thought of as a two dimensional analogue of the alkali metal cluster cations observed in zeolite lattices.[55]

The intercalation chemistry of many other layered compounds can also be described by the general reaction scheme outlined above. In the remainder of this chapter, some additional examples of redox chemistry are discussed together with other types of intercalation reactions which depend on special features of the host lattice structure or composition. In the next two sections, the intercalation chemistry of the metal phosphorus trisulphides and the metal oxyhalides is described. In addition to redox chemistry, these systems provide examples of intercalation reactions that involve modification of the host lattice structure or composition.

6.3 METAL PHOSPHORUS TRISULPHIDES

6.3.1 Structures

The metal phosphorus trisulphide phases, MPS_3, form layered structures with M = Mg, V, Mn, Fe, Co, Ni, Zn, Pd and Cd, and the corresponding selenides, $MPSe_3$, form similar structures with M = Mg, Mn, Fe, Ni and Cd.[56–58] Other variations include compounds with metal vacancies ($V_{0.78}PS_3$, $In_{2/3}PS_3$ and $In_{2/3}PSe_3$) and phases with two different metals ($Cu_{1/2}Cr_{1/2}PS_3$, $Mn_{0.87}Cu_{0.26}PS_3$, $Ag_{1/2}In_{1/2}PS_3$).[12] Two distinctly different

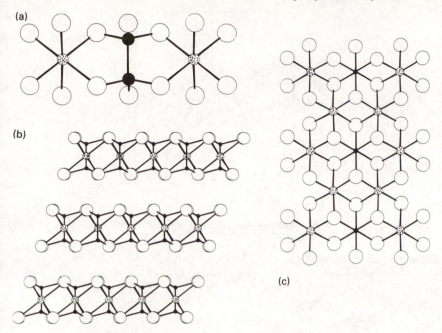

Fig. 6.12. The FePS$_3$ structure; the open circles represent sulphur, the speckled circles iron and the filled circles phosphorus. (a) The connection of the P$_2$S$_6$ and FeS$_6$ octahedra; (b) the layer stacking; (c) a single layer showing the ordered arrangement of the P$_2$S$_6$ and FeS$_6$ octahedra.

types of layer structure are known and are exemplified by monoclinic FePS$_3$ (C2/c) and rhombohedral FePSe$_3$ (R$\bar{3}$). The monoclinic FePS$_3$ structure is found for all of the first transition series sulphides and is based on a cubic close-packed anion array with alternate layers of cation sites vacant. Within a layer, the cation sites are occupied by M^{2+} cations and P$_2$ pairs, as shown in Fig. 6.12, and the anion arrangement is similar to that in CdCl$_2$. The two different octahedra are ordered so that each P$_2$S$_6$ octahedron is surrounded by six MS$_6$ octahedra and each MS$_6$ octahedron has three each of MS$_6$ and P$_2$S$_6$ neighbours (Fig. 6.12). To emphasize the structural relation to the dichalcogenides, the composition can be written M$_{2/3}$(P$_2$)$_{1/3}$S$_2$. In contrast to the sulphides, the selenide structures are based on hexagonal chalcogen packing although the unit cells are rhombohedral (R$\bar{3}$) because of M and P$_2$ order.

The crystal structures of the first transition series MPS$_3$ compounds show several interesting features.[12,59] The P–S bond distance is constant (2.032(3) Å) and the P–P distance lengthens to compensate for the different sizes of M^{2+} cations. A linear correlation between the P–P distance and the M^{2+} radius is observed, implying that weakly coupled PS$_3$ pyramids may

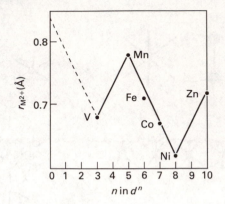

Fig. 6.13. The M^{2+} cation radius calculated from the MPS$_3$ structures (from ref. 59)

Table 6.6 *Optical band gaps of MPX$_3$ phases (eV)*

MnPS$_3$	3.0	MnPSe$_3$	2.5
FePS$_3$	1.5	FePSe$_3$	1.3
CoPS$_3$	1.4	—	—
NiPS$_3$	1.6	NiPSe$_3$	1.2
ZnPS$_3$	3.4	—	—
CdPS$_3$	3.5	—	—
In$_{2/3}$PS$_3$	3.1	In$_{2/3}$PSe$_3$	1.9

be a better description of the structure. The structural data may also be used to deduce the M^{2+} radii from the metal sulphur distances, assuming a S^{2-} radius of 1.84 Å. The results are shown in Fig. 6.13 and are a good illustration of the influence of ligand field stabilization on effective ionic radii.

Two aspects concerning the physical properties of the MPS$_3$ host lattices are worth noting briefly. First, the materials are semiconductors or insulators as indicated by the optical band gaps (Table 6.6). Second, the transition metal systems have magnetic properties characteristic of high spin octahedral cations. At low temperatures, antiferromagnetic order is observed but the susceptibility maxima are broad, indicative of the strong two-dimensional nature of the magnetic interactions.

6.3.2 Intercalation chemistry

The MPX$_3$ compounds show two quite distinct kinds of intercalation chemistry. Redox intercalation reactions analogous to those of the dichalcogenides are observed for the lower band gap materials MPS$_3$(Fe,Ni) and FePSe$_3$. The second reaction type is unique to the MPX$_3$ compounds and involves an ion exchange process whereby loss of M cations from the MPX_3 layers into the solution phase maintains charge balance on inter-calation of cations between the layers. Vacant sites in the MPX_3 layers are formed and the intercalation compounds have compositions represented by $G_x^+[M_{1-x/2}\square_{x/2}PX_3]$. Each reaction type will be discussed in turn.

6.3.2.1 Redox intercalation reactions

The redox intercalation chemistry of the MPX_3 compounds has been much less studied than that of the dichalcogenides. Most work has focused on reactions of the MPS_3 compounds with lithium because of interest in the application of the materials as cathodes in secondary batteries.[60-63] The Li_xNiPS_3 system has been the most studied for battery applications and is the best understood example. The reaction of $NiPS_3$ with excess *n*-butyl lithium in hexane leads to compositions Li_xNiPS_3 with $x > 3$, much greater than the maximum value of $x = 1.5$ which corresponds to complete occupa-tion of the interlayer sites. The X-ray powder diffraction data indicate that degradation of the host lattice occurs under these conditions and that lithium sulphide is formed. Reactions of $NiPS_3$ with n-butyl lithium, which keep $x \leqslant 1.2$, or controlled electrochemical intercalation can be carried out without destruction of the layers. Interestingly, no lattice expansion is observed, in contrast to Li_xTiS_2 (Fig. 6.6), though intercalation can be demonstrated by the formation of a hydrated phase with an increased interlayer separation on exposure to water vapour.

The absence of an expansion in the anhydrous phase is a consequence of the interlayer distance in the pure host. This distance is found from structural studies to be 3.24(2) Å for all of the first transition series MPS_3 compounds and consequently no expansion is expected based on a comparison with the results for the corresponding $LiMS_2$.[59]

Electrochemical data for Li_xNiPS_3 show two distinct regions with $0.0 < x < 0.5$ and $0.5 < x < 1.5$ (Fig. 6.14). Electrochemical reduction beyond $x = 1.5$ leads to irreversible behaviour but the shape of the discharge data indicates single phase behaviour up to $x = 0.5$. The phosphorus (^{31}P) NMR data show a single resonance in this composition range, consistent with this inter-pretation. Above $x = 0.5$, the electrochemical data are harder to interpret but the appearance of a second ^{31}P resonance in this composition region suggests two phase behaviour.

Reactions of several other MPX_3 compounds with lithium have also been studied. In general, redox behaviour is observed for those systems which have low optical band gaps (Table 6.6), though the type of host acceptor levels involved in the reduction process is not completely certain. Almost all possibilities have been suggested including M^{2+} 4s and 4p levels, the s antibonding levels of the P–P bonds and the S 3d levels. A recent theoretical study provides strong arguments for metal 3d acceptor levels, though there remain difficulties in explaining some of the observed magnetic properties.[64]

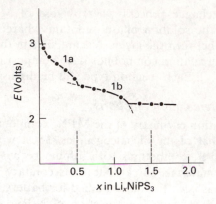

Fig. 6.14. Open circuit voltage data for the electrochemical intercalation of lithium into $NiPS_3$ (from ref. 6).

The only other examples of redox reactions which have been studied in any detail are those of the metallocenes and related organometallic molecules and these are discussed in the next section.

6.3.2.2   *Intercalation of MPX$_3$ compounds by ion exchange*[65–69]

Both $FePS_3$ and $NiPS_3$ react with toluene solutions of cobaltocene to give intercalation compounds which are similar in composition and interlayer separation to the corresponding phases formed by the transition metal dichalcogenides. However, in studies of the direct intercalation reaction of metallocenes with some wide band gap host lattices ($MnPS_3$, $ZnPS_3$, $CdPS_3$; Table 6.6), similar intercalation compounds were formed by reaction with solutions of metallocene salts, for example with cobaltocenium iodide. This reaction must occur by a different mechanism and is now known to proceed by ion exchange with Mn^{2+} cations from the layers themselves to form compounds of general stoichiometry $G_{2x}[Mn_{1-x}\square_x PS_3] \cdot yH_2O$. Several examples of intercalation compounds prepared by direct exchange of the cations within the layer are shown in Table 6.7. After the initial reaction, additional compositions can be made by cation exchange with the interlayer cations. A particularly interesting example is the exchange of $Ni(H_2O)_6^{2+}$ with $K_{0.40}[Mn_{0.80}PS_3] \cdot 0.9H_2O$.[70] Complete exchange of the potassium ions followed by dehydration gives the compound $Ni_{0.28}Mn_{0.72}PS_3$ which has identical lattice constants and magnetic properties to the phase prepared by direct reaction of the elements at high temperature. The experiment demonstrates that the ion exchange process with the layer cations is reversible. The MPS_3 lattices differ in their relative abilities to form vacant sites in the layers by direct ion exchange reactions. Ion exchange reactions

Table 6.7   *Intercalation compounds of $MnPS_3$ prepared by direct ion exchange* $G_{2x}[Mn_{1-x}\square_xPS_3] \cdot yH_2O$[66,69]

G	x	y	Δd (Å)[a]
$Co(\eta\text{-}C_5H_5)_2^+$	0.16	0.3	5.32
$Cr(\eta\text{-}C_6H_6)_2^+$	0.14	0.3	5.76
K^+	0.19	0.9	2.87
NH_4^+	0.19	1.0	2.88
Rb^+	0.19	0.9	2.90
Cs^+	0.19	0.9	3.07
$(CH_3)_4N^+$	0.15	0.9	4.95
$(C_2H_5)_4N^+$	0.14	0.3	4.68
$CH_3(CH_2)_7NH_3^+$	0.10	0.4	3.88
$C_5H_5NH^+$	0.14	0.7	3.15
$CH_3(CH_3(CH_2)_7)_3N^+$	0.11	0.3	11.95
Na^+	0.5	4.0	5.60

[a] Increase in layer spacing ($MnPS_3$, $d = 6.5$ Å).

readily occur for Mn, Cd, and Zn. In contrast, $NiPS_3$ does not undergo ion exchange and $FePS_3$ does so with difficulty and only in the presence of a complexing agent like ethylenediaminetetraacetic acid (EDTA). A correlation has been observed between the ease of ion exchange and the magnitude of the divalent metal thermal parameter determined from the crystal structure analyses[59]; $NiPS_3$ (0.74 Å^2) and $FePS_3$ (1.11 Å^2) with the lowest thermal parameters are apparently the most difficult to exchange. It is also worth noting that the P–P bond length in the $P_2S_6^{4-}$ anion adjust to accommodate the size of the specific M^{2+} cation, suggesting that opening up the MS_6 octahedra to permit M^{2+} migration is an easy process.

In summary, the MPS_3 compounds can undergo two distinct types of interlayer reactions; one requires electron transfer to the layers and the other results in ion transfer from the layers into the reactant solution. Both reactions are at least partially reversible and involve simultaneous intercalation of cationic species into the interlayer spaces. The balance between the two different reaction pathways appears to depend on the band gap of the particular solid. Ion transfer reactions are easier in the larger band gap compounds (for example, with M = Zn, Cd, and Mn) where the layers appear to behave as M^{2+} cations weakly coupled to $P_2S_6^{2-}$ units. The MPS_3 lattices behave very differently from the dichalcogenides where the strong covalent interactions between formally M^{4+} cations and S^{2-} anions precludes any ion transfer chemistry involving the layers.

6.4   TRANSITION METAL OXYHALIDES

6.4.1   Structure

A small number of metal chalcogenohalides have layered structures. Four main structure types (FeOCl, AlOCl, SmSI and PbFCl) are known, but only the FeOCl structure is known to undergo topotactic redox reactions. The orthorhombic FeOCl (Pmnm, $a = 3.7729$ Å, $b = 7.9104$ Å, $c = 3.3026$ Å) structure is shown in Fig. 6.15 and is characterized by double layers of distorted octahedra which share edges. Other compounds with the FeOCl structure are known and include MOCl (M = Ti, V, Cr) and InOX (X = Cl, Br, I). An idealized view of the structure is shown in two projections along the two orthogonal directions in the plane of the layer (Fig. 6.15). Each iron atom is coordinated to four oxide anions and two chlorides, with the two chlorides in *cis* octahedral positions on the outside of the layer. The arrangement of octahedra within a layer is similar to that found in γ-FeOOH and γ-AlOOH, and in some ternary oxides, for example β-NaMnO$_2$.

The redox intercalation chemistry of FeOCl with inorganic, organic, and organometallic guest cations is very similar to that of the dichalcogenides. However, FeOCl also shows a different types of interlayer reactivity not observed in the systems discussed earlier. Irreversible topochemical substitution reactions are observed where the outer halide layers of the FeOCl lattice are replaced by other groups. These reactions have been referred to as grafting reactions[71] and they offer some interesting possibilities for the synthesis of new materials. The second part of this section is devoted to this kind of reaction, but first some examples of the redox chemistry are outlined.

6.4.2   Redox intercalation reactions of FeOCl

6.4.2.1   *Inorganic cations*

Little information on simple cation intercalation of FeOCl is available and, as in the case of the MPS$_3$ hosts, most of it concerns lithium because of interest in the application of FeOCl as a cathode material. Some degree of reversible intercalation occurs in electrochemical or chemical reactions with n-butyl lithium to form Li$_x$FeOCl.[72] However, the compounds are not very crystalline, easily degrade and are not structurally characterized. More is known about solvated cation systems and these present some new features.[73,74] In aqueous systems, hydrated alkali intercalation compounds can be prepared using chemical reducing agents, or, alternatively, non-aqueous solvated compounds can be prepared electrochemically. The experiments in aqueous systems are complicated by the tendency of FeOCl to hydrolyse topochemically to γ-FeOOH, but this problem can be minimized by the use of single crystals. The alkali metal systems, M$_{0.14}$(H$_2$O)$_y$FeOCl

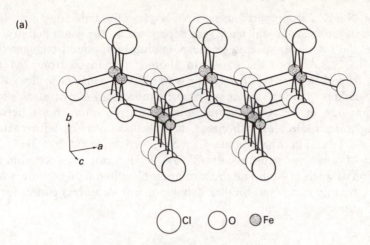

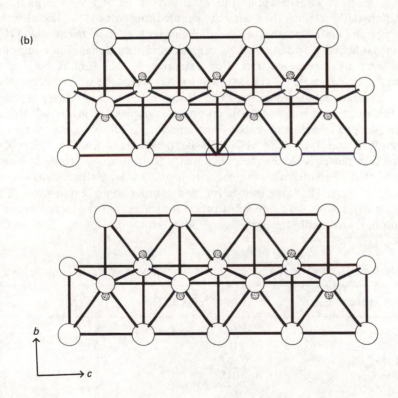

Fig. 6.15. The structure of FeOCl (a) represented as metal, oxygen, and halogen bonds (top) and (b) as anion polyhedra (bottom).

(M = Li, Na, K, Cs), prepared using $Fe(CN)_6^{4-}$ as the reducing agent, show large interlayer spacings which are very dependent on the water activity. For example, the interlayer spacing in the potassium compound changes from 11.47 Å to 22.8 Å when the contacting liquid is changed from 3M KCl solution to pure water. This behaviour is in marked contrast to that of the dichalcogenides where the spacings primarily reflect cation hydration effects. The behaviour of FeOCl indicates much stronger interactions between solvating water molecules with either the chloride layers or with partially hydrolysed layers in which some Cl^- has been replaced by OH^-. The addition of some reducing agents ($TiCl_3$, N_2H_4) can produce complete exfoliation of single crystals and the formation of colloidal dispersions which, however, rapidly hydrolyse. Similar behaviour was described earlier for the dichalcogenides.

6.4.2.2   *Organic and organometallic cations*

Much more is known about the reactions of FeOCl with organic and organometallic cations than with the simple inorganics.[75-78] The chemistry is generally similar to that of the dichalcogenides, but, because FeOCl is more oxidizing, a wider range of organic and organometallic compounds can form by direct reaction. For example, FeOCl, unlike any of the dichalcogenides or MPS_3 phases, will oxidize ferrocene to ferrocenium and form an intercalation compound. The stoichiometry and interlayer expansion for this phase and some other metallocene examples are given in Table 6.8. Even though the lattice expansions are large, all of these phases are crystalline and their unit cells can be determined from powder X-ray diffraction patterns. A change in symmetry from primitive to body-centred shows that the intercalation reaction produces a relative translation of adjacent layers. The layer translation and comparison of expansions (Table 6.8) for derivatives with ring substitution indicates that the metallocenes are accommodated with the rings perpendicular to the layers. The reaction

Table 6.8   *Organometallic intercalation compounds of FeOCl*

Organometallic guest	x in G_x [FeOCl]	Δb (Å)[a]	Ref.
$CoCp_2$	0.16	4.94	75
$FeCp_2$	0.16	5.13	75
$Fe(EtMe_4Cp)_2$	0.16	7.55	76
$Fe(Me_5Cp)_2$	0.075	7.1	78
$NiCp_2$	0.15	5.19	77
$CrCp_2$	0.14	5.19	77

[a] Interlayer expansion relative to FeOCl ($b = 7.91$ Å).

stoichiometry, $G_{1/6}FeOCl$, is close to that calculated for close-packed metallocene cations in this model.

The nature of the electron transfer process which occurs in the formation of $[(\eta\text{-}C_5H_5)_2Fe]_{1/6}FeOCl$ has been studied in detail by Mössbauer spectroscopy.[79] At temperatures in the range 77–100 K the Mössbauer spectrum shows resonances characteristic of the ferrocenium cation, two types of lattice Fe^{3+} which differ in the number of Fe^{2+} neighbours, and lattice Fe^{2+}. Quantitative measurements confirm that intercalation occurs with complete electron transfer from the guest to host lattice. At ambient temperature, only a single quadrupole split doublet is observed as a consequence of rapid electron hopping between iron atoms, both within layers and between them, via the guest molecule. Consistent with this observation, the electrical conductivity of the intercalation compound is substantially higher than that of the host lattice at 293 K.

Intercalation compounds are also formed by reaction of FeOCl with organic compounds which are good electron donors.[80] Several systems have been synthesized with organic molecules such as tetrathiofulvalene (TTF), tetrathionaphthalene (TTN) and tetrathiotetracene (TTT). The organic intercalation compounds are crystalline and the measured interlayer separations suggest orientations for the organic molecules in the interlayer space. In the examples above, orientations were proposed with the organic rings tilted with respect to the FeOCl layers in a way which permits the sulphur atoms to key into the hole formed by four layer chlorides. The observed stoichiometries are consistent with near close packing of the guests. Part of the interest in these systems stems from the possibility of obtaining very high conductivities via properly oriented organic guests. However, no substantial enhancement of the electrical conductivity has yet been observed.

Reactions with Lewis bases are the other major class of redox intercalation reactions which have been explored with the FeOCl lattice. Reactions of FeOCl with ammonia and alkylamines are among the earliest reported intercalation reactions of compounds with neutral layers.[81] Reactions with pyridine and substituted pyridines are also known and have been extensively investigated.[82–84] For example, pyridine reacts with FeOCl to form two compounds, (pyridine)$_{1/4}$FeOCl and (pyridine)$_{1/3}$FeOCl, with interlayer expansions of 5.35 and 5.53 Å, respectively. The pyridine intercalation compounds are of particular interest because they show large increases in electrical conductivity relative to the host lattice. The resistivity of (pyridine)$_{1/3}$FeOCl is 10 ohm-cm, much lower than that of FeOCl (10^7ohm-cm).

Mössbauer spectroscopy of several Lewis base intercalation compounds gives results which are similar to those for the ferrocenium compound. An Fe^{2+} resonance is observed at low temperatures which can be used to estimate the charge transfer. Measurements on several different compounds gave values in the range 0.07 to 0.14 e/FeOCl. Infrared studies of the two pyridine intercalation compounds indicates the presence of pyridinum cations and 4,4'-bipyridine, in addition to neutral pyridine molecules, suggesting a similar coupling mechanism to that proposed for the corresponding reaction with the dichalcogenides.

6.4.2.3   Substitution reactions

Topochemical substitution reactions, sometimes called grafting reactions, are an especially interesting aspect of FeOCl chemistry. The earliest examples of this type of substitution were observed in reactions of ammonia and methylamine with FeOCl at ambient temperature.[81,85] Reaction occurs with two moles of ammonia or methylamine according to the equation:

$$FeOCl + 2RNH_2 \longrightarrow FeONHR + RNH_3Cl \ (R = H \ or \ CH_3). \quad (6.5)$$

The products of the substitution reaction are X-ray amorphous and the main evidence for their formation comes from analysis and infrared measurements. Similar irreversible substitution reactions have also been observed in heating long chain alkylamine intercalation compounds of FeOCl above 110°C.[86] It is worth noting that the hydrolysis of FeOCl to γ-FeOOH (lepidocrocite) is also an example of a topochemical substitution of Cl^- by OH^-.

Substitution reactions with alkoxide ions lead to more-crystalline phases. The first example of this type was reported by Koizumi and coworkers[87] who found that reaction of (4-aminopyridine)$_{1/4}$FeOCl with methanol at 100°C led to the formation of a crystalline phase, $FeOOCH_3$, with an increased interlayer separation. Methanol does not react directly with FeOCl, but sodium methoxide does and provides a simpler synthetic route.[88] The reaction of the aminopyridine intercalation compound of FeOCl with ethylene glycol also gives a crystalline phase of composition $FeO(O_2C_2H_4)_{1/2}$ with an interlayer spacing of 10.89 Å.[89] The interlayer separation is too large for a structure in which ethylene glycolate bridges between two layers and a structure in which both ends of the glycolate bind to the same layer appears more likely.

Recently, Villieras and coworkers[71] have studied reactions of the general type:

$$KR + FeOCl \longrightarrow KCl + FeOR \quad (6.6)$$

$$K_2R + 2FeOCl \longrightarrow 2KCl + (FeO)_2R \quad (6.7)$$

with aliphatic and aromatic alkoxides and acids. Two compounds, $FeOOCH_3$, were readily formed by reaction at 60°C of FeOCl with acetone solutions of potassium methoxide and acetate, respectively. In contrast, reactions under similar conditions with potassium phenoxide and benzoate were much slower, and pure products were not obtained. The cross-linking reaction is not successful using potassium oxalate, which gives a compound analogous to the ethylene glycol phase described above. However, with the more rigid and longer molecules, *para*-hydroxy benzoic acid (I) and dimercaptothio-diazole (II), cross-linking was achieved. In both cases, the reaction was found to occur in two steps; an initial intercalation step which caused an expansion of the FeOCl interlayer distance, and a subsequent second substitution step leading to layer cross-linking. The final interlayer separations were found to be 11.33 Å for the *para*-hydroxybenzoate compound and 9.21 Å for the bridging thiazole.

$$^-O - \langle \bigcirc \rangle - CO_2^-$$

I

$$
\begin{array}{ccc}
N & \!\!-\!\!-\!\! & N \\
\| & & \| \\
C & & C \\
\diagup & \diagdown \quad \diagup & \diagdown \\
^-S & S & S^-
\end{array}
$$

II

In summary, FeOCl can undergo interlayer reactions which are irreversible and lead to replacement of Cl^- by other anionic groups without change in the oxidation state of Fe. The scope of this kind of layer derivatization has yet to be fully explored for both FeOCl and other layered oxyhalides, but offers the prospect of some interesting new systems. The compounds are examples of a growing class of layered compounds with alternating organic and inorganic layers where the organic group is attached to the inorganic layer by a covalent bond. Other examples include MoO_3–pyridine (below) and the organophosphonates of zirconium, $Zr(RPO_3)_2$, and vanadium, $VORPO_3$.[90]

6.5 LAYERED TRANSITION METAL OXIDES

Simple layered structures with true van der Waals gaps are found only for the oxides of high oxidation state transition metal cations which form strong multiple covalent bonds to oxygen, for example, MoO_3, V_2O_5 and $MOXO_4$ (M = V, Nb, Ta and X = P, As). In the absence of this strong covalent bonding, oxide layer structures are destabilized by electrostatic repulsions between layers of negatively charged oxide ions, which are not offset by sufficiently strong attractice van der Waals forces. Transition metal oxide layers may be stabilized when the electrostatic repulsion is reduced by the presence of interlayer cations or hydrogen bonding. The three types of situations which are encountered are illustrated schematically in Fig. 6.16. Oxides of type I show the same general type of topotactic reduction chemistry found for the dichalcogenides, whereas the reaction chemistry of type II oxides involves mainly ion exchange and oxidative deintercalation.

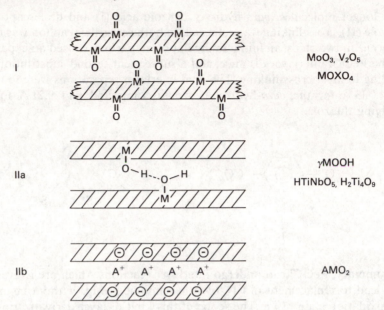

Fig. 6.16. Schematic representation of the major classes of layered oxide systems.

6.5.1 High oxidation state transition metal oxides

In this section intercalation reactions of the compounds MoO_3, V_2O_5, and $VOPO_4 \cdot 2H_2O$ are discussed. Each compound contains a transition metal ion in a strongly distorted octahedral environment, arising from $d\pi$–$p\pi$ covalent bonding of oxygen to the high oxidation state transition metal cation. The transition metal cation coordinations of the three oxides are shown in Fig. 6.17. In each case, the shortest $M = O$ distance is perpendicular to the layers.

6.5.1.1 Molybdenum trioxide

The structure of molybdenum trioxide is shown in Fig. 6.18. The layers of double octahedra are anisotropic and idealized projections along *a* and *c* in the orthorhombic unit cell are shown. Along the *c* axis, the octahedra share *cis* edges whereas along the *a* direction they share corners. The octahedra are far from ideal (Fig. 6.17) and have two short distances *cis* to each other, one pointing into the interlayer space and one along *c*. Individual layers in MoO_3 are separated by a true van der Waals gap and consequently the oxide shows the full range of topotactic redox chemistry outlined for the dichalcogenides. Molybdenum trioxide is insulating and white in colour when fully oxidized, but reductive intercalation produces both a large increase in electronic conductivity and a dramatic change in colour to metallic blue-black, even for small amounts of reduction.

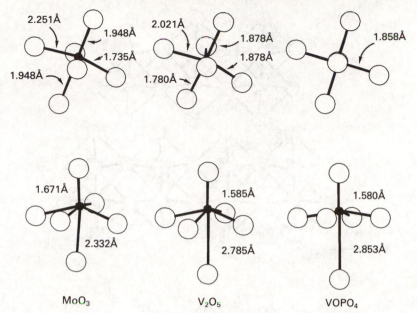

Fig. 6.17. Local transition metal coordination in layered transition metal oxides looking approximately down on a layer (top) and perpendicular to a layer (bottom).

The intercalation of MoO_3 by hydrogen is the most extensively investigated of the redox reactions. Chemical reduction of MoO_3 in aqueous acid was first shown by Glemser and coworkers[91] to result in the formation of a series of compounds which could be formulated as H_xMoO_3.[92–94] Subsequent studies have shown that the hydrogen bronzes can be prepared by electrochemical reduction in acid solution and by chemical reduction ('hydrogen spillover') with hydrogen and a noble metal (e.g. platinum) catalyst.[95,96] Four distinct hydrogen bronze phases have been established: blue I $(0.23 < x < 0.40)$, blue II $(0.85 < x < 1.04)$, red $(1.55 < x < 1.72)$, and green $(x = 1.99)$.[95] Hydrogen insertion occurs with minimal rearrangement of the MoO_3 structure, as shown in powder neutron diffraction studies of $D_{0.36}MoO_3$ and $D_{1.68}MoO_3$.[97] In the structure of $D_{0.36}MoO_3$ the deuterium atoms are attached to bridging oxygen atoms as –OD groups in intralayer sites and are not involved in hydrogen bonding between layers. In contrast, in the structure of $D_{1.68}MoO_3$ the deuteriums are attached to the terminal oxygens to form –OD$_2$ groups projecting into the interlayer space. Measurements of proton mobilities by NMR show rapid proton motion in $H_{1.68}MoO_3$ $(D = 10^{-8} \text{ cm}^2 \text{ sec}^{-1})$, but slower motion in $H_{0.36}MoO_3$ $(D = 10^{-11} \text{ cm}^2 \text{ sec}^{-1})$, consistent with the different structures of the two compounds. All of the H_xMoO_3 phases show metallic conductivity and weak temperature-independent paramagnetism.

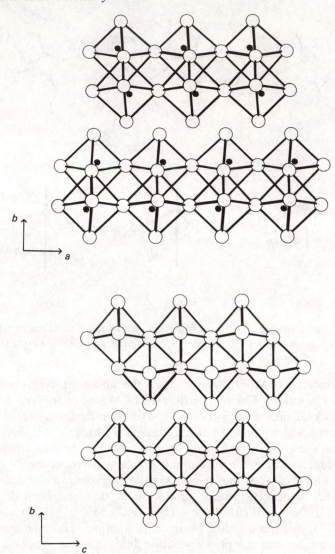

Fig. 6.18. Two views of the double layer structure of MoO_3 showing the directions of edge- and corner-sharing octahedra.

The hydrogen intercalation compounds are Brønsted acids and react with Lewis bases (L) to form intercalation compounds of general composition $L_yH_xMoO_3$. As an example, $H_{0.5}MoO_3$ reacts with pyridine to give a phase of composition (pyridine)$_{0.3}H_{0.5}MoO_3$ with an interlayer separation increased by 5.84 Å.[98] Only some of the interlayer pyridine molecules are protonated and these pyridinum cations are stabilized by hydrogen bonding to neutral pyridine molecules. Such an arrangement was proposed earlier for the pyridine dichalcogenide compounds. Unlike the dichalcogenides, however, when water is rigorously excluded a thermal redox intercalation reaction of MoO_3 with pyridine does not occur. Instead, the structure rearranges to form the layer compound MoO_3–pyridine which is pale green and not reduced.[99]

In other respects, the intercalation chemistry of MoO_3 closely resembles that of the dichalcogenides. Reduction in neutral aqueous electrolytes leads to the formation of hydrated cation intercalation compounds which can exchange ions or solvent molecules easily. Differences in behaviour in comparison to the dichalcogenides are related to the relative importance of hydrogen bonding with the oxide layers and the high stability of the hydrogen phases. For example, the compound $Na_{0.5}(H_2O)_yMoO_3$ is converted on washing with water to form $Na_{0.5-x}H_x(H_2O)MoO_3$. Treatment with dilute acid gives $H_{0.5}(H_2O)_yMoO_3$, which dehydrates irreversibly to form the phase $H_{0.5}MoO_3$.[95] Cation exchange reactions in aqueous solution are rapid at room temperature and interlayer separations are found to depend on the hydration energy of the interlayer cation, but there is no simple correlation between layer spacing and cation charge-to-radius ratio, as observed for the dichalcogenides.

6.5.1.2 Vanadium pentoxide

A projection of the structure of V_2O_5 is shown in Fig. 6.19 with the vanadium–oxygen coordination represented as octahedral and square pyramidal. In the square pyramidal representation, the structure is apparently layered. However, the interlayer interactions are sufficiently strong that in most respects V_2O_5 behaves as a three-dimensional lattice and intercalates only the smallest cations, H^+ and Li^+.

On reaction with lithium at ambient temperature, V_2O_5 forms a series of phases, $Li_xV_2O_5$ with $x < 2.0$.[33,100,101] When $x < 1.0$, the compounds are crystalline and electrochemical measurements show three single-phase regions with $x < 0.2$, $0.35 \leqslant x \leqslant 0.5$ and $0.9 \leqslant x \leqslant 1.0$. All three phases have structures related to V_2O_5. The compounds with lowest lithium content is isostructural with the corresponding high temperature phase (α), but the high temperature structures at the other compositions (β' and γ) are quite different.[102] Vanadium pentoxide is sufficiently oxidizing to react with acetonitrile solutions of lithium iodide. The results are similar to those obtained electrochemically though there are some differences in the details of the phase boundaries. The phase boundaries and the temperatures of the transformations to the high temperature phases are shown in Fig. 6.20. More strongly reducing reagents such as n-butyl lithium give reduction beyond a composition corresponding to $x = 1.0$, but the products are poorly crystalline and not well characterized.

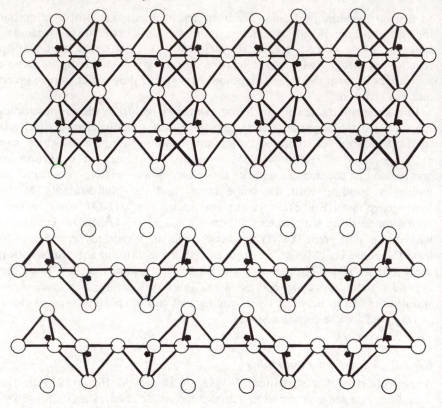

Fig. 6.19. Part of the V_2O_5 structure represented as distorted anion octahedra (top) and square pyramids (bottom).

Hydrogen intercalation compounds of V_2O_5 can be prepared by the techniques described for MoO_3.[103,104] Hydrogen 'spillover' gives the composition $H_xV_2O_5$ ($x \approx 3.8$) but, unlike the high hydrogen content molybdenum bronzes, the vanadium compounds are X-ray amorphous and semiconducting. Electrochemical synthesis using a solid proton conducting electrolyte[105] indicates that three distinct phases in the $H_xV_2O_5$ system exist with compositions $x \leqslant 0.5$, $1.3 < x < 2.3$, and $3.0 < x < 3.8$. The lowest hydrogen content phase can be obtained in crystalline form using the spillover route and by limiting the amount of hydrogen and subsequent annealing at 50°C. Infrared data for this phase and its deuterium analogue ($H_{0.33}V_2O_5$ and $D_{0.39}V_2O_5$) indicate the presence of an –OH group, while inelastic neutron diffraction studies of the higher hydrogen content phase, $H_{2.5}V_2O_5$, indicate that the bulk of the hydrogen is present as $–OH_2$ groups. The structures of the vanadium hydrogen bronzes thus appear to be closely analogous to the corresponding MoO_3 phases.

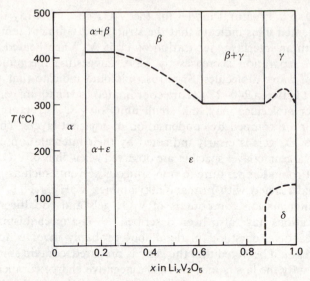

Fig. 6.20. Stability diagram for $Li_xV_2O_5$ phases showing the transformations from the intercalation compounds prepared at low temperature to the high temperature equilibrium phases (from ref. 101).

As mentioned above, V_2O_5 behaves more like a three-dimensional structure than a layered compound and attempts to intercalate large cations are either unsuccessful or lead to structural rearrangement. In contrast, vanadium pentoxide when prepared in gel form behaves much more like a layered phase in its intercalation chemistry.[106] The vanadium pentoxide gels have been known for many years[107] but have received recent attention because of their electronic properties and because they can readily be cast into thin films. One convenient synthetic procedure is to prepare decavanadic acid by ion exchange of an aqueous sodium vanadate solution on an ion exchange resin. The resulting decavanadic acid solution polymerizes spontaneously in a few hours at room temperature to form a red viscous gel which typically contains about 1 per cent of V (IV). Removal of excess water at room temperature leads to the formation of a solid gel with an approximate composition $V_2O_5 \cdot 1.6H_2O$. On heating, the gels lose water in distinct steps.

$$V_2O_5 \cdot 1.6H_2O \xrightarrow[<150°C]{} V_2O_50.5H_2O \xrightarrow[240-270°C]{} V_2O_50.1H_2O \xrightarrow[>300°C]{} V_2O_5$$

$$(6.8)$$

Complete removal of water above 300°C leads to recrystallization with the formation of orthorhombic V_2O_5. A variety of techniques, including X-ray diffraction, electron microscopy, and vibrational spectroscopy, have been used to develop a structural model for the V_2O_5 gels.[108] X-ray diffraction studies of oriented films indicate that the structure at ambient temperatures is layered with an interlayer separation of 11.55 Å. When heated to 150°C the interlayer separation decreases by 2.8 Å, suggesting the removal of a single layer of water molecules. Spectroscopic data indicate that the water molecules lost between 240–270°C are coordinated to vanadium centres. The residual water molecules may link small units of V_2O_5 structure together (see below) or are trapped by condensation of adjacent layers. The layered structure of V_2O_5 gels is clearly indicated by their intercalation chemistry. For example, large interlayer spacings are observed when films of $V_2O_5 \cdot 1.6H_2O$ are soaked at room temperature in non-aqueous solvents such as propylene carbonate and reacted with primary alkylamines, $C_nH_{2n+1}NH_2$ ($n = 1$ to 18).[109,110] Intercalation compounds of V_2O_5 gels and cobalticenium and ferricenium cations have also been described.[111] The mechanisms of these reactions are not understood in detail but probably involve host lattice reduction. It should be noted that the gels, as prepared, contain some V (IV), and consequently the layers may possess a negative charge compensated by interlayer H_3O^+. To what extent further reduction is involved in the above reactions is not known. However, reduced gels can be prepared and are stable up to a vanadium (IV) content of 20 per cent. They have a larger interlayer spacing and water content ($2.5H_2O$), and presumably contain interlayer cations. Overall, the general structural and reaction chemistry of V_2O_5 gels has a much stronger parallel to that of crystalline $VOPO_4$ than to V_2O_5. The intercalation chemistry of $VOPO_4$ is discussed in the next section.

6.5.1.3   Vanadium phosphate

Vanadium phosphate shows a wide range of intercalation reactions which parallel in many respects the chemistry of the covalent oxide systems described earlier.[112] Several differences in behaviour are interesting to discuss and some of its intercalation chemistry serves to clarify V_2O_5 gel reactions.

The α form of $VOPO_4$ is a layered compound and is one member of a series of complex oxides of general formula $MOXO_4$ (M = V, Nb, Ta, Mo; X = P, As, S) which are isostructural. The tetragonal layer structure of these compounds is made up of distorted MO_6 octahedra and XO_4 tetrahedra which are linked by corner-sharing oxygen atoms. Each octahedron is joined to four tetrahedra as shown in Fig. 6.21. The layers are joined along the c axis by corner-sharing the remaining two *trans* vertices of the octahedron. Although, the oxygen atoms in each MO_6 unit form a nearly perfect octahedron, the M atoms are displaced off-centre along the c axis as shown in Fig. 6.17 for $VOPO_4$. The vanadium coordination is consequently better represented as square pyramidal and is similar to the coordination in V_2O_5. An important difference between V_2O_5 and $VOPO_4$ is the number of interlayer interactions of the type V=O---V per unit area which is reduced

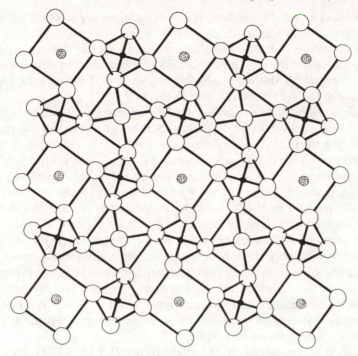

Fig. 6.21. Projection of the $VOPO_4$ structure perpendicular to a layer. The open circles are oxygen, filled circles phosphorus, and speckled circles vanadium. The VO_5 square pyramidals are alternately oriented up and down relative to the page.

in $VOPO_4$ by the presence of the phosphate groups. Consequently, the interlayer interactions in $VOPO_4$ are much weaker and $VOPO_4$ behaves in its chemistry like a two-dimensional system. An immediate indication of the layered nature of the structure is the ease with which it forms mono- and dihydrates. Structural studies of the dihydrate show that one water molecule is coordinated to the vanadium atom and that the second is located in the interlayer space. On dehydration, the two distinct H_2O molecules are lost in successive steps. The process is reversible and is closely analogous to the transformations between the hydrated phases in V_2O_5 gels. The intercalation and deintercalation of water between $VOPO_4$ and $VOPO_4 \cdot H_2O$ can be viewed as a special type of intercalation reaction in which coordination of the guest species to a metal centre provides the driving force. Similar reactions have been described for larger donor ligands such as straight chain alcohols and for stronger Lewis bases such as pyridine and 4,4'-bipyridine.[112] Reaction of either $VOPO_4$ or $VOPO_4 \cdot 2H_2O$ with pyridine leads to the formation of the compound, $VOPO_4 \cdot$ pyridine, which has all of the vanadium sites in the metal oxide layer coordinated by pyridine. Only very small amounts (<1 per cent) of V (IV) and pyH^+ are detected. Similar results were obtained with 4,4-bipyridine, but the stoichiometry is $VOPO_4$(bipyridine)$_{0.5}$ and comparison of its interlayer separation with those of other pyridine derivatives indicates that adjacent layers are cross-linked or 'pillared' by the bifunctional ligand. These coordination intercalation compounds are generally similar to the corresponding pyridine and substituted pyridine derivatives of the type $MoO_3 \cdot L$ and $WO_3 \cdot L$.[99]

Vanadium phosphate dihydrate readily undergoes redox intercalation reactions with alkali-metal and alkaline-earth-metal cations in the presence of a reducing agent.[13] The oxidizing power of V^{5+} is such that only mild reducing agents, for example aqueous or aqueous alcohol solutions of metal iodides, are required. The reactions are analogous to those used to prepare lithium intercalation compounds of V_2O_5; iodide is oxidized to iodine, V (V) is partially reduced to V (IV), and the metal cation present in solution is inserted between the layers. An important difference between $VOPO_4 \cdot 2H_2O$ and the other oxide lattices discussed above is that the layers have low electronic conductivity. Redox intercalation generally requires both ionic transport of the guest species in the interlayer region and electronic transport in the host lattice. Consequently, it has been proposed that the reductant, I^-, itself is intercalated into the interlayer space of $VOPO_4 \cdot 2H_2O$ and subsequently expelled in oxidized form. A similar mechanism has been proposed for intercalation of α-$RuCl_3$.[113]

The phase relationships in the systems $M_x VOPO_4 \cdot yH_2O$ have been surveyed for the alkali and alkaline earth cations and a small number of divalent transition metal cations, and determined in more detail for M = Na and Cs. All of the intercalation compounds show complicated phase relations and each system has several phases with different degrees of non-stoichiometry. The interlayer spacings versus composition for the $Na_x VOPO_4 \cdot 2H_2O$ phases are shown in Fig. 6.22 as an example of the type of behaviour observed. In $VOPO_4 \cdot 2H_2O$, the layers are held together by relatively weak hydrogen bonding. On intercalation, the layers become negatively charged and are brought closer together by electrostatic interaction with the interlayer cations. The result is that intercalation of cations actually produces a reduction in the interlayer separation rather than the expansion that is usually observed.

6.5.2 Layered AMO_2 oxides

The layered oxides, AMO_2, comprise alternating layers of edge-shared MO_6 octahedra and layers of alkali metal cations, and are structurally similar to the alkali–transition metal dichalcogenides. Non-stoichiometric phases, $A_x MO_2$, can be synthesized at high temperatures by substitution of a tetravalent cation of a different metal or by partial oxidation of the trivalent M cation. In some cases, metal or oxygen vacancies have also been observed. In general, stable phases cannot be synthesized at high temperature for $x < 0.5$.

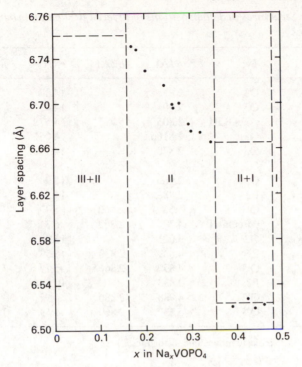

Fig. 6.22. Layer spacing versus composition for sodium intercalation compounds of $VOPO_4 \cdot 2H_2O$.

The lithium and sodium transition metal oxide phases prepared at high temperature that have been investigated for reactivity at lower temperatures are listed in Table 6.9. The structures of these oxides, like those of the alkali metal dichalcogenides, depend on the alkali cation content and size, and on the ionicity of the M–O bond, but there are important differences. For example, the transition metal is octahedrally coordinated in almost all cases and the pure MO_2 phases cannot be prepared even at low temperature because some alkali metal content is necessary to stabilize the layer structure. Staging is not found because the interlayers cannot be completely emptied of cations.

The main layer structure types can be written in the same stacking notation developed earlier. Three main sequences are found:[114,115]

aC(b)AcB(a)CbA(c)Ba	Type Ia or O3
aC(b,a)CaB(c,a)Ba	P2
aB(a)BcA(c)AbC(a)Ca	Type Ib or P3

As in the earlier section, capital letters are used to indicate anionic sites, lower case letters cation positions, and parentheses have been used to

Table 6.9 *Structures and compositions of layered lithium and sodium transition metal oxides*

Compound	Type[a]	a (Å)	b (Å)	c (Å)	β (°)	Ref.
$LiVO_2$	O3	2.83		14.87		121
$LiCrO_2$	O3	2.90		14.43		123
$LiMnO_2$	γ-FeOOH	2.805	5.757	4.572		116
$LiCoO_2$	O3	2.8166		14.052		119
$LiNiO_2$	O3	2.878		14.190		122
$NaTiO_2$	O3					124
$NaCrO_2$	O3	2.960		15.94		128
$Na_{0.7}MnO_{2.25}$	P2	2.888		11.24		125
α-$NaMnO_2$	O′3	5.63	2.860	5.77	112.9	125
β-$NaMnO_2$	γ-FeOOH	4.77	2.852	6.31		125
α-$NaFeO_2$	O3	3.025		16.09		123
$Na_{0.60}CoO_{1.92}$	P′3	4.839	2.830	16.53		126
$Na_{0.77}CoO_{1.96}$	O′3	4.890	2.866	5.77	111.28	126
$Na_{0.71}CoO_2$	P2	2.833		10.82		126
α-$NaCoO_2$	O′3	4.988	2.880	6.17	122.6	126
$NaNiO_2$	O′3	5.33	2.860	5.59	110.5	127

[a] Primes (e.g., O′3) indicate monoclinic distortions.

indicate alkali cation layers. The designations O3, P2, and P3 indicate the nature of the alkali coordination, octahedral (O) or trigonal prismatic (P), and the number of layers per unit cell. The O3 and P3 structures are identical to the Ia and Ib structures shown in Fig. 6.5 for the alkali metal dichalcogenides. The P2 structure (Fig. 6.23(a)) has two sets of interlayer trigonal prismatic sites which are not equivalent; one set is offset from the octahedra in the layers above and below and is energetically more favourable. The P2 structure is not observed for the intercalated dichalcogenides, but has been

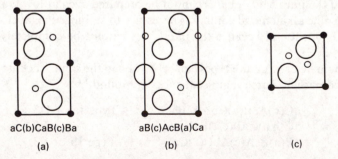

aC(b)CaB(c)Ba   aB(c)AcB(a)Ca

(a)   (b)   (c)

Fig. 6.23. Projections of the stacking sequences of some alkali metal oxide systems (see text).

reported for $NaTiS_2$ and $NaTiSe_2$ prepared at high temperatures. The $LiTiS_2$ structure is not observed in the oxide systems, presumably because of unfavourable electrostatic repulsions between metal ions in chains of face-shared octahedra along the hexagonal c axis. One oxide, Li_2MnO_2, which was prepared by reaction of n-butyl lithium with the spinel $LiMn_2O_4$, does, however, have a structure analogous to the intercalated dichalcogenide Li_2VSe_2 (Fig. 6.23(c)) with the Li^+ cations in tetrahedral sites.[116]

6.5.2.1 *Intercalation–deintercalation chemistry*

Most of the reaction chemistry of the alkali–transition metal oxides involves oxidation of the host lattice electrochemically or with chemical reagents such as iodine or bromine in acetonitrile or chloroform. The chemical reactions are analogous to the oxidation of $LiVS_2$ to form VS_2 using I_2 in CH_3CN, described above, though generally the transition metal oxide systems are much harder to oxidize than the alkali disulphides. In electrochemical oxidation, the applied cell voltages are limited only by electrolyte stability, whereas chemical oxidations depend on the particular redox couple chosen. The I_2/I^- and Br_2/Br^- couples in acetonitrile have been estimated as 3.3 and 3.7 volts with respect to Li^+/Li and 2.94 and 3.34 volts with respect to Na^+/Na. Not all of the alkali metal cations can be removed by either synthetic method without irreversible structural rearrangement. In lithium systems, the formation of the three-dimensional spinel structure in preference to the layer structures is particularly favourable. Correspondingly, lithium intercalation of oxide spinels is quite facile and has been studied in its own right by several groups.[117,118]

6.5.2.2 *Deintercalation of lithium*

Deintercalation of lithium from $LiMO_2$ (M = Cr, V, Ni, Co) has been investigated. From the standpoint of applications of these materials as cathodes in lithium cells, $LiCoO_2$ is the most interesting in that it has the widest range of non-stoichiometry and rapid lithium diffusion.[119,120] The electrochemical data are shown in Fig. 6.24 with respect to the Li^+/Li couple. In the composition range shown, the structure is unchanged apart from an increase in the c axis spacing from 14.08 Å ($x = 1.0$) to 14.42 Å ($x = 0.49$). The contraction in the c axis with increasing lithium content is in the opposite sense to the behaviour of Li_xTiS_2 and reflects the greater electrostatic repulsion between oxide layers. The unusually wide range of non-stoichiometry observed for Li_xCoO_2 is thought to result from displacements of Co^{4+} cations producing dipole–dipole couplings between adjacent oxide layers and alleviating the electrostatic repulsion effects. Measurements of lithium mobility in Li_xCoO_2 give comparable diffusion coefficients to Li in Li_xTiS_2 ($D \approx 10^{-8}$ cm^2 sec^{-1}).[120]

The behaviour of $LiVO_2$ on delithiation illustrates the more complex behaviour that is often observed in layered oxides.[121] On removal of lithium,

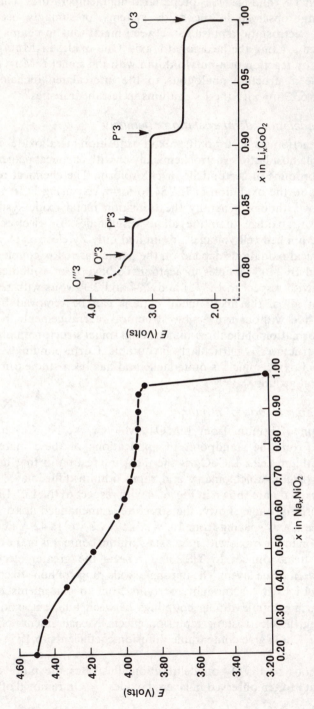

Fig. 6.24. Electrochemical data for alkali metal removal from $NaNiO_2$ (left—from ref. 120) and from $LiCoO_2$ (right—from ref. 128).

the layer structure is preserved up to a composition $Li_{0.67}VO_2$, but further lithium removal leads to two phase behaviour until another single phase is formed at $Li_{0.3}VO_2$. Analysis of the powder X-ray diffraction intensities for the last compound shows that vanadium atoms from the VO_2 layers migrate into the vacant sites between the layers created by lithium removal. The product of this process (which presumably occurs slowly in the two phase region) can be written $Li_x V_{1/3}[V_{2/3} \square_{1/3}]O_2$. The structure corresponds to the octahedral part of the spinel lattice (atacamite framework). Mild heat treatment of the two phase system, $Li_{0.5}VO_2$, leads to the formation of a cubic spinel phase.

Electrochemical deintercalation of $Li_x NiO_2$, which occurs for the composition range $0.4 < x < 1$, shows very complex behaviour but does not apparently involve migration of nickel into the interlayer sites.[122] However, the composition $Li_{0.5}NiO_2$ also transforms above 150°C to a spinel structure analogous to the vanadium system. $LiCrO_2$ shows little evidence for deintercalation.[123]

6.5.2.3 Deintercalation of sodium

Deintercalation of sodium from the host lattices $Na_x MO_2$ (M = Ti, Cr, Mn, Co, and Ni) can be achieved by electrochemical oxidation. With the exception of the titanium compound, high applied voltages are required in order to effect oxidation of the trivalent transition metal to the $4+$ oxidation state. The titanium system, in contrast, is relatively easy to oxidize, but the layer structure can only be preserved to a composition of $Na_{0.71}TiO_2$.[124] Further oxidation produces irreversible changes which may be due to titanium migration into interlayer sites.

In general, the electrochemical behaviour of these systems is very complicated. They show only limited ranges of solid solution but large number of intermediate phases. It is not possible to discuss here details for all of the systems that have been studied, but some general features can be illustrated. Three situations can be distinguished: oxidation starting from the fully reduced O3 structure, the behaviour of P2 phases, and reactions of phases prepared at high temperature with different alkali contents but similar structures. The first type of behaviour is illustrated by the open circuit voltage/composition data for $NaNiO_2$ shown in Fig. 6.24. The first point to note is that the maximum amount of oxidation corresponds to a composition $Na_{0.79}NiO_2$ and this low level of alkali removal is typical for electrochemical oxidation of the fully reduced phases. Higher levels of sodium removal ($x = 0.5$) have, however, been reported in chemical oxidation using bromine as the reagent.[123] The second point to note is that oxidation proceeds through a series of step phases which involve interconversions between O3 and P3 type structures differing only by a translation of one layer relative to another. In the nickel system, Ni^{3+} is a low spin ion and all of the structures (indicated by the primes in Fig. 6.24) are distorted to monoclinic symmetry as a consequence of the Jahn–Teller effect. The interlayer separation increases with sodium removal in a manner analogous to that observed in $LiCoO_2$.

Electrochemical reaction of compounds with P2 structures gives somewhat different behaviour. For example $Na_{0.7}MnO_{2.25}$ and $Na_{0.7}CoO_{1.96}$ can be both oxidized and reduced over similar composition ranges: $0.45 < x < 0.85$ and $0.46 < x < 0.83$ for the manganese and cobalt compounds, respectively.[125,126] Again the electrochemical behaviour is complex and indicates the existence of several discrete compositions, but the P2 structural arrangement is always preserved. This is not unexpected since conversion to the other layer structures cannot occur by simple layer translation but requires a 60° rotation of one slab relative to the next. Finally, it should be noted that differences in the high temperature structures determine the operable range for intercalation–deintercalation even when the compositions are similar. Thus $NaCoO_2$ (O3) can be oxidized only to $Na_{0.82}CoO_2$ (P'3), whereas the compound $Na_{0.83}CoO_2$ (P2), prepared at high temperature, can be further oxidized to $Na_{0.46}CoO_2$ (P2). Clearly more work is required to understand the subtleties involved in ambient temperature transformations between these complex metastable phases.

6.5.2.4 Ion exchange behaviour

As expected from the general electrochemical and structural features of the layered oxides, they undergo ion exchange reactions under mild conditions leading to the formation of metastable oxides.[124] For example, $LiCrO_2$ can be synthesized by ion exchange of $NaCrO_2$ in molten lithium nitrate. Ion exchange of $Na_{0.6}CoO_2$ (P3) and $Na_{0.7}CoO_2$ (P2) by lithium ions in methanol at 64°C occurs, but is accompanied by some solvent oxidation since the products have compositions close to $LiCoO_2$. The exchange of the P3 structure gives the stable high temperature O3 phase whereas exchange of the P2 compound gives the unusual two layer O2 structure shown in Fig. 6.23(b). An interesting recent example[127] is the formation of a new crystalline nickel hydroxide by hydrolysis of $NaNiO_2$ in aqueous ammonium chloride followed by reduction with sodium sulphite solution. The reactions are all topotactic and can be summarized as

$$NaNiO_2(O'3) \xrightarrow[NH_4Cl]{hydrolysis} \gamma\text{-}NiOOH(P3) \xrightarrow[Na_2SO_3]{reduction} \alpha^*\text{-}Ni(OH)_2(P3)$$

$$\xrightarrow[solution]{KOH} \beta\text{-}Ni(OH)_2(O1)$$

The final step (treatment with KOH solution) removes the interlayer water and forms the stable hydroxide structure. The successive transformations, O3 to P3 to O1, require only equal layer translations along the $(11\bar{2}0)$ direction (see Fig. 6.25).

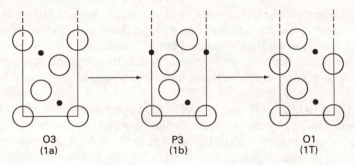

Fig. 6.25. Structural transformations observed in the conversion of NaNiO$_2$ to Ni(OH)$_2$ (adapted from ref. 127).

6.6 SUMMARY AND CONCLUSION

The reaction chemistry of several different classes of layered oxides (AMO$_2$, MoO$_3$, V$_2$O$_5$, VOPO$_4$, and FeOCl) and chalcogenides (MX$_2$ and MPS$_3$) has been surveyed to provide an overview of synthesis and characterization methods and of the factors that determine reaction types, reaction stoichiometries, product structures, and phase transitions in the formation of intercalation compounds. The main emphasis has been on electron transfer reactions in two-dimensional host lattices, but some examples have been given with the same solids to illustrate other kinds of reactivity, including Lewis and Brønsted acid–base reactions, anion substitution, and ion exchange. The different kinds of reaction that have been considered are classified in Table 6.10.

In a general way, it can be seen from these examples that the reactions expected for layered solids are similar to those observed in homogeneous solution though some differences are to be anticipated, particularly for the reaction kinetics. In the heterogeneous case, the reaction rates are constrained by the diffusion of the molecules or ions in the solid state, by kinetic barriers associated with deformation of the solid lattice at the reaction front, and,

Table 6.10 *Reactions of layered compounds*

Reduction	TiS$_2$ + n-C$_4$H$_9$Li	$\longrightarrow$ LiTiS$_2$ + C$_8$H$_{18}$
Oxidation	LiVS$_2$ + 1/2I$_2$	$\longrightarrow$ VS$_2$ + LiI
Substitution	FeOCl + KOCH$_3$	$\longrightarrow$ FeOOCH$_3$ + KCl
Lewis acid/base	VOPO$_4$ + C$_5$H$_5$N	$\longrightarrow$ VOPO$_4$·C$_5$H$_5$N
Brønsted acid/base	H$_x$MoO$_3$ + RNH$_2$	$\longrightarrow$ (RNH$_3$)$_x$(RNH$_2$)$_y$MoO$_3$
Ion exchange	NaNiO$_2$ + H$_3$O$^+$	$\longrightarrow$ NiOOH + Na(H$_2$O)$_n^+$
Ion exchange	MnPS$_3$ + Co(η-C$_5$H$_5$)$_2^+$	$\longrightarrow$ Mn$_{0.87}$PS$_3$[Co(η-C$_5$H$_5$)$_2$]$_{0.36}$

in three-dimensional solids, by the fixed size of the vacant lattice sites and their connectivity in the structure. In spite of these restrictions, by suitable choice of reaction conditions many types of reactions with a wide range of solids can be carried out. A particularly effective way of enhancing the kinetics is to minimize diffusion distances by the use of very small uniform particles.

In addition to inorganic reactions of layered compounds, organic chemistry is also observed to occur in interlamellar spaces. The oxidative coupling reactions of pyridine by TaS_2, discussed in Section 6.2.4, are one example. However, in the sheet silicates and in other layered systems not discussed in this chapter, organic reactions have been studied much more extensively. The addition of methanol to iso-butylene to form methyl *tert*-butyl ether (MTBE) on Cu^{2+} exchanged montmorillonite (J. M. Thomas in ref. 1) and the formation of $Zr(O_3POCH_2CH_2OH)_2$ by reaction of $Zr(O_3POH)_2$ with CH_2CH_2O (G. Alberti and U. Costantino in ref. 1) illustrate the possibilities for catalytic and synthetic chemistry.

Several interesting and relatively new topics, not covered or only briefly touched on, are worth noting as areas that are likely to see future development. Layer structures in which an organic fragment is attached to a layer through a covalent bond (derivatized layers) were discussed with some examples for FeOCl. Other lattices can also be synthesized with this same general feature. For example, the organophosphonates of zirconium ($Zr(O_3PR)_2$) and vanadium ($VO(O_3PR)H_2O$) have both been synthesized with a wide range of organic groups (R) and shown to have interesting interlamellar reactivity. Both classes of compound offer some new possibilities for selective sorption and catalysis. Modification of $Ni(CN)_2$ by coordination of organic bases to form compounds of stoichiometry $Ni(CH_3NH_2)_2Ni(CN)_4$ (the Hoffman clathrates)[3] is a further illustration of a derivatized layer structure. The organic part is attached to the layer in this case by coordination to a metal centre in the same way as in the similar compounds $VOPO_4 \cdot$ pyridine and $MoO_3 \cdot$ pyridine. Large inorganic species (pillars) may be introduced between the layers as an alternative to organic molecules, if a synthetic goal is to form compounds with greater thermal stability. Ion exchange of the clay, montmorillonite with the large oxycation, $Al_{13}O_4(OH)_{24}^{7+}$, is perhaps the best current example. Calcination, after the ion exchange, eliminates the hydroxyls and attaches an aluminium oxide fragment to the silicate surface of the clay. The resulting pillared clay has substantial permanent microporosity and is comparable in many ways to a zeolite. The ability of the montmorillonite clay to expand its interlayer separation in water to a value comparable to the diameter of the hydrated aluminium cation enhances the ion exchange kinetics in the first step of the reaction.

Ion exchange reactions are also of interest from a different perspective. Many alkali metal, early transition metal, ternary oxides have been synthesized with layered and framework structures and shown to exchange alkali metal ions for protons in aqueous acids. The products of these reactions are solid Brønsted acids with acidities which reflect both the chemical composition and structure. In some cases, different structures with the same composition can be synthesized. For example, $HNbO_3$ with a perovskite structure is formed by proton exchange of $LiNbO_3$, whereas the same composition synthesized from $TlNbO_3$ has the pyrochlore framework. Dehydroxylation of the proton-exchanged oxides may lead to the formation of new metastable oxide structures (e.g. the synthesis of $TiO_2(B)$ from $H_2Ti_4O_9$).

In conclusion, the use of reactions carried out on preformed solids for the modification and control of physical and chemical properties and for the synthesis of new types of materials continues to be a major direction for research in this area. Further work directed towards design of the host lattice structure and reactivity offers the prospect of controlling the arrangement of the guest molecules or ions in the solid structure and thereby the ability to promote specific chemical reaction pathways or to enhance particular physical processes.

6.7 GENERAL REFERENCES

1. *Intercalation chemistry* (ed. M. S. Whittingham and A. J. Jacobson), Academic Press, New York (1982).
2. *Physics and chemistry of materials with layer structures*, Reidel, Holland: (1) Vol. 2, *Crystallography and crystal chemistry of materials with layered structures* (ed. F. Levy), (1976); (2) Vol. 5, *Structural chemistry of layer-type phases*, F. Hulliger (ed. F. Levy), (1976); (3) Vo. 6, *Intercalated layered materials* (ed. F. Levy), (1979): (*i*) *Intercalation in layered transition metal dichalcogenides*, G. V. Subba Rao and M. W. Shafer, pp. 99–199; (*ii*) *Alkali metal intercalation compounds of transition metal chalcogenides: TX_2, TX_3 and TX_4 chalcogenides*, J. Rouxel, pp. 201–250.
3. *Inclusion compounds* (ed. J. L. Atwood, J. E. D. Davies, and D. D. MacNicol), Academic Press, New York (1984): (1) *Intercalation compounds*, R. Schöllhorn, Chapter 7, pp. 249–349; (2) *The Hoffmann-type and related inclusion compounds*, Chapter 2, pp. 29–57.
4. *Treatise on solid state chemistry* (ed. N. B. Hannay), Plenum, New York (1976): Vol. 3, *Inclusion compounds*, F. R. Gamble and T. H. Geballe, Chapter 2, pp. 89–166.
5. *Chemistry of intercalation compounds: metal guests in chalcogenide hosts*, M. S. Whittingham, *Prog. Sol. State Chem.*, **12**, 41–99 (1978).
6. *Recent progress in intercalation chemistry: alkali metals in chalcogenide host structures*, J. Rouxel, *Rev. Inorg. Chem.*, **1**, 245–279 (1979); *Low dimensional chalcogenides as secondary cathodic materials: some geometric and electronic aspects*, J. Rouxel and R. Brec, *Ann. Rev. Mater. Sci.*, **16**, 137–162 (1986).
7. *Reversible topotactic redox reactions of solids by electron/ion transfer*, R. Schöllhorn, *Angew. Chem.*, **92**, 1015–1035 (1980); *Angew. Chem. Int. Ed.*, **19**, 983–1003 (1980).
8. *Extended linear chain compounds* (ed. J. S. Miller), Plenum, New York (1982): Vol. 2, *An added dimension—two-dimensional analogs of one dimensional materials*, M. B. Dines and M. Marrocco, Chapter 1, pp. 1–57.

9. *Electron/ion transfer reactions of solids with different lattice dimensionality*, R. Schöllhorn, *Pure Appl. Chem.*, **56**, 1739–1752 (1984).
10. *Interaction of alkylamines with different types of layered compounds*, G. Lagaly, *Solid State Ionics*, **22**, 43–51 (1986).
11. *Graphite intercalation compounds*, H. Selig and L. B. Ebert, *Adv. Inorg. Radiochem.*, **23**, 281–327 (1980).
12. *Review on structural and chemical properties of transition metal phosphorus trisulfides* MPS_3, R. Brec, *Solid State Ionics*, **22**, 3–30 (1986).
13. *Redox intercalation reactions of* $VOPO_4 \cdot 2H_2O$ *with mono- and divalent cations*, A. J. Jacobson, J. W. Johnson, J. F. Brody, J. C. Scanlon, and J. T. Lewandowski, *Inorg. Chem.*, **24**, 1782–1787 (1985) and references therein.
14. *Soft chemistry: the derivatives of two dimensional silicon*, M. Fouletier and M. Armand, *Rev. Chim. Min.*, **21**, 468–475 (1984).
15. *Clay mineralogy*, R. E. Grim, 2nd edn., McGraw-Hill, New York (1968).
16. *Structural chemistry of silicates*, F. Liebau, Springer-Verlag (1985).
17. *Layered* $K_2Ti_4O_9$ *and the open metastable* $TiO_2(B)$ *structure*, M. Tournoux, R. Marchand, and L. Brohan, *Prog. Sol. State Chem.*, **17**, 33–52 (1986).
18. *Interlayer chemistry between thick transition–metal oxide layers: synthesis and intercalation reactions of* $K[Ca_2Na_{n-3}Nb_nO_{3n+1}]$ $(5 < n < 7)$, A. J. Jacobson, J. W. Johnson, and J. T. Lewandowski, *Inorg. Chem.*, **24**, 3727–3729 (1985) and references therein.
19. *Inorganic ion exchange materials*, A. Clearfield, CRC Press (1982).
20. *Synthesis of anionic clay minerals (mixed metal hydroxides, hydrotalcite)*, *Solid State Ionics*, **22**, 135–141 (1986) and references therein.
21. *Energy conversion and storage using insertion materials*, G. Betz and H. Tributsch, *Prog. Sol. State Chem.*, **16**, 195–290 (1985).

6.8   REFERENCES

22. Chianelli, R. R., *J. Crystal Growth*, **34**, 329 (1976) and Chianelli, R. R., Scanlon, J. C., and Rao, B. M. L., *J. Sol. State Chem.*, **29**, 323 (1979).
23. Michel, C. and Raveau, B., *Rev. Chim. Min.*, **21**, 407 (1984).
24. Brown, B. W. and Beernsten, D. J., *Acta Cryst.*, **18**, 31 (1965).
25. Rüdorff, W., *Chimia*, **19**, 489 (1965).
26. Murphy, D. W. and Christian, P. A., *Science*, **205**, 651 (1979).
27. Dines, M. B., *Mater. Res. Bull.*, **10**, 287 (1975); Whittingham, M. S. and Dines, M. B., *J. Electrochem. Soc.*, **124**, 1387 (1977).
28. Omloo, W. P. F. A. and Jellinek, F., *J. Less Comm. Met.*, **20**, 121 (1970).
29. Van Laar, B. and Ijdo, D. J. W., *J. Sol. State Chem.* **3**, 590 (1971).
30. Schöllhorn, R. and Lerf, A., *Z. Naturforsch.*, **29b**, 804 (1974).
31. Murphy, D. W., Cros, C., DiSalvo, F. J., and Waszczak, J. V., *Inorg. Chem.*, **16**, 3027 (1977).
32. van Bruggen, C. F., Haange, R. J., Wiegers, G. A., and de Boer, D. K. G., *Physica*, **99B**, 166 (1980).
33. Whittingham, M. S., *J. Electrochem. Soc.*, **123**, 315 (1976).
34. Thompson, A. H., *J. Electrochem. Soc.*, **126**, 608 (1979).
35. Safran, S., *Phys. Rev. Lett.*, **44**, 937 (1980).

36. Hibma, T., *J. Sol. State Chem.*, **34**, 97 (1980).
37. Schöllhorn, R. and Weiss, A., *Z. Naturforsch.*, **28b**, 711 (1973).
38. Schöllhorn, R. and Meyer, H., *Mater. Res. Bull.*, **9**, 1237 (1974).
39. Lerf, A. and Schöllhorn, R., *Inorg. Chem.*, **16**, 2950 (1977).
40. Schöllhorn, R., Sick, E., and Lerf, A., *Mater. Res. Bull.*, **10**, 1005 (1975).
41. Johnston, D. C., *Mater. Res. Bull.*, **17**, 13 (1982).
42. Boos-Alberink, A. J. A., Haange, R. J., and Wiegers, G. A., *J. Less Comm. Met.*, **63**, 69 (1979).
43. Hendricks, S. and Teller, E., *J. Chem. Phys.*, **10**, 147 (1942).
44. Riekel, C., Reznik, H. G., and Schöllhorn, R., *J. Sol. State Chem.*, **34**, 253 (1980).
45. Murphy, D. W. and Hull, G. W., *J. Chem. Phys.*, **62**, 973 (1975).
46. Nazar, L. F. and Jacobson, A. J., *J. C. S. Chem. Commun.*, 570 (1986).
47. Weiss, A. and Ruthardt, R., *Z. Naturforsch.*, **24b**, 355 (1969) and **24b**, 1066 (1969).
48. Gamble, F. R., DiSalvo, F. J., Klemm, R. A., and Geballe, T. H., *Science*, **168**, 568 (1970).
49. Gamble, F. R. and Silbernagel, B. G., *J. Chem. Phys.*, **63**, 2544 (1975); Silbernagel, B. G., Dines, M. B., Gamble, F. R., Gebhard, L. A., and Whittingham, M. S., *J. Chem. Phys.*, **65**, 1906 (1976).
50. Schöllhorn, R. and Zagefka, H. D., *Angew. Chem.*, **89**, 193 (1977); *Angew. Chem. Int. Ed.*, **16**, 199 (1977).
51. McKelvy, M. J. and Glaunsinger, W. S., *Solid State Ionics*, **25**, 287 (1987).
52. Schöllhorn, R., Zagefka, H. D., Butz, T., and Lerf, A., *Mater. Res. Bull.*, **14**, 369 (1979).
53. Dines, M. B., *Science*, **188**, 1210 (1975).
54. Clement, R. P., Davies, W. B., Ford, K. A., Green, M. L. H., and Jacobson, A. J., *Inorg. Chem.*, **17**, 2754 (1978).
55. Westphal, U. and Geismar, G., *Z. Anorg. Allg. Chem.*, **508**, 165 (1984).
56. Klingen, W., Ott, R., and Hahn, H., *Z. Anorg. Allg. Chem.*, **396**, 271 (1973).
57. Klingen, W., Eulenberger, G., and Hahn, H., *Z. Anorg. Allg. Chem.*, **401**, 97 (1973).
58. Ouvrard, G., Brec, R., and Rouxel, J., *Mater. Res. Bull.*, **20**, 1181 (1985).
59. Brec, R., Ouvrard, G., and Rouxel, J., *Mater. Res. Bull.*, **20**, 1257 (1985).
60. Le Méhauté, A., Ouvrard, G., Brec, R., and Rouxel, J., *Mater. Res. Bull.*, **12**, 1191 (1977).
61. Thompson, A. H. and Whittingham, M. S., *Mater. Res. Bull.*, **12**, 741 (1977).
62. Brec, R., Schleich, D. M., Ouvrard, G., Louisy, A., and Rouxel, J., *Inorg. Chem.*, **18**, 1814 (1979).
63. Brec, R., Ouvrard, G., Louisy, A., Rouxel, J., and Le Méhauté, A., *Solid State Ionics*, **6**, 185 (1982).
64. Whangbo, M.-H., Brec, R., Ouvrard, G., and Rouxel, J., *Inorg. Chem.*, **24**, 2459 (1985).
65. Clement, R. and Green, M. L. H., *J. Chem. Soc. Dalton*, 1566 (1979).
66. Clement, R., *J. C. S. Chem. Commun.*, 647 (1980).
67. Clement, R., *J. Am. Chem. Soc.*, **103**, 6998 (1981).
68. Clement, R., Garnier, O., and Jegoudez, J., *Inorg. Chem.*, **25**, 1404 (1986).
69. Clement, R., Audiere, J. P., and Renard, J. P., *Rev. Chim. Min.*, **19**, 560 (1982).
70. Clement, R. and Michalowicz, A., *Rev. Chim. Min.*, **21**, 426 (1984).
71. Villieras, J., Chiron, R., Palvadeau, P., and Venien, J. P., *Rev. Chim. Min.*, **22**, 209 (1985).

72. Palvadeau, P., Coic, L., Rouxel, J., and Portier, J., *Mater. Res. Bull.*, **13**, 221 (1978).

73. Meyer, H., Weiss, A., and Besenhard, J. O., *Mater. Res. Bull.*, **13**, 913 (1978).

74. Weiss, A. and Sick, E., *Z. Naturforsch.*, **33b**, 1087 (1978).

75. Halbert, T. R. and Scanlon, J., *Mater. Res. Bull.*, **14**, 415 (1979).

76. Halbert, T. R., Johnston, D. C., McCandlish, L. E., Thompson, A. H., Scanlon, J. C., and Dumesic, J. A., *Physica*, **99B**, 128 (1980).

77. Schafer-Stahl, H. and Abele, R., *Angew. Chem. Int. Ed. Engl.*, **19**, 477 (1980).

78. Stahl, H., *Inorg. Nucl. Chem. Lett.*, **16**, 271 (1980).

79. Palvadeau, P., Coic, L., Rouxel, J., Menil, F., and Fournes, L., *Mater. Res. Bull.*, **16**, 1055 (1981).

80. Averill, B. A. and Kauzlarich, S. M., *Mol. Cryst. Liq. Cryst.*, **107**, 55 (1984).

81. Hagenmuller, P., Portier, J., and Barbe, B., *Vietnam. Chim. Acta*, 59 (1966).

82. Kanamaru, F., Shimada, M., Koizumi, M., Takano, M., and Takada, T., *J. Sol. State Chem.*, **7**, 297 (1973).

83. Kikkawa, S., Kanamaru, F., and Koizumi, M., *Bull. Chem. Soc. Jpn.*, **52**, 963 (1979).

84. Eckert, H. and Herber, R. H., *J. Chem. Phys.*, **80**, 4526 (1984).

85. Hagenmuller, P., Rouxel, J., and Portier, J., *C. R. Acad. Sci. Paris*, **254**, 2000 (1962).

86. Weiss, A. and Choy, J. H., *Z. Naturforsch.*, **39b**, 1193 (1984).

87. Kikkawa, S., Kanamaru, F., and Koizumi, M., *Inorg. Chem.*, **15**, 2195 (1976).

88. Son, S., Kikkawa, S., Kanamaru, F., and Koizumi, M., *Inorg. Chem.*, **19**, 262 (1980).

89. Kikkawa, S., Kanamaru, F., and Koizumi, M., *Inorg. Chem.*, **19**, 259 (1980).

90. Johnson, J. W., Jacobson, A. J., Brody, J. F., and Lewandowski, J. T., *Inorg. Chem.*, **23**, 3842 (1984).

91. Glemser, O. and Hutz, G., *Z. Anorg. Allg. Chem.*, **264**, 17 (1951).

92. Glemser, O., Hutz, G., and Meyer, G., *Z. Anorg. Allg. Chem.*, **285**, 173 (1956).

93. Kihlborg, L., Hägerström, G., and Ronnquist, A., *Acta Chem. Scand.*, **15**, 1187 (1961).

94. Wilhelmi, K. A., *Acta Chem. Scand.*, **23**, 419 (1969).

95. Birtill, J. J. and Dickens, P. G., *Mater. Res. Bull.*, **13**, 311 (1978).

96. Schöllhorn, R., Kuhlmann, R., and Besenhard, J. O., *Mater. Res. Bull.*, **11**, 83 (1976).

97. Dickens, P. G., Crouch-Baker, S., and Weller, M. T., *Solid State Ionics*, **18/19**, 89 (1986).

98. Schöllhorn, R., Schulte-Nölle, T., and Steinhoff, G., *J. Less Comm. Met.*, **71**, 71 (1980).

99. Johnson, J. W., Jacobson, A. J., Rich, S. M., and Brody, J. F., *J. Am. Chem. Soc.*, **103**, 5246 (1981).

100. Dickens, P. G., French, S. J., Hight, A. T., and Pye, M. F., *Mater. Res. Bull.*, **14**, 1295 (1979).

101. Murphy, D. W., Christian, P. A., DiSalvo, F. J., and Waszczak, J. V., *Inorg. Chem.*, **18**, 280 (1979).

102. Hagenmuller, P., Galy, J., Pouchard, M., and Casalot, A., *Mater. Res. Bull.*, **1**, 45 (1966).

103. Dickens, P. G., Chippindale, A. M., Hibble, S. J., and Lancaster, P., *Mater. Res. Bull.*, **19**, 319 (1984).
104. Tinet, D. and Fripiat, J. J., *Rev. Chim. Min.*, **19**, 612 (1982).
105. Tinet, D., Legay, M. H., Gatineau, L., and Fripiat, J. J., *J. Phys. Chem.*, **90**, 948 (1986).
106. Livage, J. and Lemerle, J., *Ann. Rev. Mat. Sci.*, **12**, 103 (1982).
107. Watson, J. H. L., Heller, W., and Wojtowicz, W., *Science*, **109**, 274 (1949).
108. Legendre, J.-J. and Livage, J., *J. Coll. Int. Sci.*, **94**, 75 (1983).
109. Aldebert, P., Baffier, N., Gharbi, N., and Livage, J., *Mater. Res. Bull.*, **16**, 949 (1981).
110. Bouhaouss, A. and Aldebert, P., *Mater. Res. Bull.*, **18**, 1247 (1983).
111. Aldebert, P. and Paul-Boncour, V., *Mater. Res. Bull.*, **18**, 1263 (1983).
112. Johnson, J. W., Jacobson, A. J., Brody, J. F., and Rich, S. M., *Inorg. Chem.*, **21**, 3820 (1982).
113. Schöllhorn, R., Steffen, R., and Wagner, K., *Angew. Chem. Int. Ed.*, **22**, 555 (1983).
114. Delmas, C., Fouassier, C., and Hagenmuller, P., *Mater. Res. Bull.*, **11**, 1483 (1976).
115. Delmas, C., Braconnier, J.-J., Fouassier, C., and Hagenmuller, P., *Z. Naturforsch.*, **36b**, 1368 (1981).
116. David, W. I. F., Goodenough, J. B., Thackeray, M. M., and Thomas, M. G. S. R., *Rev. Chim. Min.*, **20**, 636 (1983).
117. Thackeray, M. M., Baker, S. D., Adendorff, K. T., and Goodenough, J. B., *Solid State Ionics*, **17**, 175 (1985).
118. Chen, C. J., Greenblatt, M., and Waszczak, J. V., *Solid State Ionics*, **18/19**, 838 (1986).
119. Mizushima, K., Jones, P. C., Wiseman, P. J., and Goodenough, J. B., *Mater. Res. Bull.*, **15**, 783 (1980).
120. Thomas, M. G. S. R., Bruce, P. G., and Goodenough, *Solid State Ionics*, **17**, 13 (1985).
121. de Picciotto, L. A. and Thackeray, M. M., *Mater. Res. Bull.*, **20**, 187 (1985).
122. Thomas, M. G. S. R., David, W. I. F., Goodenough, J. B., and Groves, P., *Mater. Res. Bull.*, **20**, 1137 (1985).
123. Miyazaki, S., Kikkawa, S., and Koizumi, M., *Syn. Met.*, **6**, 211 (1983).
124. Delmas, C., Braconnier, J.-J., Maazaz, A., and Hagenmuller, P., *Rev. Chim. Min.*, **19**, 343 (1982).
125. Mendiboure, A., Delmas, C., and Hagenmuller, P., *J. Sol. State Chem.*, **57**, 323 (1985).
126. Delmas, C., Braconnier, J.-J., Fouassier, C., and Hagenmuller, P., *Solid State Ionics*, **3/4**, 165 (1981).
127. Braconnier, J. J., Delmas, C., Fouassier, C., Figlarz, M., Beaudouin, B., and Hagenmuller, P., *Rev. Chim. Min.*, **21**, 496 (1984).
128. Braconnier, J. J., Delmas, C., and Hagenmuller, P., *Mater. Res. Bull.*, **17**, 993 (1982).

7 Zeolites

John M. Newsam

7.1 INTRODUCTION

Zeolites are special. They form a fascinating class of minerals. They are widely used in ion-exchange applications, offer unique properties as sorbents and molecular sieves, and they play a dominant role in heterogeneous catalysis.

The properties of zeolites derive directly from the particular characteristics of their crystal structures and the broad field of solid state chemistry provides few other examples where the interplay between local structure and bulk macroscopic properties can be so directly observed. Academic and industrial interest in zeolites continues to grow steadily. This reflects the expanding scope of zeolite synthetic procedures, and further developments in commercial zeolite utilization. In addition, many of the solid state characterization tools that have been developed over the past two decades or so[1] are directly applicable to zeolite structural problems. The deeper understanding of zeolite chemistry that such techniques promise affords the hope of more direct control of zeolite synthesis, and of developing predictive capabilities for zeolite selection and optimization for particularly desired applications.

7.2 STRUCTURES

The term zeolite was coined in 1756 by the Swedish mineralogist A. F. Cronstedt who observed that the mineral stilbite frothed and gave off steam when heated in a blowpipe flame. The name comes from the Greek, $\zeta\epsilon\iota\nu$—boil and $\lambda\iota\theta os$—stone, and describes a particular class of minerals related to feldspars and feldspathoids. These are tectoaluminosilicates. That is, they have crystal structures that are constructed from TO_4 tetrahedra (T = tetrahedral species, Si, Al, etc.), each apical oxygen atom of which is shared with an adjacent tetrahedron (Figs. 7.1–7.3). Tectosilicates thus always have a framework metal-to-oxygen atom ratio of 2. Examples of tectosilicates include the feldspars,[2,3] feldspathoids[3] and the various structures adopted by SiO_2 itself.[4] Zeolites are, by definition, distinguished in having more open structures that will reversibly sorb and desorb water (or larger molecules), and that contain large non-framework cations that can readily be exchanged.

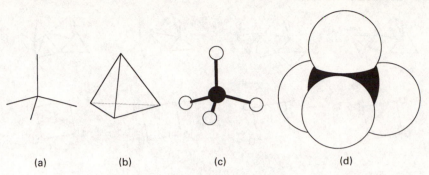

Fig. 7.1. A TO$_4$ tetrahedron, the primary building block of all zeolite structures, is variously drawn in stick (a), or ball and stick representation (c), using approximate van der Waals radii (Si 1.4 Å, O 1.35 Å—*d*), or as a tetrahedron (b), with lines connecting the apical oxygen atoms.

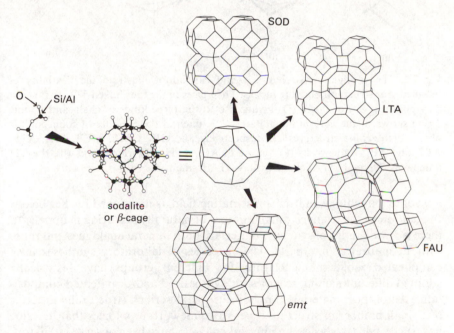

Fig. 7.2. The figurative construction of four different zeolite frameworks that contain sodalite or β-cages (truncated octahedra). A pair of TO$_4$ tetrahedra sharing one vertex is linked into a single sodalite cage. In a less cluttered representation, the oxygen atoms are omitted and the cage is drawn as straight lines connecting the tetrahedral (T) sites (with hidden lines removed). The sodalite cage unit is found in the SOD, LTA, and FAU frameworks. The *emt* framework is a hexagonal variant of FAU that occurs in EMC-2 and to a limited extent in zeolite ZSM-20[21].

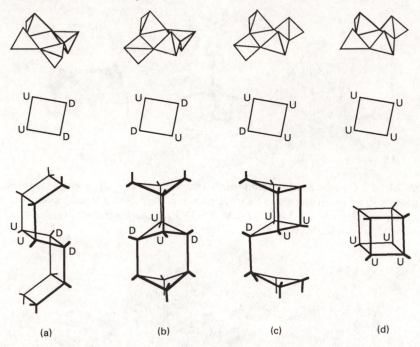

Fig. 7.3. The configurations that a 4-ring can adopt illustrate the flexibility of tetrahedra as structural building blocks. The apices of the four linked TO_4 tetrahedra may point up (U) or down (D) giving rise to the three kinds of chain and double 4-ring shown at the bottom (after Barrer[106]). Each of these modes of connection is observed in zeolite structures[9] in, respectively, the GIS, GME, MER, PHI, ATF, ATT, and APC frameworks (a), the AEL, AFI, and APD frameworks (b), the STI framework (c), and the AFY, AST, and LTA frameworks (d).

Zeolite mineralogy, in itself a fascinating field, is mentioned further below. In addition to the relatively small number of naturally occurring materials,[5–8] there is a wide range of synthetic materials, both mineral analogues and those without natural counterparts. The successes of laboratory syntheses have complicated zeolite nomenclature, for different groups have historically adopted different naming schemes for synthesized and/or patented compositions. All of these materials can, however, be described structurally in terms of a small number of structure types or framework topologies that describe how successive tetrahedra are interlinked (~ 85 have been observed to date).[9] Each observed framework topology is assigned a unique framework code (Table 7.1). Thus, for example, zeolites A (also called 3A, 4A, 5A depending on the non-framework cation composition), alpha, ZK-4, N-A, and SAPO-42 all share the LTA framework (Fig. 7.2). A material with the zeolite A framework has not yet been found to occur naturally. The mineral faujasite and synthetic zeolites X, Y, CSZ-3, and SAPO-37 adopt the FAU framework

(Fig. 7.2). (The framework code is subsequently given in parentheses following each material name; in the cases of the more recent structure determinations for which framework codes have not yet been assigned, an arbitrary three letter designation is given in lower case italics.) The framework code defines the topology, the extended manner in which the TO_4 tetrahedra are interconnected in three dimensions. The structural details, such as the bond lengths and angles for the framework components and the distribution of non-framework species, depend on the composition and conditions.

The individual TO_4 tetrahedra in zeolite structures are generally close to regular, but the shared oxygen linkage can accommodate a wide range of T–O–T bond angles from $\sim 125°$ to $\sim 180°$. The tetrahedron can, therefore, be combined readily to form a variety of different structures, ranging from amorphous materials, through glasses and condensed aluminosilicates such as the feldspathoids, to the more open zeolites (Figs 7.2 and 7.3). Zeolite framework structures have pores that vary in shape, size, and dimensionality (Figs 7.2, 7.4–7.6 and Table 7.1). The fact that zeolites are crystalline (their extended structures are produced by regular translational repeats of the unit cell), with microporosity that is an intrinsic characteristic of the crystal structure, distinguishes them from several other microporous materials such as carbon molecular sieves, silica gel, and certain categories of pillared clay.[10,11] As we see below, the strict regularity of the pore structure enables high selectivities to be achieved in both catalysis and sorption processes.

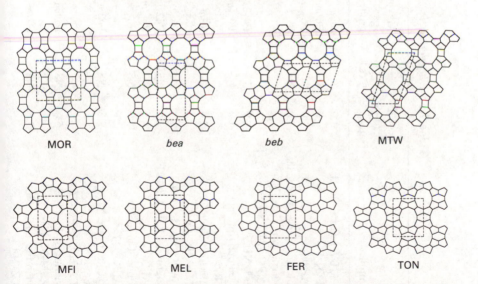

Fig. 7.4. Projections along principal crystallographic directions of eight zeolite structures that contain 5-ring units. The structures, drawn as straight lines connecting adjacent T-sites, in each case are based on reported crystal structures. The unit cell outlines are indicated by the dashed lines.

Table 7.1 *Characteristics of selected zeolite frameworks*

Code	Examples[a]	Typical Si:(T–Si) ratio	Occ.[b]	Maximal space group[c]	SBU[d]	FD[e]	N_T[f]	Pore structure[g]
ABW	Li–A(BW), Cs[SiAlO$_4$]	1.0	S	$Imam$	4, 6, 8	19.0	8	**8** 3.4 × 3.8*
AFI	AlPO$_4$-5	1.0	S	$P6/mcc$	4, 6	17.5	24	**12** 7.3*
AFS	MAPSO-46	1.0	S	$P6_3/mcm$	6≡1	13.7	56	**12** 6.3* $\leftrightarrow$ **8** 4.0 × 4.0**
ANA	analcime, leucite, pollucite, viseite, wairakite, Na-B AlPO$_4$-24, Cs$_2$[FeSi$_5$O$_{12}$]	2.0	NS	$Ia\bar{3}d$	4, 6, 6-2	18.6	48	**8** distorted
bea	beta[h], NU-2[h]	10.0	S	$P4_122$	5-1 + 4	15.5	64	**12** 7.3 × 6.0***
CAN	cancrinite, tiptopite, ECR-5	1.0	NS	$P6_3/mmc$	6	16.7	12	**12** 5.9*
CHA	chabazite, Linde D, Linde R, ZK-14, SAPO-34, MeAPO-47	2.0	S	$R\bar{3}m$	6, 6-6	14.6	36	**8** 3.8 × 3.8**
EDI	edingtonite, K-F, Linde F	1.5	NS	$P\bar{4}2_1m$	4≡1	16.6	10	**8** 2.8 × 3.8** $\leftrightarrow$ **8** variable
emt	ZSM-20[i]	4.5	S	$P6_3/mmc$	6-6, 6-2	12.7	192	{**12** 7.6 $\leftrightarrow$ **12** 7.6 × 5.7}***
ERI	erionite, Linde T[j], AlPO$_4$-17	3.0	NS	$P6_3/mmc$	6	15.6	36	**8** 3.6 × 5.1***
FAU	faujasite, Linde X, Linde Y, LZ-210, SAPO-37	2.5	NS	$Fd\bar{3}m$	6-6, 6-2	12.7	192	**12** 7.4***
FER	ferrierite, Sr-D, FU-9, ZSM-35, ISI-6	5.0	NS	$Immm$	5-1	17.7	36	**10** 4.2 × 5.4* $\leftrightarrow$ **8** 3.5 × 4.8*
HEU	heulandite, clinoptilolite, LZ-219	3.5	NS	$C2/m$	4-4≡1	17.0	36	**8** 2.6 × 4.7* $\leftrightarrow$ {**10** 3.0 × 7.6* + **8** 3.3 × 4.6*}

Code	Type species[a]		Occurrence[b]	Symmetry[c]	SBU[d]	FD[e]	nT[f]	Channels[g]
KFI	ZK-5, Ba-P, Ba-Q	2.0	S	$Im\bar{3}m$	6-6, 4, 8, 6-2	14.7	96	**8** 3.9*** | 3.9***
LTA	Linde A, ZK-4, N-A, alpha, ZK-21, ZK-22, SAPO-42	1.0	S	$Pm\bar{3}m$	4-4, 8, 6-2	12.9	24	**8** 4.1***
LTL	Linde L, K(Ba)-G(L) ECR-3, perlialite	3.0	NS	$P6/mmm$	6	16.4	36	**12** 7.1*
MAZ	mazzite, omega, ZSM-4	3.0	NS	$P6_3/mmc$	5-1, 4	16.1	36	**12** 7.4* ↔ **8** 3.4 × 5.6*
MEL	ZSM-11	>30.0	S	$I\bar{4}m2$	5-1	17.7	96	**10** 5.3 × 5.4***
MFI	ZSM-5, silicalite, AMS-1B, NU-4	>15.0	S	$Pnma$	5-1	17.9	96	{**10** 5.3 × 5.6 ↔ **10** 5.1 × 5.5}***
MOR	mordenite, ptilolite Zeolon, Na-D	5.0	NS	$Cmcm$	5-1	17.2	48	**12** 6.7 × 7.0* ↔ **8** 2.6 × 5.7*
MTN	ZSM-39, Dodecasil-3C	∞	S	$Fd\bar{3}m$	5 + 5-1	18.7	136	**6**
MTW	ZSM-12, CZH-5, NU-13	>40.0	S	$C2/m$	5-1 + 4	19.4	28	**12** 5.5 × 5.9*
NAT	natrolite, mesolite, scolecite	1.5	NS	$I4_1/amd$	4=1	17.8	40	**8** 2.6 × 3.9* ↔ **8** variable*
OFF	offretite, TMA-O, Linde T[j]	3.5	NS	$P\bar{6}m2$	6	15.5	18	**12** 6.7* ↔ **8** 3.6 × 4.9**
RHO	rho, pahasapaite	3.0	NS	$Im\bar{3}m$	8-8, 6, 6-2	14.3	48	**8** 3.6*** | 3.6***
SOD	sodalite, ultramarine, nosean, tugtupite, AlPO$_4$-20	1.0	NS	$Im\bar{3}m$	6, 4, 6-2	17.2	12	**6**
TON	Theta-1, Nu-10, KZ-2, ISI-1, ZSM-22	>30.0	S	$Cmcm$	6, 5-1	19.7	24	**10** 4.4 × 5.5*
VFI	VPI-5, AlPO$_4$, AlPO$_4$-54, MCM-9	1.0	S	$P6_3/mcm$	4-2	14.2	36	**18** 11.2*

[a] Type species on which framework code is based is given first.
[b] Occurrence; N = natural mineral, S = synthetic, NS = both.
[c] Highest symmetry for the framework type; symmetries actually adopted by example materials may be lower.
[d] Secondary building unit (see Fig. 7.8). Frequently more than one is appropriate, and only the most useful are given (see ref. 9).
[e] Framework density in T-atoms per 1000 Å³.
[f] Number of T-atoms in the (highest symmetry) unit cell.
[g] Nomenclature of Meier and Olson.[9] Bold numbers indicate number of T (or O) atoms in the defining ring. Approximate aperture free diameters are then given for the type species in Å, the number of asterisks indicating if the channel system is 1-, 2-, or 3-dimensional. For more than one channel ↔ (or |) indicates whether (or not) channels interconnect.
[h] Structure comprises bea–beb intergrowths.
[i] Structure comprises FAU–emt intergrowths.
[j] Structure comprises ERI–OFF intergrowths.

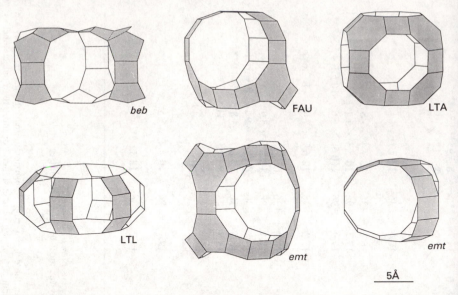

Fig. 7.5. Representations of the primary pore systems of several zeolites, drawn as straight lines connecting the adjacent T-sites that define the pore's perimeter. The pores taken are taken from representative crystal structures and are drawn to the same scale.

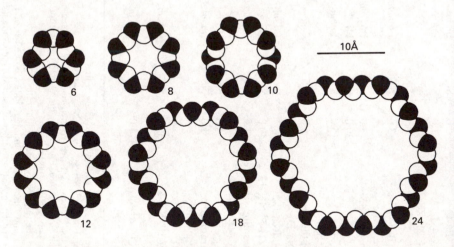

Fig. 7.6. Aperture dimensions depend primarily on the number of O or T atoms in the ring. The illustrated apertures (drawn using approximate van der Waals radii of T 1.4 Å, O 1.35 Å) are taken from representative crystal structures of SOD (6-ring), LTA (8), MFI (10), and FAU (12) framework materials. The 18 and 24-ring apertures are clipped from model structures derived respectively from the 81(1) and 81(2) nets[108] with atomic coordinates optimized by distance least squares[83] assuming $AlPO_4$ compositions.

Further, in marked contrast to most other catalytic systems in which activity is restricted to external surfaces, in zeolites all atoms comprising the extended three-dimensional structure are within a few angstroms of the internal surface and can influence or contribute to processes occurring there. Effective surface areas of $300-700 \, m^2 \, g^{-1}$ are typical for zeolites, and, for usual crystallite sizes of $\sim 0.1-5 \, \mu m$, imply that more than 98 per cent of the total surface area is internal.

The aperture dimensions which control entry into the internal pore volume are determined primarily by the number of T-atoms and oxygen atoms in the ring which defines them (Figs 7.6 and 7.7). Effective aperture sizes range from $\sim 4 \, \text{Å}$ for 8-ring structures (such as those of zeolites A and ZK-4 (LTA), rho (RHO), ZK-5 (KFI), through $\sim 5.4 \, \text{Å}$ for 10-rings (ZSM-5 (MFI), Theta-1 (TON), $\sim 7.4 \, \text{Å}$ for 12-rings (zeolites X (FAU), L (LTL), ZSM-12 (MTW), beta (*bea/beb*)) to $\sim 10.2 \, \text{Å}$ for 18-rings (VPI-5 (VFI), an $AlPO_4$ composition providing, at present, a single and very recent example). Hypothetical frameworks with larger apertures such as 24-rings can also be constructed (Fig. 7.6). The absence, until recently, of zeolites with larger than 12-ring pore systems (with aperture dimensions of $\sim 7.4 \, \text{Å}$) prompted investigations of alternative classes of microporous materials with larger pore systems, such as pillared clays.[10,11]

Although all zeolite structures have as primary building units TO_4 tetrahedra, similarities and relationships between the various full structure types become apparent only when somewhat larger units are considered. Most of the observed frameworks can be constructed from a single type of secondary building unit (SBU) from amongst those illustrated in Fig. 7.8. The group of fibrous natural zeolites, edingtonite (EDI), thomsonite (THO), natrolite (NAT), mesolite (NAT), and scolecite (NAT), share the $4 = 1$ SBU (Figs 7.8 and 7.9). Most of the high silica zeolites produced by direct synthesis have an SBU based on the 5-ring (typical examples as shown in projection in Fig. 7.4). It is attractive to consider a role for silicate or aluminosilicate anions with structures based on the SBUs of Fig. 7.8 during the process of zeolite synthesis, and there is good evidence for the solution occurrence of a wide range of silicate anions including, for example, those with the 4, 4–4 and 6–6 structures.[12] Our understanding of the details of zeolite synthesis on a molecular level[13] remains, however, limited, preventing the use of SBUs for other than categorizing and understanding the topologies of zeolite structures once formed.

There are other convenient ways of discussing zeolite framework structures. The structures in Fig. 7.2 are illustrated as constructed from a near-regular archimedean polyhedron, the truncated octahedron. A pentagonal dodecahedral unit is seen in the MTN and MEP frameworks.[9] The observed occurrences of these various types of cages leads naturally to a consideration of other ways in which they might be interlinked, leading to new hypothetical framework structures. Consideration, for example, of the ways in which

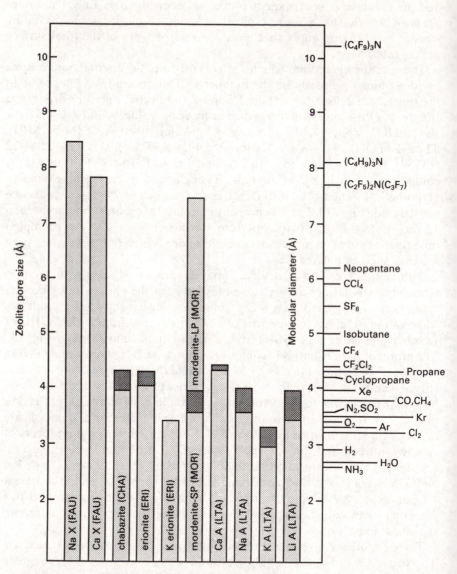

Fig. 7.7. Kinetic diameters for selected molecular species (right) contrasted with the effective pore sizes of representative zeolites (left). The darker shadings indicate increases that accompany a temperature rise 77–420 K. The differences between potassium (3A), sodium (4A), and calcium (5A) zeolite A reflect the presence of non-framework cations in partially window blocking sites (after Breck[50]).

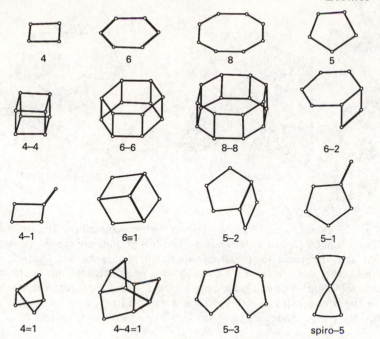

Fig. 7.8. The various zeolite structures can be classified according to their secondary building units (containing four or more tetrahedra), the more common of which are illustrated here (after Meier and Olson[9]). The LTA framework (Fig. 7.2), for example, can be constructed entirely by linking 4–4 units (and see Table 7.1).

cubo-octahedra (the shape of the LTA supercage shown in Fig. 7.5) can be interconnected[14,15] leads to a family of structures that includes the LTA, KFI, and RHO frameworks. The description of the RHO-framework as a hypothetical structure type in fact preceded both the synthesis and structural characterization of the type species, aluminosilicate zeolite rho (RHO),[16] and the discovery of the natural mineral pahasapaite (RHO), a beryllophosphate with the same framework topology.[17]

Many zeolite structures, when viewed in projection along one or more crystallographic directions, appear as two-dimensional 3-connected nets (see, for example Fig. 7.4). Any 2-D 3-connected net comprised solely of even membered rings can be converted to a 3-D 4-connected net simply by alternation of Up and Down linkages.[18] Other modes of inter-sheet connection are generally possible, and the families of 2-D 3-connected nets thus provide one basis for generating hypothetical zeolite structures. The character of the stacking of sheets along one axis is a good means of classifying many of the known zeolite structure types. The ABC-6 family of structures (Fig. 7.10),[15] for example, shares the 6-ring as an SBU (Fig. 7.8). In developing a 3-D 4-connected structure from these units, the next 6-ring can stack

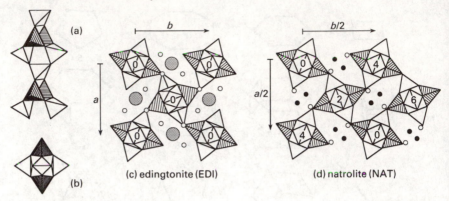

Fig. 7.9. The chains formed from 4=1 units that are common to the structures of the fibrous zeolites (drawn as tetrahedra—two linked units are shown, viewed from the side (a), and from the top (b); aluminate tetrahedra are shaded). Different modes of connection of these chains are seen in the structures of edingtonite (EDI) (c) and natrolite (NAT) (d). The unit cells, heights of the central tetrahedron in unit of $c/8$, and locations of non-framework species are indicated (open circles—water, closed circles—Na^+, large shaded circles—Ba^{2+}) (after Gottardi and Galli[7]).

vertically above the first (labelled A) to generate a hexagonal prism (AA sequence), or it can be offset, AB. The third layer can then be positioned above the first, AAA or ABA, the second, AAB or ABB, or neither (ABC). The next, fourth layer must then stack over one of the A, B, or C positions. In terms of 3-dimensional connectivity, there is no restriction on the character of the ABC . . . stacking sequence, although certain arrangements such as the double hexagonal prism, AAA, have never been observed. Nearly 20 per cent of the (to date) observed zeolite framework structures fall into this category,[9] including those of sodalite (SOD; with a cubic ABC . . . sequence—Figure 7.2; the 6-ring stacking direction, $[111]_{cubic}$, is oblique to the page), cancrinite (CAN; with AB . . .) and chabazite (CHA; AABBCC . . .) (Fig. 7.10). The sequences of erionite (ERI—AABAAC) and offretite (OFF—AABAAB; a double repeat is given for comparison) are similar and, perhaps reflecting this similarity, intergrowths between the two structure types are common.[19] Various intergrowths between other members of the ABC-6 family of structures have also been reported.[20]

Other families of zeolite structures also provide examples of different structures (observed or hypothetical) that differ only in the mode of stacking along one direction. Given the conditions of kinetic control under which zeolites are synthesized (discussed below), the relatively common occurrence of stacking disorder is unsurprising. Figure 7.2 illustrates how changing the operation (an inversion) that relates successive sheets along $[111]_{cubic}$ in the FAU framework to a mirror operation gives rise to a new structure type,

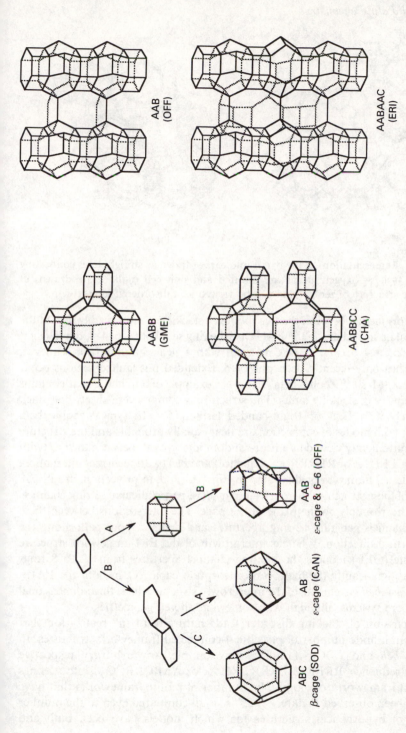

Fig. 7.10. Schematic illustrating how different modes of stacking of 6-ring SBUs in superposition or offset in one of two senses give rise to a series of structure types. This ABC-6 family includes the framework structures of sodalite (SOD), cancrinite (CAN), gmelenite (GME), chabazite (CHA), offretite (OFF), and erionite (ERI).

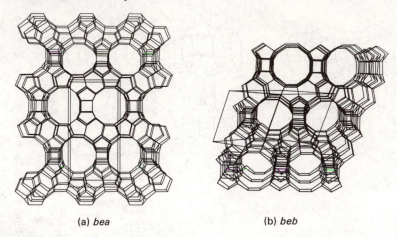

(a) *bea*   (b) *beb*

Fig. 7.11. Representations of the two frameworks (drawn as straight lines connecting adjacent T-sites—oxygen atoms are omitted and unit cell outlines are drawn) of which the structure of zeolite beta can be viewed as a disordered intergrowth.[22, 23]

emt. The distinction between the two modes of stacking is most readily seen at the hexagonal prisms where, in the FAU case the ring sequence is 6–4–4, but in *emt* it is 4–4–4 or 6–4–6 (Fig. 7.2). A material with a near fault-free *emt* sequence, EMC-2, has only recently been observed. Extended but faulted regions occur in zeolite ZSM-20.[21] Zeolite beta is another example of a material that contains a high density of stacking faults. The structure is comprised of sheets that stack successively in a Left- or Right-handed fashion.[22, 23] In typical zeolite beta materials both modes of connection are near equally probable and the structure is then conveniently viewed as a near-random intergrowth between the *bea* (with a pure LLLLL . . . or RRRRR . . . sequence) and *beb* (with recurrent alternation, RLRLRL . . .) frameworks illustrated in Fig. 7.11 and, in projection, in Fig. 7.4. In both of these structures there is a similar set of perpendicular 12-ring channels running horizontally, in the plane of the page. The intersections between these sets of channels generate 12-ring apertures and define a pore path along the third, vertical direction. Whereas intergrowth of the ERI-framework structure in offretite (OFF) reduces the limiting channel apertures from 12 to 8 rings (reducing significantly the hydrocarbon sorption capacity), in both the FAU–*emt* and *bea–beb* systems, all of the intergrowth structures have three-dimensional 12-ring pore systems, although with somewhat altered geometries.

   A discussion of stacking disorder leads naturally to the realization that there is an infinite number of possible 4-connected framework structures. In the *bea–beb* and ABC-6 families, for example, any arbitrary respective stacking sequence, RRLRLLLRRRLRL . . . or AABCABCC . . . etc., defines a distinct framework topology. The number of zeolite frameworks that have actually been observed to date, ∼ 85,[9] is small compared even to the number (∼ 800) of hypothetical structures for which models have been built and topologies categorized. This data base of models provides a first point of reference when considering structural models for a new material (utilized, to effect, in structure solutions of zeolites rho (RHO),[16] ZSM-20 (FAU/*emt*),[21] and VPI-5 (VFI)[24]). When, as more frequently happens, the new material does not conform with a known theoretical model, other structure solution techniques must be brought to bear.

7.3 COMPOSITIONS

An appreciation of zeolite structure is the entry point into the structural chemistry of these systems, for it is on the basis of structural characteristics that the zeolite family is defined, and it is structure that determines their utility. Compositionally, we can write a general formula for the aluminosilicate zeolites as

$$M_{x/m}^{m+} \ \cdot \ [Si_{1-x}Al_xO_2] \ \cdot \ nH_2O$$

Non-framework cations   Framework   Sorbed phase

The framework (the basis for all of the preceding discussion) has a metal-to-oxygen atom ratio of 2, with, as constituents, tetrahedra of net composition $[SiO_2]$ and $[AlO_2]^-$. The framework is anionic, with a net negative charge equal to the number of aluminium atoms in framework T-positions. A corresponding number of non-framework cations is required for charge compensation. These species are incorporated during synthesis, and there is a subtle balance between aluminium concentration, basicity, base character, and other factors that determine the framework structure and composition of the crystallized zeolite. The non-framework cations are usually sited in, or have access to, the pore system and can readily be exchanged by treatment in a suitable salt solution or molten salt (this property leads to a range of commercial applications). The third structural component is represented by the sorbed phase, water in the above formulation. The aqueous (or, in some cases discussed below, organic) component is, one suspects, a key component during zeolite formation, where it forms an integral part of the developing structure. Following crystallization, the sorbed phase can be liberated or extracted from the sample at higher temperature or under reduced external pressure, without destruction of the aluminosilicate framework. The structural integrity of zeolites following dehydration thus distinguishes them from classical salt hydrates, where loss of the water of crystallization is accompanied by gross structural change. The three facets of zeolite structure—the framework, the non-framework cations, and the sorbed phases—are each of interest and each plays a role of varying importance in applications.

The framework composition can show considerable variability. The range of accessible Si:Al ratios, which can be adjusted either by changes in synthesis conditions or by post-synthesis modification, varies from one framework to another, but is always greater than 1.0. Loewenstein[25] observed that most of the known aluminosilicates had Si:Al $\geqslant$ 1.0, and rationalized this (and other observations) in terms of an Al–O–Al linkage avoidance rule. On a simple level Loewenstein's rule can be understood in terms of the net negative charge of an $[AlO_2]^-$ tetrahedron in a tectoalumino-silicate structure. Adjacent $[AlO_2]^-$ tetrahedra would involve adjacent negative charges and would hence be unfavourable. Although Loewenstein's rule is uniformly obeyed in zeolite systems, there are exceptions, most notably in the SOD framework, where condensed (non-zeolite) aluminate forms, such as $Ca_4[Al_6O_{12}]SO_4$ (SOD)[26] are known, and bicchulite (SOD), $Ca_4[Si_2Al_4O_{12}](OH)_4$, can be prepared hydrothermally.[27] Certain general-izations follow from Loewenstein's rule. At lower Si:Al ratios ($< \sim 1.4$), Si–Al alternation must develop long-range coherence. As a result, the unit cell repeat distance of zeolite A (LTA) with Si:Al = 1.0, for example, is not the 12.3 Å repeat between successive sodalite cage centres seen in Fig. 7.2, but rather double this distance in each direction. There is also, for example, a lower Si:Al limit of significantly greater than 1.0 accessible in structures containing 5-rings (such as those shown in Fig. 7.4). An odd-membered ring cannot accommodate strict T-site alternation.

A zeolite's chemical characteristics depend on the framework composition. At higher Si:Al ratios, the non-framework cation complement is reduced and, structurally, there is a trend towards a prevalence of 5-membered rings. The higher silica materials have increased hydrothermal stabilities, their acid site density is decreased (but with a concomitant increase in the individual acid site strengths), and their sorptive characteristics are hydrophobic or organophilic rather than hydrophilic. The limiting Si:Al ratio corresponds to a pure SiO_2 composition and several open silicate frameworks have been described, including high silica sodalite (SOD),[28] silicalite (MFI)[29] (the SiO_2 analogue of zeolite ZSM-5 (MFI)), melanophlogite (MEP)[30], and ZSM-39 (MTN)[31] (the analogues respectively of the type I and II cubic clathrate hydrates). Adhering to a strict definition of zeolites, the open silica polymorphs are better termed porosils or clathrasils, depending on whether or not their pore spaces are accessible.[32]

In addition to Si–Al replacement, a range of T-atom heterosubstitutions are possible. Many gallosilicate zeolites have been described, with properties generally similar to their aluminosilicate counterparts.[33,34] No natural gallosilicate zeolites have, however, yet been found. Aluminogermanates and gallogermanate analogues are known, although currently less well explored.[35] A wide range of other aluminium replacements, either total or partial, have also been described. Phosphorus[35,36] and boron substitution[34,37,38] in several aluminosilicate zeolites have been reported. Substitutions of, for example, B, P, Fe, Ni, Sn, and Ti in the high silica zeolite ZSM-5 have been claimed (although definitive evidence for framework cation site location can be difficult to obtain). Transition metal substitutions into several lower Si:Al ratio zeolites have also been explored. Notably, pure ferri- and borosilicate sodalites (SOD) have been prepared and characterized.[39,40]

Investigations at Union Carbide in the early 1980s yielded a new family of microporous molecular sieves, the aluminophosphates.[41,42] These materials generally also form 4-connected frameworks, with alternating Al and P on the tetrahedral sites, and several of the structure types observed are common with those adopted by aluminosilicate zeolites.[9,43] Although by classical definition not strictly zeolites (which, as above, should contain Si and Al on the T-sites), these new molecular sieves are naturally included here. A formula for the $AlPO_4$-*n* family can be written as

$$[AlPO_4] \cdot y R \cdot n H_2O$$

The framework can be considered to be comprised of $[AlO_2]^-$ and $[PO_2]^+$ units, strictly alternating, and the framework is therefore neutral. The Al–P alternation prohibits the occurrence of 5-ring units, and the $AlPO_4$-*n* structures are constructed predominantly from 4- and 6-ring units, although the categorization of structures is complicated by the occasional adoption by aluminium of 5- or 6-coordination.[43] The neutrality of the framework and the absence, therefore, of non-framework cations implies that the $AlPO_4$-*n* materials will be organophilic and non-acidic. Cation exchange capacity and the possibility of acidity can, however, be introduced by partial substitution, generally for phosphorus as

$$M_{x/m}^{m+} \cdot [Si_x P_{1-x} AlO_4] \cdot y R \cdot n H_2O$$

This substitution by silicon gives rise to a family of silicoaluminophosphates (SAPO-*n*) that includes novel structure types in addition to those that are common with those of aluminosilicate zeolites and $AlPO_4$-*n* materials.[34,44] Incorporation of a range of other metal cations in place of, or in addition to, silicon has also been reported, the materials being termed, by analogy, MeAPO-*n* and MeAPSO-*n*, respectively, where Me can be any one of at least 13 elements, Li (monovalent), Be, Mg, Co, Fe, Mn, or Zn (divalent), B, Ga, or Fe (trivalent), or As (pentavalent).[45] Although the structural chemistries of the large number of materials included in these families have as yet been little explored, some principles have emerged, including the extension of Loewenstein's rule to P–O–P, P–O–Si, Me–O–Al, and Me–O–Me linkage avoidance.[46] Several new structure types have been found and more will likely emerge as these phase fields are further studied.

7.4 GEOLOGICAL OCCURRENCE

A variety of zeolites occur naturally.[5–8] The large and beautiful zeolite mineral specimens in collections are generally extracted from basaltic vugs and cavities in which zeolites are common, although on a small scale. Deposits occur on a much larger scale in sedimentary tuff deposits in which they have been produced by recrystallization of source material (volcanic ash) under the action of mineralizing solutions. This crystallization process has evolved over geological time scales, although the composition of the recrystallized material, the nature of co-occurring phases, and the depth

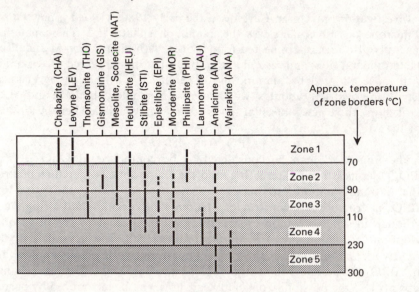

Fig. 7.12. Zoning of zeolites found in geothermal areas in Iceland. The distribution of individual species within the zones and the approximate zone boundary temperatures are indicated (after Kristmannsdottir and Tomasson[109]).

and/or presumed temperature of the deposit, can provide insight into the manner in which zeolite crystallization has occurred. In active thermal areas, where correlations between occurrence and temperature/composition are probably simpler (Fig. 7.12), occurrence patterns are consistent with those observed under controlled laboratory conditions. The ranges of physical and chemical conditions generating natural zeolites have been relatively restricted. There are, for example, no known natural occurrences of highly siliceous zeolites, such as members of the pentasil (ZSM-5 (MFI)–ZSM-11 (MEL)) family, or of other systems that are most readily produced in the presence of organic templates (see below).

For many zeolites, natural basaltic deposits provide the only source of large single crystals. Thus laboratory syntheses, although capable of accessing a much broader range of compositions, conditions, and structure types, only rarely yield large specimens. Mineral collections typically contain a variety of zeolite structure types including stilbite (STI), harmotome (PHI), various sodalites (in its better known form, ultramarine) (SOD), heulandite (HEU), analcime (ANA), pollucite (ANA), and edingtonite (EDI). Occurrences of faujasite (FAU) and laumontite (LAU) are rare,[7] and there are zeolites, such as mazzite (MAZ),[47] pahasapaite (RHO),[17] and perlialite (LTL),[48,49] for which only one or two occurrences are known. In one sense this is unsurprising. Laboratory syntheses demonstrate that the formation of certain materials can be critically dependent upon the chemical composition of the crystallizing medium.

Relatively large scale, mineable sedimentary occurrences of zeolites enable their use in a variety of applications,[6, 7, 50, 51] predominantly in ion-exchange or sorbent capacities. Natural clinoptilolite (HEU) has been used for extracting radioactive ^{137}Cs and ^{90}Sr from high- and low-level wastes of nuclear facilities (including use at the Three Mile Island site), and natural mordenite (MOR) has been studied in Sweden as a decontaminating animal feed supplement following the Chernobyl disaster. Clinoptilolite (HEU) is widely used for water purification (for extracting ammonium ions) and, particularly in Japan, as an animal feed supplement (the presence of the zeolite can improve growth rates and reduce the water content and malodour of excrement). Horticultural and aquacultural uses have also been explored. Natural zeolites have been used as fillers in paper and construction products, as polishing agents in toothpaste, and in medical recycle–dialysis systems. The relatively low cost of natural zeolites makes their use as sorbents attractive, such as in O_2–N_2 separations (see below), SO_2 and/or CO extraction from flue gases, or simply as dessicants. Catalytic applications have also been considered, although the presence of impurities generally limits the usefulness of the natural materials.

7.5 ZEOLITE SYNTHESIS

7.5.1 Aluminosilicate gels

Other than the time scale, the conditions that have given rise to natural zeolite formation can be reproduced in the laboratory. R. M. Barrer demonstrated in the 1940s and 50s that a series of zeolitic materials could be synthesized under hydrothermal conditions.[35, 50] Such syntheses entail the combination of an alumina component, a silica component, and inorganic base(s). This mixture forms a gel or viscous liquid which is allowed to crystallize, usually under autogenous pressure, for a period of between a few hours to several weeks at temperatures between $\sim 60°C$ and $\sim 200°C$. In a typical zeolite A (LTA) synthesis,[35, 50, 53] for example, hydrated alumina ($Al_2O_3 \cdot 3H_2O$) is dissolved in a concentrated solution of NaOH (~ 20 N). This (cooled) solution is then blended with a 1 N solution of sodium metasilicate ($Na_2SiO_3 \cdot 9H_2O$) to give a thick white gel (of typical relative composition $2.1Na_2O \cdot Al_2O_3 \cdot 2.1SiO_2 \cdot 60.0H_2O$) which is then loaded into a plastic or Teflon bottle and crystallized at $\sim 90°C$ over ~ 6 h. The optimal crystallization time is somewhat variable, and samples are therefore extracted during the crystallization and examined by powder X-ray diffraction (PXD). Initially, no sharp peaks are seen in the PXD profile indicating that only amorphous components are present. After an induction period of some 3 h, sharp peaks from crystalline zeolite A (LTA) appear, and increase steadily in intensity with time until a maximum (plateau) is reached, at which point the mixture is filtered, washed, and dried. If crystallization is allowed to continue beyond this point, the solid phase undergoes reconversion to basic

Fig. 7.13. SEM pictures of samples of zeolite ZK-4 (LTA)—(a), zeolite X (FAU)—(b), high silica sodalite (SOD)—(c) and silicalite (MFI)—(d) [courtesy J. P. Verduijn, E. W. Corcoran, S. B. Rice, and J. L. Pizzulli]. In each case the longest lower bar indicates 10 µm.

sodalite (SOD), a more condensed structure (Fig. 7.2). A micrograph of cubic zeolite A crystals produced by such a crystallization is shown in Fig. 7.13.

This synthesis illustrates Ostwald's rule of successive transformations (and see Fig. 7.15 below), which states that in the formation of polymorphs from vapour, liquid, or solution, the first polymorph to appear is the least thermodynamically stable and it is successively replaced by the more stable forms. Zeolite A (LTA) is produced under kinetic control and, subsequently, the more thermodynamically stable product, basic sodalite (SOD), appears. Kinetic control implies that nucleation phenomena will be important, and the courses of zeolite crystallizations are generally sensitive to slight changes in the chemical and physical environment (Fig. 7.14, Table 7.2). Thus, if the hydrated alumina in the above preparation is replaced by a less reactive aluminium source such as $\alpha\text{-Al}_2\text{O}_3$, although crystalline products are formed, zeolite A (LTA) is, at best, a minority component. Changes in the relative $\text{Na}_2\text{O–SiO}_2\text{–Al}_2\text{O}_3\text{–H}_2\text{O}$ gel composition and/or crystallization conditions give rise to a wide range of possible products from this system, including LTA, FAU, SOD, MOR, GIS, ANA, and CHA/GMA framework materials.[35]

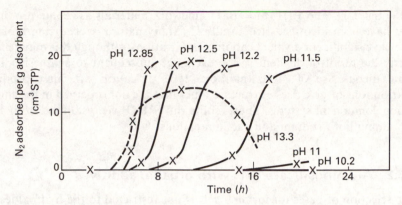

Fig. 7.14. The dependence on pH of the yield of crystalline mordenite at 300°C (after Barrer[35]).

Table 7.2 *Factors affecting the nature, composition, and quality of zeolite crystallization products*

Gel (or solution) composition
pH
Nature of starting ingredients
Temperature
Time
Pressure
Cold age treatment
Seeding conditions
Degree of agitation

Clearly, this diversity implies the possibility of simultaneous crystallization of more than one phase, and multiphasic products are indeed common. The products that are observed as a function of each of the system variables (Table 7.2) define crystallization fields for the various possible products and, generally, conditions that optimize the purity and yield of a desired material.

Use of a different inorganic base gives rise to a new range of possible products. The $M_2O-SiO_2-Al_2O_3-H_2O$; M = Li crystallization field is less rich in microporous materials, but two relatively compact zeolites, Li-A(BW) (ABW) and Li-H (structure unknown) are found.[35] In the M = K system, new structure types appear, including LTL, EDI, and PHI framework materials.[35] Rubidium and caesium hydroxides also give different product distributions, although no new large-pore frameworks are observed. Crystallizations from gels containing Group IIa cations yield ANA, THO, EPI, PHI, and MOR (M = Ca), FER, ANA, CAN, YUG, GME, CHA, HEU, and MOR

(M = Sr), LTL and PHI (M = Ba) framework materials as zeolite products that have been identified structurally.[35] (Many naturally occurring zeolites also have significant divalent cation concentrations, although this may reflect partly the result of divalent cation exchange subsequent to synthesis.) The simultaneous use of more than one type of cation can alter product distributions or give rise to structure types that are not observed in the single cation component systems. Thus, zeolite rho (RHO) is crystallized from gels containing both sodium and caesium cations.[16]

7.5.2   Aluminosilicate gels with organic additives

The selection of bases for forming gels is not restricted to the hydroxides of the alkali metals and alkaline earths. Barrer and Denny in 1961 isolated new materials on using tetramethylammonium (TMA) hydroxide as the base component.[54] Thus, in a typical zeolite A synthesis in which the sodium hydroxide is replaced in part by TMAOH, higher silica forms of zeolite A, termed alpha, N-A or ZK-4 (LTA), and of sodalite (SOD)[54,55] are produced. Zeolites omega (MAZ) or ZSM-4 (MAZ) (Fig. 7.15), (Na, TMA)-E (EAB), and O (OFF) are also crystallized from similar gel compositions. The organic base may have a number of different roles.[56] It alters the chemical and physical character of the crystallizing gel. The addition of TMAOH to sodium silicate solutions, for example, dramatically increases the population of the octameric, double 4-ring silicate anion (Figs 7.3 and 7.8).[12] Additionally, the organic cation is generally observed to be occluded within the crystallized zeolite, resulting in different framework compositions (a sodalite cage, can, for example, accommodate only a single TMA^+ cation (Fig. 7.16); the

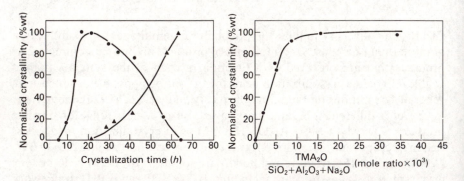

Fig. 7.15. The crystallization of zeolite ZSM-4 (MAZ) from $TMA_2O \cdot Na_2O \cdot Al_2O_3 \cdot SiO_2 \cdot H_2O$ composition gels which requires a limiting TMA_2O concentration (right), and which occurs progressively following initial crystallization of a Y-type zeolite (FAU), illustrating Ostwald's rule (left—triangles ZSM-4 (MAZ), circles zeolite Y (FAU)) (after Barrer[35]).

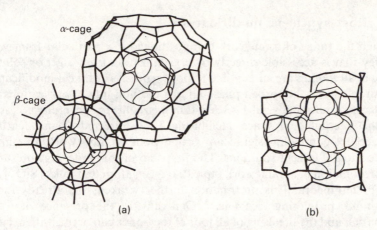

α-cage

β-cage

(a)   (b)

Fig. 7.16. Illustration of tetramethylammonium cations occluded within the sodalite (β) and supercage (α) of the zeolite ZK-4 (LTA) structure (a), and of a tetrapropyl-ammonium cation housed at the channel intersections in ZSM-5 (MFI) (b).

framework aluminium content required for charge balance is thus limited, resulting in higher framework Si:Al ratios) and leading to the concept of a 'templating' effect in which the bulky organic species provides a basis around which the developing zeolite cages form.[56]

The TMA$^+$ cation fits relatively snugly into the sodalite cages of sodalite (SOD), ZK-4 (LTA), and SAPO-37 (FAU) (Figs 7.2 and 7.16), and into the gmelinite cages of zeolites omega (MAZ) and O (OFF). Crystallization of these materials is promoted in TMA-containing gels (Fig. 7.15). Studies of zeolite crystallizations using organic bases led in 1967 to the synthesis of zeolite beta[57] (*bea/beb*), the first in a long series of high silica zeolites, including ZSM-5 (MFI), ZSM-8 (MFI), ZSM-11 (MEL), ZSM-12 (MTW), ZSM-18 (MEI), and ZSM-20 (*emt*/FAU). Tetrapropylammonium (TPA) hydroxide is used in the synthesis of ZSM-5 (MFI), and the occluded TPA$^+$ cation is observed at a concentration of one per channel intersection, in a configuration that has a 'C$_4$H$_9$ arm' directed along each of the channel openings (Fig. 7.16).[58,59] Although this conformation is consistent with the templating model, the model cannot be exclusive because ZSM-5 (MFI) can also be synthesized in the absence of organic additives.[56]

Neutral organic additives, such *n*-propylamine have been used successfully in ZSM-5 (MFI) syntheses, and a series of microporous silicas (porosils or clathrasils) can be crystallized hydrothermally in the presence of neutral organic components. A large number of organic bases and amines, as well as linear polyelectrolytes, alcohols, ketones, etc., have been tested in various zeolite syntheses.[56] There is, however, considerable scope for further experimentation, given the number of parameters which are variable in a typical synthesis (Table 7.2).

7.5.3   Post-synthesis modification

Although the range of framework compositions for a particular framework topology that is accessible directly by crystallization from a gel or solution is limited, all zeolites are susceptible to post-synthesis chemical modification. On a simple level, the sorbed phase and/or the non-framework cations can be replaced. Additionally, all T-sites within the zeolite are generally accessible from the internal pore space, facilitating T-site directed chemical attack. Most zeolites containing significant proportions of framework aluminium dissolve rapidly in acid solutions. The first step of this process is presumed to be protonation of a framework (apical) oxygen atom that links $[SiO_2]$ and $[AlO_2]^-$ tetrahedra. This protonation further weakens the already longer Al–O bond, promoting cleavage.[60] Generally in the presence of water, protonation and the breaking of all four of the bonds can occur, with excision of an $AlO_4H_n^{x-}$ species from the framework. The isolated defect that is formed (decorated by surrounding hydroxyl groups—a hydroxyl 'nest') may undergo no further modification, or it may, particularly at higher concentrations (i.e. for initially low Si:Al framework ratios), be implicated in gross structure collapse. Additionally, it affords a potential site for T-atom reinsertion.

T-atom reinsertion is both intriguing from a mechanistic standpoint, and important technologically. An ammonium exchanged zeolite Y (FAU) with a Si:Al ratio of ~ 3, when subjected to repeated steaming, acid extraction, and ammonium exchange treatments, is converted to a highly crystalline material which has a framework composition close to SiO_2 and which shows only a small number of point defects. The zeolite framework has effectively been completely annealed or recrystallized, without, at any stage, destruction of the basic FAU framework structure. The dealumination process, by increasing framework Si:Al ratios, improves the zeolite's hydrothermal stability and alters its chemical characteristics. Additionally, it can facilitate the application of certain characterization tools, such as ^{29}Si nuclear magnetic resonance (n.m.r.—mentioned further below), for which interpretations are complicated in the presence of framework aluminium. Aluminium extraction can also be achieved by mild acids, or by other reagents such as ethylenediaminetetraacetic acid (EDTA), $SiCl_4$, or $(NH_4)_2SiF_6$. The latter two reagents presumably permit direct silicon insertion, without the necessity of silicon transport from other regions of the crystallite. Aluminium reinsertion, and the introduction of other cations (such as B, Ga, and Fe) via dealumination chemistry, has also been discussed.[61,62]

7.5.4   Other phase fields

Initial laboratory syntheses in the 1940s and 1950s yielded a series of low Si:Al ratio zeolites such as A (LTA) and X (FAU), supplemented in the late 1960s and 1970s by high silica zeolites such as beta (*bea/beb*), ZSM-5 (MFI), and ZSM-12 (MTW). The 1980s have witnessed a huge increase in the number of known zeolites and related crystalline microporous materials, including new porosils and clathrasils, the $ALPO_4$-n, SAPO-n, MeAPO-n, and MeAPSO-n families, and extensive T-atom hetero-substitutions in aluminosilicate zeolites. Gallosilicates,[33,34] aluminogermanates and gallo-germanates,[35] with framework structures similar to those of the analogous aluminosilicates, can often be made, for example, by suitable replacement of the alumina and or silica component in the synthesis medium. Novel structure types are also observed. Boron incorporation is achieved by using H_3BO_3 or $B(OC_2H_5)_3$ as an additive during synthesis.[34,37,38] Similarly, recent reports describe the introduction of transition metal cations into T-site positions by the addition of suitable soluble metal salts to the crystallizing medium.[34,39,63] In many instances, such as that of $FeCl_3$ additions, an insoluble hydroxide is precipitated in the alkaline medium, but it remains an integral and dispersed part of the gel, apparently facilitating incorporation into the crystallizing zeolite.

Some aluminosilicate phases related to zeolites can be prepared by conventional solid state synthesis routes, namely solid–solid reactions at higher temperatures. A range of materials with the sodalite framework structure (SOD) can be produced by combining appropriate proportions of the oxides or salts above $\sim 800°C$.[35] Similar syntheses of the condensed materials $Cs[AlSiO_4]$,[64] $Cs_2[FeSi_5O_{12}]$,[65] $Cs_{0.35}[Al_{0.35}Si_{2.65}O_6]$,[66] (with, respectively, the ABW, ANA, and BIK frameworks), and of $Cs[AlSi_5O_{12}]$[67] (which also has a 4-connected framework structure) have been reported.

The $ALPO_4$-n aluminophosphates are crystallized at 100–300°C over a period of between 2 h and several weeks from gels formed by adding an organic templating agent to a mixture of an active form of alumina (typically CATAPAL—approx. $Al_2O_3 \cdot 1.5H_2O$), orthophosphoric acid (H_3PO_4), and water. The pH of such gels (5–8) is notably much lower than those typical in aluminosilicate zeolite syntheses (11–14). The character of the synthesis products is determined largely by the nature of the organic additive, although the degree of specificity associated with a particular templating species varies considerably. Thus the formation of $ALPO_4$-5 (AFI) is promoted by a large number of neutral organic amines,[34,41,42] with the apparent constraint only that the pore (channel) system be appropriately filled. Other structure types, such as $ALPO_4$-33 (ATT) and $ALPO_4$-20 (SOD), have been produced using only a single type of template (in these cases TMAOH). A notable new addition to this family is the 18-ring $ALPO_4$ material, VPI-5[24] (VFI), produced using di-n-propylamine or tetrabutylammonium hydroxide. For SAPO-n materials, synthesis conditions are similar, with silica sol typically used as the silica source.[44] The MeAPO-n and McAPSO-n families are produced by appropriate metal salt additions[45,46]; the more acidic pHs reduce complications associated with insoluble hydroxide precipitation.

Other microporous materials that are frequently compared with zeolites include inclusion complexes of urea or cyclodextrins, amorphous silica, aluminas, microporous carbons, and pillared layered materials such as clays. Clays are silicate minerals, with crystal structures comprised of layers (made up of sheets of tetrahedrally coordinated silicon/aluminium, and one or more sheets of octahedrally coordinated cations such as Al^{3+}, Mg^{2+}, Fe^{2+}, or Fe^{3+}). The absence of strong layer–layer binding permits easy cleavage parallel to the sheets, either mechanically (mica or talc are well-known examples that have no net layer charge) or by swelling in an appropriate aqueous medium. The ready exchange of the interlayer cations permits introduction of large polyoxocations (such as $Al_{13}O_4(OH)_{24}(H_2O)_{12}^{7+}$) which on calcination above $400°C$ dehydrate to oxide pillars that separate the clay layers. The resulting pillared clays[10,11] can have high surface areas, $< 200 \, m^2 \, g^{-1}$, and relatively large effective aperture dimensions, $\geqslant 8 \, Å$, although they are structurally less regular than zeolites and prove troublesome to produce in large quantities.

The basic structural building blocks of zeolites are tetrahedra. A large number of open framework structures can also be constructed from combinations of MO_6 octahedra and TO_4 tetrahedra. Several cage or channel structures have already been reported; many others are probably awaiting discovery.

7.5.5 Crystallization mechanisms

The process of zeolite crystallization on a molecular scale remains little understood,[13,34,35,50] although the development of tools for characterizing the synthesis medium and the crystallized product(s) is contributing steadily to our knowledge. A typical synthesis involves initial gel formation, slow dissolution of gel components into the aqueous component, nucleation (from either solution or the gel), crystal growth, and then further evolution, perhaps by redissolution of this first phase and nucleation and growth of a second crystalline phase (Fig. 7.15), etc. Silicate gels and solutions generally support equilibria between large numbers of distinct silicate anions (in various states of condensation). Such anions can, under favourable conditions, be extracted chemically,[35] or probed directly by ^{29}Si n.m.r. techniques.[12] However, it has, as yet, proven impossible to measure the size or molecular character of the species that initially nucleate the crystallizing zeolite, or the species that dominate the crystal growth process. The n.m.r. techniques view a solution-average of configurations within a relatively small length scale, $< \sim 6 \, Å$, and scattering or microscopy techniques probe length scales larger than $\sim 40 \, Å$. Studies of molecular aggregation in the intermediate size regime are experimentally taxing, although techniques such as Raman, infrared, or inelastic neutron scattering spectroscopies have potential for contributing insight in this area.

The Edisonian approach (of 'try it and see') to zeolite synthesis has been extremely productive. The belief, however, is that a detailed understanding of the molecular mechanisms of zeolite crystallization will yield opportunities for producing, and even tailoring the characteristics of, new materials. A large number of relatively simple, hypothetical 4-connected frameworks are, in terms of geometric constraints, plausible. Materials with many of these structure types will probably prove accessible by appropriate chemical and physical control of the process of structure formation.

7.6 ZEOLITE CHARACTERIZATION

Adequate characterization of zeolites and related materials requires definition of chemical composition and of structure across a range of length scales. On a local scale the local geometries of framework cations (T), the coordination of non-framework species, and the environments of acid sites or of catalytically active metal centres, are of interest. The pore architecture and the framework connectivity and geometry are generally evident in the range ~ 5–20 Å and, on longer length scales, the uniformity of composition and the structural regularity are important, for these may also influence the efficiencies of processes that rely on zeolite structural properties. A variety of chemical and physical techniques are therefore useful, and zeolite characterization has, in many instances, provided key incentives for technique development.

7.6.1 Chemical composition

Zeolite elemental compositions are generally analysed by microprobe, by inductively coupled plasma emission spectroscopy (ICPES), or by atomic absorption spectroscopy. Electron microprobe analyses measure the intensity of X-ray fluorescence induced by the beam in the electron microscope as a function of energy. The energy of the emitted X-ray photon is an elemental characteristic, and elements with atomic numbers down to ~ 8 (oxygen) can be analysed quantitatively. The scanning transmission electron microscope (STEM) also offers an effective spatial resolution of the order of 100 Å, permitting full elemental analyses of discrete microcrystallites or the tracking of compositional variations (such as Si:Al ratios) across larger particles. In a typical ICPES analysis, the sample is dissolved in acid, fused with an appropriate flux such as lithium tetrafluoroborate, or injected directly into the ICPES spectrometer where atomic emission is excited by inductively coupling to a plasma. Conventional atomic absorption spectroscopy is generally applied to acid solutions of the zeolite. Zeolite bulk chemical analyses yield averaged sample compositions and do not, for example, distinguish between zeolite framework constituents and detrital or second phase material.

All primary zeolite constituents are amenable to study by solid state nuclear magnetic resonance (n.m.r.).[1,12] Proper calibration of n.m.r. signal intensities provides, in principle, one analysis method. More usefully, chemical shift variations can enable resolution of different elemental environments. As a result of the deshielding influence of aluminium, the ^{29}Si n.m.r. spectrum of, for example, a typical zeolite X (FAU) consists of up to five peaks (separated by about 4.5 ppm) that correspond to the differing possible levels of aluminium substitution, $n = 0$, 1, 2, 3, or 4, in the four first neighbour T-sites.[1,12] The n.m.r. experiment is quantitative, and (when Loewenstein's rule is obeyed) each Al atom is counted four times, enabling the framework Si:Al ratio, R, to be calculated as $R = 4\Sigma I_n/\Sigma n I_n$; for $n = 0$–4. This analysis is environment-sensitive, discriminating between framework and detrital or second phase aluminium-containing components. Similarly, although by Loewenstein's rule each framework aluminium atom has only silicon atoms in its first shell, and although quantitative assessments of spectra from the quadrupolar ^{27}Al nucleus are less straightforward, ^{27}Al n.m.r. can be used to analyse the relative amounts of aluminium in framework and extra-framework environments.

7.6.2 Pore architecture

Measurements of sorption behaviour[68] indicate the character of a zeolite's pore structure. In the as-synthesized state, the pore space is fully occupied by water, by neutral organic (or inorganic) components, or by inorganic or organic cations. The neutral and organic components can be driven off thermally, generally leaving an intact open framework. The temperature required to drive off a given species depends on the strength of its interaction with the zeolite, and thermogravimetric analyses (t.g.a) (which track weight change as a function of temperature) can therefore be revealing. Water is generally lost in the range ~ 90–$\sim 200°C$; the ammonium cation is decomposed to a proton and NH_3 at ~ 180–$\sim 300°C$, and organic template cations are usually decomposed and evolved at ~ 375–$\sim 550°C$. The capacity of the calcined zeolite for water (or for other small molecules such as N_2) indicates the total accessible internal pore volume (which in the most open structure of zeolite Y represents up to some 50 per cent of the total crystalline volume), or the framework density (the number of T-atoms per 1000 Å^3). The capacities for larger molecules measure the void volume accessible through the limiting apertures (Fig. 7.6) for the particular sorbate size (Fig. 7.7), permitting further categorization of the zeolite's pore system. These types of measurements are usually made as an adsorption isotherm which tracks uptake as a function of partial sorbate pressure (measured gravimetrically at equilibrium on a sensitive balance). At low partial pressures, p, the loading level, q, varies linearly (Henry's law), $q = Kp$, and the temperature dependence of the equilibrium constant yields the isosteric heat, the heat evolved on sorption at infinite dilution (where sorbate–sorbate interactions are absent). The remainder of the isotherm generally has the Brunauer type I form,[68]

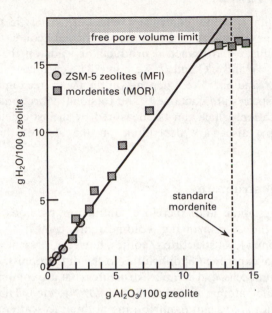

Fig. 7.17. Water sorption capacities (at 25°C, $P/P_0 = 0.042$) of mordenite and ZSM-5 zeolites scale with framework aluminium content and indicate an effective capacity of 4 water molecules per aluminium atom (after Weisz[99]).

with steadily increasing loading as a function of sorbate pressure up to a plateau of complete pore filling. The presence of significantly different types of sorption site, or of dominating sorbate–sorbate interactions, are often evidenced by deviations from this type I behaviour.[68]

Water interacts strongly with zeolites of low Si:Al ratios. Heats of sorption of 20–30 kcal mol^{-1} at low and ~ 17 kcal mol^{-1} at high loadings are typical (adsorption heats for *n*-alkanes in sodium zeolite X (FAU) are, for comparison, some 3.5 kcal mol^{-1} per carbon atom number[50]). For higher framework Si:Al ratios, measured equilibrium water uptake varies with framework charge, and, for example, the capacities of mordenites (MOR) scale directly with framework aluminium content (Fig. 7.17). The strength of the water–zeolite interaction (that gives rise to the use of lower Si:Al ratio zeolites such as A (LTA) and X (FAU) as desiccants) derives primarily from the coordination by water of non-framework cations. Recent neutron diffraction studies in which water sorbate locations and orientations have been measured[69,70] indicate near optimal distances to non-framework cations, but generally weak hydrogen-bonded linkages between the water molecules and framework oxygen atoms.

Several other techniques provide data on zeolite pore architectures, including analyses of certain photochemical reactions, low-pressure pore size analysis (measuring pore size distributions $> \sim 5$ Å based on interpretation of N_2 or Ar uptake at low controlled pressures), catalytic test reactions (such as *n*-decane hydrocracking over the Pt-loaded zeolite, for which the distribution of linear and branched products depends on the cage dimensions and shape[71]), and ^{13}C and ^{129}Xe n.m.r. The chemical shifts of ^{13}C in TMA$^+$ cations[72] and of ^{129}Xe[73] depend on the mean free dimensions within which these species are contained. The torsional vibrational frequencies of the TMA$^+$ cation, which can be measured by inelastic neutron scattering spectroscopy, also vary depending on the size of the entrapping cage.[74]

7.6.3 Crystal structure

Zeolites are crystalline and, therefore, diffraction methods (which were introduced in the accompanying volume[1]) are central to attempts to characterize atomic-level structure. Zeolites, however, present a particularly challenging set of experimental difficulties to the diffractionist. Zeolites have relatively large unit cells, and complex structures that are comprised of atoms of low and similar atomic numbers (Si, Al, O, Na, etc.). The pore spaces within the zeolites are vacant (contributing nothing to scattered intensities) or filled with species which are typically highly mobile and, therefore, diffuse on the time scale of the diffraction experiment. Synthetic zeolites are almost invariably microcrystalline, and problems of phase impurity, inhomogeneity, and defects or disorder of various types are common. In addition, local orderings, such as certain non-framework cation configurations, or local framework aluminium distributions that are not reproduced by the lattice translational periodicity, are probed only in an averaged way by diffraction experiments.

Tabulating some of the technical difficulties in this way highlights approaches that can be taken towards their solution.[75] Certain structural characteristics can be probed by conventional single-crystal X-ray or neutron diffraction studies of natural mineral samples.[69] Careful manipulation of synthesis conditions has allowed the formation of larger crystals ($\leqslant 100\,\mu$m; $1\,\mu$m = 0.001 mm) of a small number of synthetic zeolites, such as A (LTA), X (FAU), and ZSM-5 (MFI), although particle sizes $< \sim 5\,\mu$m are more typical. Conventional single-crystal X-ray diffraction would require zeolite crystals larger than $\sim 50\,\mu$m, and zeolite structural studies have, therefore, relied heavily on powder X-ray diffraction methods.

Electron diffraction has been applied frequently to determining the unit cell sizes and space groups of new zeolite structures. Although zeolites are prone to radiation damage in the electron beam, when conditions do permit they can be interesting subjects. Zeolite pore systems give good electron scattering contrasts, yielding interpretable lattice images even under moderate resolution. Pore stacking arrangements, twinning, intergrowths, and other forms of stacking disorder can be directly observed.[20-22,76,77] Under high resolution conditions, smaller structural features such as 6- and 5-rings can also become evident, aiding structure solution.[22,78]

Different zeolite materials frequently have similar chemical compositions but differ structurally. The positions and intensities of peaks in the powder X-ray diffraction (PXD) pattern are therefore a basis for phase identification and definition. The identity or novelty of a new material is deduced by comparison against a reference data base of PXD patterns published in the open or patent literature, measured from standard samples, or calculated based on known structural models.[79] The PXD pattern provides an indication of sample 'crystallinity' based on the measured peak intensities or widths of standard reflections, and a quantitative measure of the relative amounts of different phases when more than one is present. There is also a tradition of using the PXD pattern quantitatively in zeolite structure solution and refinement.[80] Refinement (optimization of approximate atomic coordinates based on least-squares minimization of the difference between observed and calculated PXD patterns) can be based on integrated peak intensities or, more commonly, on Rietveld analysis[1,81] of the full diffraction profile. This type of curve-fitting procedure allows pattern decomposition (when the peak intensities are taken as independent variables), or, as is more usual, in direct optimization of a trial structural model.[81]

Structure solution, the initial derivation of an approximate framework model, however, is only rarely possible directly. The Patterson and direct methods (*ab initio*) techniques that can be applied in a near-routine fashion to single-crystal data[1] require a reasonably precise measurement of each unique structure factor. Peak overlap, exacerbated by sample contributions to peak widths, prevents such measurements for most polycrystalline zeolites, and structure solutions have generally been approached by building physical models. The expanding data base of hypothetical structures[82] provides a first point of reference in evaluating the structure of a new material. Known zeolite structures satisfy a well-defined set of geometrical constraints. The TO_4 terahedra are always close to regular, and observed T–O distances, T–O–T angles (or T–T distances), and O–O distances fall within well-defined bounds. The approximate atomic coordinates of a framework model can be adjusted so as to optimize the degree to which these distance constraints are satisfied. The distance least squares (DLS) technique was first developed specifically for zeolite applications in 1969,[83] and DLS refinement is now a standard step in framework model evaluation. Similar distance constraints can be used to supplement the measured diffracted intensities in refinement, enabling even complex zeolite structures to be tackled by Rietveld methods.[80]

Synchrotron X-ray facilities offer significant advantages in diffraction studies of zeolites and related materials.[75,84,85] PXD patterns can usefully be recorded at significantly higher ($\geqslant 4 \times$) instrumental resolution, facilitating phase identification, allowing more complicated structures to be refined using Rietveld methods, and enhancing the possibilities for *ab initio* structure solution. Recent examples include a first refinement of the structure of zeolite ZSM-11 (MEL),[86] and the solution of the framework structure of the clathrasil sigma-2 (SGT), largely by direct methods based on decomposed peak intensities.[87] Improved resolution also permits more information to be extracted from the powder diffraction peak profiles themselves, such as that relating to the stacking disorder in zeolites such as beta (*bea/beb*) and ZSM-20 (*emt/*FAU). The brightness of the synchrotron X-ray source (the combination of intensity and low intrinsic divergence) permits conventional measurements to be made on much smaller particles. Feasibility studies have demonstrated the viability of diffraction experiments on single microcrystals in the micron size regime,[88] and improvements in the handling and alignment of crystallites of this size make initial setup, at least, relatively straight-forward.[85] The synchrotron source also promises advances in other areas. The intensity and white character of the source will potentially allow zeolite structure refinements based on single-crystal Laue diffraction data accumulated on a sub-second time-scale, and sensitivity to specific elements in powder or single-crystal studies may be greatly enhanced by exploiting anomalous scattering effects.[85]

X-ray diffraction, however, is not well suited for several zeolite structural questions, such as distinguishing Si, Al, and/or P, locating H^+ or Li^+, determining water or hydrocarbon sorbate configurations, or measurements under non-ambient conditions. In these areas neutron diffraction is attractive, although as both atomic neutron scattering cross-sections and the fluxes available at today's neutron facilities are relatively low, large samples are required. Studies of synthetic zeolites, therefore, necessitate the use of powder diffraction and, generally, Rietveld structure refinement techniques. The first zeolite powder neutron diffraction (PND) study appeared in 1977,[89] although initial studies were restricted to high symmetry, cubic materials. Recent applications[70] include lower symmetry materials and more complicated structures. Work on a series of gallosilicates has provided data on the effects of framework cation substitution. Several studies have yielded information about non-framework (detrital) aluminate species generated by framework dealumination. The complete hydrated structure of a polycrystalline sample of zeolite Li-A(BW) (ABW) was determined and a small number of complete studies of zeolite–hydrocarbon sorbate systems have also been described.[70] These direct measurements of the preferred sites and orientations of hydro-carbon sorbates are invaluable in the context of attempts to improve computer modelling methods for zeolite–hydrocarbon systems.

Molecular graphics[90] and computer modelling techniques[91,92] are being applied broadly to studies of zeolite structures and structural chemistries. Construction, modification, and evaluation of zeolite framework models is potentially much quicker on the computer, and a variety of solids modelling tools such as distance least squares, molecular mechanics, or total energy minimization techniques[1,92] can then be applied directly to interesting structures. Quantum-mechanical approaches are restricted at present mainly to investigations of small, model clusters,[93] but they can provide insight into, for example, the nature of zeolite Brønsted acidity. Relatively simple treatments of the interaction between hydrocarbon sorbate molecules and the zeolite (in which the interaction energy is computed as the sum over all relevant atom–atom interactions, each generally being phrased in terms of modified Lennard-Jones potentials[1]), permit estimates of energy minimum configurations, isoteric heats of sorption, and diffusion pathways.[90–94] The path of the hydrocarbon in the zeolite can be traced graphically, or by using Monte Carlo or Molecular Dynamics approaches similar to those applied to studies of protein structure and function. Simulation results can then be compared with experiment, potentially enabling improvement of the modelling methods. Ultimately, it is hoped to simulate the key properties of zeolites by combining a sophisticated computer modelling system with an improved ability to measure both sorption properties and the locations and motions of hydrocarbon sorbates.

7.6.4 Local site environments

Probes of local site environments that have been applied to zeolite problems include extended X-ray absorption fine structure (EXAFS), Mössbauer spectroscopy (of ^{57}Fe, ^{151}Eu, or ^{119}Sn containing materials), far infrared spectroscopy (far-IR),[95] electron spin resonance, and electron spin-echo spectrometry,[96] and n.m.r. Conventional infrared and Raman spectrocopies[1] are sensitive to local structural features, but interpretations are generally limited to relatively simple dynamical models of framework vibrational modes. Spectral features in the far IR region (~ 40–350 cm^{-1}) are associated with non-framework cation motion and can suggest coordination geometries, and indicate changes in relative site populations with changing composition or conditions.[95] EXAFS has only limited usefulness as a probe of framework cation environments (which are exclusively 4-coordinate, with near tetrahedral geometry), but can yield coordination numbers and geometries of catalytically active metal centres. Solid state n.m.r. measurements made with magic angle sampling spinning (which averages to zero the chemical shift and dipolar interaction anisotropies) provide, as above, data on the chemical composition of the first coordination shell(s) of the target nucleus.[1,12] Additionally, ^{29}Si (and ^{27}Al) chemical shifts, δ, depend on the local geometry, there being an approximately linear correlation between δ and the mean first neighbour Si–T distance (or Si–O–T angle).[12] Most non-framework species are also accessible by n.m.r. and detailed studies of the locations and relaxational characteristics of the cations ^{7}Li$^+$ and ^{23}Na$^+$ in zeolites A (LTA) and X (FAU) have, for example, been described. Water, hydrocarbon sorbates, and templating species have been studied by ^{1}H, ^{13}C, and ^{2}H (deuterium) n.m.r. Interpretation of the measured quadrupolar splittings of deuterium nuclei in hydrocarbon sorbates can permit the character of sorbate molecular motions to be directly inferred.

7.7   ZEOLITES AS ION EXCHANGERS

Zeolites, by definition, have a capacity for partial or complete exchange of their non-framework cation complement by treatment with an appropriate salt solution or molten salt. The character of the ion-exchange equilibrium between a zeolite and a contacting solution or salt (reflected in the pattern of the ion-exchange isotherm—Fig. 7.18) depends on several factors, including the character of the cations in question and the possible coordination environments for non-framework species provided by the zeolite structure. The maximum ion exchange capacity is determined by the framework Si:Al ratio, although the actual capacity may be lower if a proportion of the cations

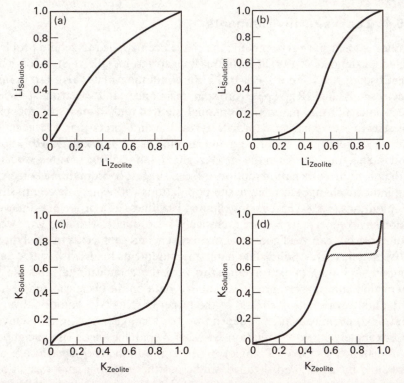

Fig. 7.18. Four characteristic zeolite ion exchange isotherms, for, respectively, Na–Li in sodalite hydrate (SOD) (a), Na–Li in cancrinite hydrate (CAN) (b), Na–K in zeolite Na-P (GIS) (c), and Na–K in zeolite K-F (EDI) (d)—the isotherm shows significant hysteresis at high zeolite K$^+$ contents (after Barrer[35]).

are sited within small, inaccessible cages. Effective exchange rates are generally very fast. Ion exchange can typically be achieved at room temperature by pouring an aqueous salt solution through the wet zeolite on a Buchner funnel. A homogeneous sample of, for example, Na/Li zeolite A (LTA) can be made merely by tumbling together hydrated powder samples of the Na and Li forms at room temperature.

Several commercial applications of zeolites exploit their ion exchange characteristics. Most, such as radioactive wastewater treatment, water purification, and agricultural, horticultural, and medical uses rely, as introduced above, on natural mineral zeolites such as clinoptilolite (HEU) or mordenite (MOR) which are relatively inexpensive ($2-20p \, lb^{-1}$). Synthetic zeolite A (LTA) has been widely used as a partial replacement for sodium tripolyphosphate (STPP) in low phosphate detergents. Zeolite A performs as a filler and exchanges Ca^{2+} and Mg^{2+} cations out of the washing water. Subsequently, under neutral or mildly acidic conditions in the environment, the zeolite dissolves to form relatively benign silicate and aluminate species. Worldwide annual zeolite A production for detergent applications probably tops 250 000 metric tons.

7.8 ZEOLITE CATALYSTS

Synthetic zeolites related to faujasite (FAU) were developed for fluid catalytic cracking (FCC) applications in the early 1960s. Catalytic cracking is the process by which heavier components of crude oil are broken down into lower molecular weight fragments such as those that are used in gasoline.[97] This process occurs via carbonium ion chemistry and is, therefore, catalysed by solid acids. Brønsted acid sites are typically protons attached to bridging framework oxygen atoms and are introduced into the zeolite via ion exchange. For example, exchange of sodium zeolite Y (FAU) in NH_4Cl solution gives the ammonium form, which on calcination loses NH_3 to generate the protonic or acid form of the zeolite. Intrinsically, this process has complications due to the instability of the zeolite in an acid environment. Some framework dealumination almost invariably accompanies formation of the protonic form. The number of Brønsted acid sites is tied to the framework composition and at high Si:Al ratios there is a direct correlation between catalytic activity and acid site density or framework aluminium content (Fig. 7.19). The acidity of the individual sites (a measure of how readily the proton can be abstracted) depends on their environment and is highest for protons associated with only a single, isolated framework aluminium atom. The early development of zeolite-based FCC catalysts required methods of stabilizing the zeolite structure under the taxing conditions of the catalytic cracker and regenerator (Fig. 7.20). Carefully controlled dealumination permits 'ultrastable' higher Si:Al ratio materials

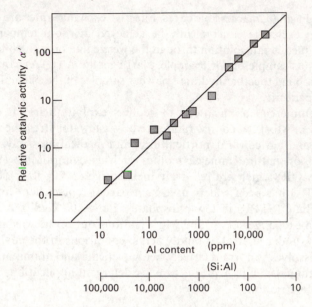

Fig. 7.19. The relative *n*-hexane cracking activity of ZSM-5 (MFI) zeolites plotted against framework aluminium content (after Haag[110]).

to be produced that have improved hydrothermal stabilities. Alternatively, exchange by a trivalent cation such as La^{3+} or Ce^{3+} imparts structural stability by virtue of its charge and location within the smaller (sodalite) cages of the zeolite (Fig. 7.2). Exchange of trivalent Ln^{3+} species can also generate Brønsted acid sites directly by cation hydrolysis, $Ln^{3+} \cdot H_2O \rightarrow Ln(OH)^{2+} + H^+$.

Fluid catalytic cracking is a complex process.[97] A typical catalyst pellet contains 5–40 per cent of a Y-type zeolite (FAU), a silica or alumina binder or matrix, and a clay filler. Pretreated oil at $\sim 300°C$ (typically vacuum gas oil extracted from the residuum of an initial atmospheric distillation of raw crude) is sprayed into the riser to be mixed with hot catalyst pellets (at $\sim 700°C$). Much of the cracking chemistry occurs within a short time scale ($< \sim 5$ sec) in the riser before entry into the reactor proper (Fig. 7.19). Large hydrocarbon molecules that have accepted one or more protons from the zeolite or binder rearrange and fragment (crack) in a variety of ways. A series of cyclones in the reactor separate the catalyst particles, which are passed to the regenerator where coke deposited on the catalyst is burnt off. The regeneration is exothermic, enabling the entire process to operate at close to an even heat balance. The FCC unit operates continuously, with new catalyst pellets being introduced constantly into the riser to compensate for losses due to degradation or attrition.

One suspects that the FCC process, even today after decades of research,

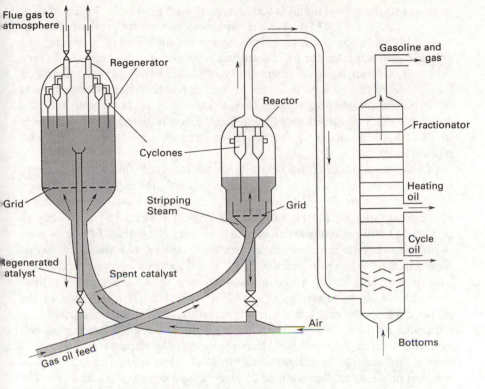

Fig. 7.20. Conceptual fluid bed catalytic cracker schematic (after Heinemann[111])

still has scope for considerable improvement. Our understanding of reaction pathways, both propitious and those leading to deactivation (coke formation), is limited. The FCC feed contains a wide range of hydrocarbon species, and catalysed cracking is complicated by the presence of variable but significant amounts of sulphur- and nitrogen-containing heterocycles and traces of heavy metals such as Ni and V. The improved performance of zeolite FCC catalysts compared to their amorphous aluminosilicate predecessors is measured by both better activity maintenance and greater product selectivities to liquid fractions such as those used in gasoline. The differing selectivities of zeolite based catalysts arise in part from reactions occurring within the zeolite pores. However, many of the initial hydrocarbon reactants are too large to enter into the zeolite, and must, therefore, undergo initial cracking on external surfaces or on the matrix. The increased lifetimes of zeolite catalysts reflect constraints on the formation of coke (more extended aromatic species) within the zeolite, and possibly an ability to regenerate active surface sites from aluminium leached from inside the framework. Perhaps reflecting a multiple role for the zeolite catalyst, recent FCC research has emphasized the use of additives to the conventional faujasite-type (FAU) materials. Such additives (e.g. zeolite ZSM-5 (MFI)) have the capability of altering product distributions and hence, for example, increasing the octane value of products. The annual catalyst market for FCC materials is several hundred million dollars. However, a better measure of the utility of zeolite catalysts is the increased value added to the process by virtue of their implementation. Reduced crude oil import demands and allayed capital expenditures, combined with the increased yields of valued products, place the accumulated worth of zeolites in FCC applications over the past two decades well in excess of \$100 billion.

FCC applications of zeolite catalysts are the most economically significant, but their desirable properties as acid catalysts are also exploited in a range of other conversions.[98–102] Xylene isomerization occurs via a carbonium ion intermediate and is catalysed by zeolite ZSM-5 (MFI). The diffusivity of *p*-xylene (1,4-dimethylbenzene) is much greater than that of the *o*- and *m*-isomers. Once formed within the zeolite, *p*-xylene can therefore escape more rapidly, enabling the reaction to be biased towards *p*-xylene production. This process illustrates a second key feature of zeolite catalysts: their potential for shape selectivity (Fig. 7.21).[98–101] The pore structure of the zeolite controls the approach of reactants to the active internal site, and, as illustrated by the xylene isomerization example (and related benzene alkylation and toluene disproportionation reactions), the departure of products from it. In addition to reactant and product selectivities, the character of reaction products can be modified by steric constraints at the active site. The relative amounts of xylene isomerization and (unwanted) bimolecular disproportionation depend on the effective pore dimension of the zeolite catalyst (Fig. 7.22). Steric constraints on reaction transition states can rule out particular reaction pathways (Fig. 7.21). In the acid-catalysed transalkylation of methylethylbenzenes at $\sim 300°C$, for example, 1-methyl-3,5-diethylbenzene represents 1/3 of the equilibrium product composition, and $\sim 1/5$ of the product over an amorphous acid catalyst, but is observed only in traces when the reaction is catalysed by H-mordenite (MOR).[101]

The conversion of alcohols to olefinic products over zeolite 5A (LTA) provides a good example of reactant selectivity (Fig. 7.23). Under conventional conditions, the secondary alcohols form the more stable carbonium ions, and, therefore, convert under acid catalysed conditions to olefins much more rapidly than the primary alcohols. Using zeolite 5A (LTA), however, only the primary alcohols can enter into the zeolite to reach the active acid sites, and these undergo relatively rapid conversion. The higher alcohols are excluded and are thus effectively inert. An important commercial process that exploits this type of reactant selectivity is catalytic dewaxing, commercially termed 'M-forming' or 'Selectoforming'. Compared to the branched isomers, the normal paraffins have poor octane numbers and higher melting points. They are thus undesirable in gasoline and contribute to wax formation in

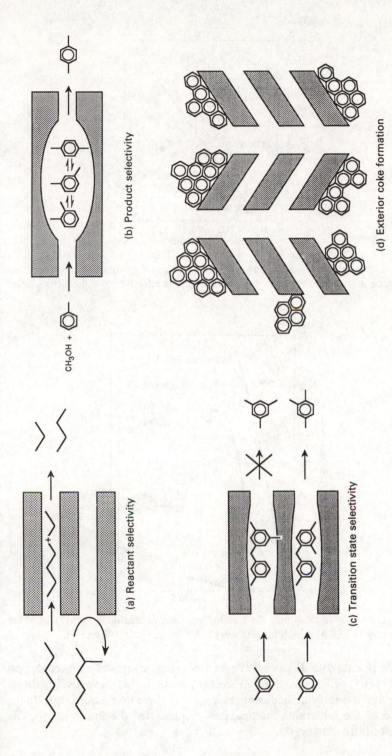

Fig. 7.21. Schematic of the types of shape selectivity exhibited by zeolite catalysts (after Csicsery[101]).

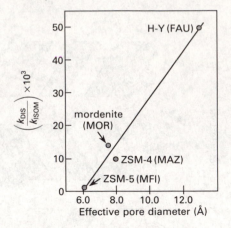

Fig. 7.22. The relative rates of the unwanted bimolecular disproportionation reaction versus xylene isomerization for zeolites with different effective pore diameters (after Weisz[99]).

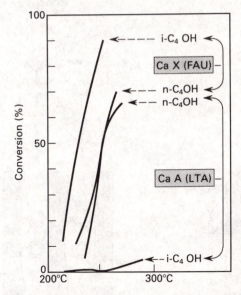

Fig. 7.23. Dehydration of *n*- and iso-butanol on calcium zeolite X (FAU) , and on calcium zeolite A (LTA) which admits only the *n*-isomer (after Weisz[99]).

diesel fuel, In catalytic dewaxing the hydrocarbon stream is contacted with a ZSM-5 (MFI) or Pt-loaded beta (*bea/beb*) zeolite, which selectively admits the linear and mono-branched isomers, cracking them into gaseous products. The value of the remaining stream, now depleted of the linear alkanes, is thus significantly enhanced.

Zeolite ZSM-5 has received wide publicity following the 1986 startup of the Motunui plant in New Zealand that converts methanol to gasoline.[103,104] The commercial viability of the evolution of CH_3OH to olefinic and ultimately aromatic products on this acid zeolite arises in part from the intracrystalline (steric) constraints that limit the production of aromatic species larger than C_9 or C_{10}, greatly reducing deactivation via coke formation (Fig. 7.21).

The preceding examples illustrate zeolite acid catalysis. Catalytically active metal species can also be introduced into a zeolite host, generally by ion exchange and subsequent hydrogen reduction. Compared to conventional metal loadings on, for example, silica or alumina, the use of a zeolite support may help to stabilize high metal dispersions, and offers the possibility of reaction pathways being dictated by shape selective constraints. A variety of applications have been studied, including reforming (conversion of alkanes to aromatics) on platinum loaded zeolites, and Fischer–Tropsch chemistry ($CO + H_2 \rightarrow$ hydrocarbons) on iron-containing systems. Zeolites have also been used in a diverse range of laboratory-scale syntheses of organic compounds.[105]

7.9   ZEOLITES AS SEPARATING AGENTS

Certain of the structural features of zeolites that are significant in catalysis are also exploited in separations. The lower Si:Al ratio zeolites have high affinities for water, and are widely used as desiccants in laboratory scale applications and, commercially, for drying gas streams. The largest recurring zeolite desiccant uses (several thousand metric tons annually) are in refrigerators and between the panes of insulating glass windows.

For larger molecules, access to the energetically favourable internal sorbates sites is controlled by the size of the pore apertures and by the presence or absence of non-framework cations in partially blocking positions (Figs 7.6 and 7.7). Zeolite A is an 8-ring zeolite (Figs 7.2 and 7.6), with a mean aperture dimension of ~ 4.7 Å. In the as-prepared sodium form, however, sodium cations occupy partially blocking sites in the 8-ring windows, reducing the free aperture dimension to approximately 4 Å. Zeolite 4A will admit water, CO_2, NH_3, etc. and has been used for removing CO_2 from natural gas. By exchange with the larger potassium cation, the degree of pore blockage can be enhanced, reducing the mean free aperture dimension to approximately 3 Å. The zeolite, now termed 3A, will admit water and ammonia, but larger molecules such as alcohols or alkanes are completely excluded, or 'sieved' out. Zeolite 3A is used for intensive drying of gas streams. Partial exchange by divalent Ca^{2+} ions removes Na^+ cations from the blocking sites, permitting the full aperture size to be realized. The zeolite (called 5A) will then admit *n*-alkanes, but exclude the branched isomers. Zeolite 5A (LTA) is used commercially for extracting *n*-alkanes (precursors to detergents) from various hydrocarbon streams. This system illustrates two important features of zeolites from a size exclusion sorbent perspective. Firstly, the aperture dimensions are reproduced in a completely regular fashion by the translational periodicity of the crystal, enabling high selectivities to be obtained. Secondly, the sieving characteristics can be fine tuned by appropriate cation exchange, or indeed, by framework modification. ZK-4 zeolites (LTA) have lower framework aluminium contents than type A zeolites (LTA). The correspondingly reduced number of non-framework cations can be accommodated at sites other than those partially blocking the 8-ring windows. When the occluded tetramethylammonium cations (Fig. 7.16) have been removed, ZK-4 (LTA) zeolites will then admit normal alkanes.

Zeolites in these types of applications have been given the term molecular sieve because of their ability to discriminate between molecules on the basis of their size and shape. A large number of such separations have been studied.[106] It should be emphasized, however, that today's applications are in one sense not strictly sieving (where the zeolite permits ongoing passage of the smaller component). Rather, the processes are cyclical. The sorbed component is taken up by the zeolite, and subsequently removed in a separate reactivation step. Reduction in the external pressure (*pressure swing*), increased temperature (*thermal swing*), or the use of a desorbent such as ammonia, have all been used commercially for this reactivation step.[68,107] True molecular sieving applications in which zeolite crystals in a membrane permit permeation of the 'sieved' component(s) are, however, conceivable and are currently under development.

Molecular sieving is a good example of how structure on an atomic scale can determine macroscopic properties, and how a detailed knowledge of structure can enable prediction of performance. A wider range of separations exploit the differing affinities for differing molecules that can all enter into the zeolite. Several factors determine the equilibrium composition of the sorbed phase within a zeolite in contact with a gas or liquid mixture. When the affinities between the zeolite and the various molecules present differ significantly, the compositions of sorbed and contacting phases will be different and the zeolite can be used to effect separations.

Nitrogen has a molecular quadrupole moment and is sorbed more strongly than oxygen by zeolites that have low Si:Al ratios (in which electric field gradients are high). A zeolite in contact with air will, therefore, preferentially sorb the nitrogen, leaving a gas enriched in oxygen. For certain applications (such as small portable units) this sorptive separation (which can also be achieved using microporous carbon absorbents) is more economically attractive than conventional low temperature distillation. The Lindox and Unox processes are commercial examples of air separations which operate under pressure swing adsorption.

For less volatile materials, and for systems in which the difference(s)

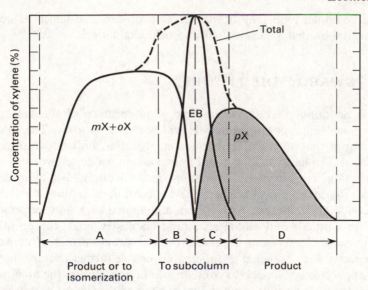

Fig. 7.24. Schematic of the emergence as a function of time of C_8 aromatic hydrocarbon isomers from a column of zeolite X or Y (FAU) adsorbent in the Asahi process (*o*X, *m*X, *p*X—*o*-, *m*-, *p*-xylene; EB—ethylbenzene) (after Ruthven[68]).

between sorption affinities is less marked, the pressure or thermal swing approaches become impractical. The selectivity of the zeolite can, however, be exploited in a chromatographic type separation. A pulse of hydrocarbon mixture (often with a suitable diluent) is introduced into a column of the zeolite sorbent. A flow of desorbent carries the mixture down the column at rates that are for each component inversely related to its affinity to the zeolite. The first component to emerge is the least strongly sorbed and so on (Fig. 7.24). The Asahi process operates on this basis, separating mixtures of the C_8 aromatic isomers (*o*-, *m*-, and *p*-xylene, and ethylbenzene) in a cyclical fashion (Fig. 7.24).[68] A cyclical process is, however, inconvenient from an engineering and process standpoint. The PAREX process, which is also used to extract *p*-xylene (a precursor for terephthalic acid in polyester production), separates the column into a series of sections.[68] The C_8 hydrocarbon mixture and desorbent inlets, and raffinate and extractate outlet positions, are successively switched by a rotary valve to successive sections of the column. Thus, although each stage operates cyclically, the composite provides for continuous operation.

Historically, the development and optimization of zeolites for hydrocarbon separations has required an Edisonian approach. Many of the factors which dictate the strengths of the interactions of sorbates with each other and with the zeolite are, however, now being explored in detail. There is, therefore, the hope that a quantitative understanding of such interactions will permit detailed computer modelling of the sorption process and hence reduce the experimental burden associated with sorbent evaluations.

7.10   TOWARDS THE FUTURE

There is no doubt that the technological importance of zeolites and of zeolite-related microporous materials will continue to grow. The rigorous steric control that such materials offer in processes, both separations and catalysis, is a quality that can only increase in value as we seek more sophistication and selectivity in management of our global hydrocarbon resources. As an empirical science, the application of zeolites has been of large economic significance. As has been highlighted here, however, research on zeolite synthesis and zeolite structural chemistry holds the promise of enabling molecular control of framework construction and of active site environments, such as will, it is anticipated, enable further improvements in zeolite applications. We seek to control with fine resolution the architecture of these crystalline solids, and to tailor their chemistries for particularly desired properties. This is the true realm of solid state chemistry.

7.11   REFERENCES

1. Cheetham, A. K. and Day, P. (eds), *Solid state chemistry: techniques*, Oxford University Press, Oxford (1987).
2. Ribbe, P. H. (ed.), *Feldspar mineralogy* (*Rev. Mineral. Vol. 2*), Mineralogical Society of America, Washington, DC (1981).
3. Brown, W. L. (ed.), *Feldspars and feldspathoids* (NATO ASI Series No. C137), D. Riedel, Dordrecht (1984).
4. Liebau, F., *Structural chemistry of silicates*, Springer-Verlag, New York (1985).
5. Mumpton, F. A. (ed.), *Mineralogy and geology of natural zeolites* (*Rev. Mineral. Vol. 4*), Mineralogical Society of America, Washington, DC (1981).
6. Sand, L. B. and Mumpton, F. A. (eds), *Natural zeolites. Occurrence, properties, use*, Pergamon Press, Oxford (1978).
7. Gottardi, G. and Galli, E., *Natural zeolites*, Springer-Verlag, New York (1985).
8. Gottardi, G. In *New developments in zeolite science and technology* (eds Murakami, Y., Iijima, A., and Ward, J. W.), pp. 41–39, Kodansha, Tokyo and Elsevier, Amsterdam (1986).
9. Meier, W. M. and Olson, D. H., *Atlas of zeolite structure types*, Butterworths, Surrey (1987).
10. Vaughan, D. E. W. In *Perspectives in molecular sieve science*, ACS Symp. Ser. No. **368** (eds Flank, W. H. and Whyte, T. E.), pp. 308–323, American Chemical Society, Washington, DC (1988).
11. Pinnavaia, T. J., *Science*, **220**, 365–371 (1983).
12. Engelhardt, G. and Michel, D., *High-resolution solid-state NMR of silicates and zeolites*, Wiley, New York (1987)

13. Occelli, M. and Robson, H. E. (eds), *Zeolite synthesis*, ACS Symp. Ser. No. **398**, American Chemical Society, Washington, DC (1989).
14. Moore, P. B. and Smith, J. V., *Mineralog. Mag.*, **35**, 1008–1014 (1964).
15. Smith, J. V. and Bennett, J. M., *Am. Mineral.*, **66**, 777–788 (1981).
16. Robson, H. E., Shoemaker, D. P., Ogilvie, R. A., and Manor, P. C. In *Molecular sieves*, ACS Adv. Chem. Ser. No. **121** (eds Meier, W. M. and Uytterhoeven, J. B.), pp. 106–115, American Chemical Society, Washington, DC (1973).
17. Rouse, R. C., Peacor, D. R., Dunn, P. J., Campbell, T. J., Roberts, W. J., Wicks, F. J., and Newbury, D., *Neues Jahrb. Miner.*, **1987**, 433–440 (1987).
18. Wells, A. F., *Three-dimensional nets and polyhedra*, Wiley, New York (1977).
19. Gard, J. A. and Tait, J. M., *Acta Cryst.* **B28**, 825–834 (1972).
20. Millward, G. R., Ramdas, S., and Thomas, J. M., *Proc. Roy. Soc. (London)*, **399**, 57–71 (1985).
21. Newsam, J. M., Treacy, M. M. J., Vaughan, D. E. W., Strohmaier, K. G., and Mortier, W. J., *J. Chem. Soc. Chem. Comm.*, 493–495 (1989).
22. Newsam, J. M., Treacy, M. M. J., DeGruyter, C. B., and Koetsier, W. T., *Proc. Roy. Soc. (London)* A**420**, 375–405 (1988).
23. Higgins, J. B., LaPierre, R. B., Schlenker, J. L., Rohrman, A. C., Wood, J. D., Kerr, G. T., and Rohrbaugh, W. J., *Zeolites*, **8**, 446–452 (1988).
24. Davis, M. E., Saldarriaga, C., Montes, C., Garces, J., and Crowder, C., *Nature (London)*, **331**, 698–699 (1988).
25. Loewenstein, W., *Am. Mineral.*, **39**, 92–96 (1954).
26. Halstead, P. E. and Moore, A. E., *J. Appl. Chem.*, **12**, 413–417 (1962).
27. Gupta, A. K. and Chatterjee, N. D., *Am. Mineral.*, **63**, 58–65 (1978).
28. Bibby, D. M. and Dale, M. B., *Nature*, **317**, 157–158 (1985).
29. Flanigen, E. M., Bennett, J. M., Grose, R. W., Cohen, J. P., Patton, R. L., Kirchner, R. L., and Smith, J. V., *Nature (London)*, **271**, 512–516 (1979).
30. Gies, H., *Zeit. Kristallogr.*, **164**, 247–257 (1983).
31. Schlenker, J. L., Dwyer, F. G., Jenkins, E. E., Rohrbaugh, W. J., and Kokotailo, G. T., *Nature (London)*, **294**, 340–342 (1981).
32. Liebau, F., *Zeolites*, **3**, 191–193 (1983).
33. Newsam, J. M. and Vaughan, D. E. W. In *New developments in zeolite science and technology* (eds Murakami, Y., Iijima, A., and Ward, J. W.), pp. 457–464, Kodansha, Tokyo and Elsevier, Amsterdam (1986).
34. Szostak, R., *Molecular sieves: principles of synthesis and identification*, Van Nostrand Rheinhold, New York (1988).
35. Barrer, R. M., *Hydrothermal chemistry of zeolites*, Academic Press, London, (1982).
36. Flanigen, E. M. and Grose, R. W. In *Molecular sieve zeolites—I*, ACS Adv. Chem. Ser. No. **101** (eds Flanigen, E. M. and Sand, L. B.), pp. 76–101, American Chemical Society, Washington, DC (1971).
37. Taramasso, M., Perego, G., and Notari, B. In *Proc. 5th Int. Conf. on Zeolites* (ed. L. V. Rees), pp. 40–48, Heyden, London (1980).
38. Meyers, B. L., Ely, S. R., Kutz, N. A., Kaduk, J. A., and Van den Bossche, E., *J. Catal.*, **91**, 352–355 (1985).
39. Szostak, R. and Thomas, T. L., *J. Chem. Soc. Chem. Comm.*, 113–114 (1986).
40. Hoelderich, W., Eur. Patent 0219, 810 (1986).
41. Wilson, S. T., Lok, B. M., Messina, C. A., Cannan, T. R., and Flanigen, E. M., *J. Am. Chem. Soc.*, **104**, 1146–1147 (1982).

42. Wilson, S. T., Lok, B. M., Messina, C. A., and Flanigen, E. M. In *Proc. 6th. Int. Conf. on Zeolites* (eds Olson, D. and Bisio, A.), pp. 97–109, Butterworths, Surrey, UK (1984).

43. Bennett, J. M., Dytrych, W. J., Pluth, J. J., Richardson, J. W., and Smith, J. V., *Zeolites*, **6**, 160–162 (1986).

44. Lok, B. M., Messina, C. A., Patton, R. L., Gajek, R. T., Cannan, T. R., and Flanigen, E. M., *J. Am. Chem. Soc.*, **106**, 6092–6093 (1984).

45. Flanigen, E. M., Lok, B. M., Patton, R. L., and Wilson, S. T. In *New developments in zeolite science and technology* (eds Murakami, Y., Iijima, A., and Ward, J. W.), pp. 103–112, Kodansha, Tokyo and Elsevier, Amsterdam (1986).

46. Flanigen, E. M., Patton, R. L., and Wilson, S. T. In *Innovation in zeolite materials science*, Stud. Surf. Sci. Catal. Vol. 37 (eds Grobet, P. J. *et al.*), pp. 13–27, Elsevier, Amsterdam (1988).

47. Galli, E., Passaglia, E., Pongiluppi, D., and Rinaldi, R., *Contrib. Miner. Petrol.*, **45**, 99–105 (1974).

48. Men'shikov, Y. P., *Zap. Vses. Mineral. O-va.*, **113**, 607–612 (1984).

49. Konev, A. A., Sapazhnikov, A. N., Afonina, G. G., Vorob'ev, E. I., Arsenyuk, M. I., and Lapides, I. L., *Zap. Vses. Mineral. O-va.*, **115**, 200–204 (1986).

50. Breck, D. W., *Zeolite molecular sieves: structure, chemistry and use*, Wiley and Sons, London (1973) (reprinted R. E. Krieger, Malabar, FL (1984)).

51. Dyer, A., *Chemistry and industry*, 241–245 (1984).

52. Manly, R. and Holmes, J., *New Scientist.*, **1989**, 39–43 (1989).

53. Milton, R. M., U.S. Patent No. 2,882,243 (1959).

54. Barrer, R. M. and Denny, P. J., *J. Chem. Soc.* (*London*), 971–982 (1961).

55. Kerr, G. T. and Kokotailo, G. T., *J. Am. Chem. Soc.*, **83**, 4675–4675 (1961).

56. Lok, B. M., Cannan, T. R., and Messina, C. A., *Zeolites*, **3**, 282–291 (1983).

57. Wadlinger, R. L., Kerr, G. T., and Rosinski, E. J., U.S. Patent No. 3,308,069 (1967).

58. Price, G. D., Pluth, J. J., Smith, J. V., Bennett, J. M., and Patton, R. L., *J. Am. Chem. Soc.*, **104**, 5971–5977 (1982).

59. van Koningsveld, H., van Bekkum, H., and Jansen, J. C., *Acta Cryst.* B**43**, 127–132 (1987).

60. Mortier, W. J. and Schoonheydt, R. A., *Progr. Solid State Chem.*, **16**, 1–125 (1985).

61. Hamdan, H., Sulikowski, B., and Klinowski, J., *J. Phys. Chem.*, **93**, 350–356 (1989).

62. Lutz, W., Lohse, U., and Fahlke, B., *Cryst. Res. Technol.*, **23**, 925–933 (1988).

63. For example, Nair, V., Szostak, R., Agrawal, P. K., and Thomas, T. L. In *Catalysis 1987*. Stud. Surf. Sci. Catal. No. **38** (ed. Ward, J. W.), pp. 209–219, Elsevier, Amsterdam (1988).

64. Klaska, R. and Jarchow, O., *Naturwiss.*, **60**, 299–299 (1973).

65. Torres-Martinez, L. M. and West, A. R., *Zeit. Kristallogr.*, **175**, 1–7 (1986).

66. Annehed, H. and Fälth, L., *Zeit. Kristallogr.*, **166**, 301–306 (1984).

67. Araki, T., *Zeit. Kristallogr.*, **152**, 207–213 (1980).

68. Ruthven, D. M., *Principles of adsorption and adsorption processes*, Wiley-Interscience, New York (1984).

69. Kvick, A., *Trans. Am. Crystallogr. Assoc.*, **22**, 97–106 (1986).

70. Newsam, J. M., *Materials science forum*, Vol. **27/28**, pp. 385–396 (1987).

71. Martens, J. A. and Jacobs, P. A., *Zeolites*, **6**, 334–348 (1986).

72. Jarman, R. H. and Melchior, M. T., *J. Chem. Soc. Chem. Commun.*, 414–415 (1984).

73. Fraissard, J., Ito, T., Springuel-Huet, M., and Demarquay, J. In *New developments in zeolite science and technology* (eds Murakami, Y., Iijima, A., and Ward, J. W.), pp. 393–400, Kodansha, Tokyo and Elsevier, Amsterdam (1986).

74. Brun, T. O., Curtiss, L. A., Iton, L. E., Kleb, R., Newsam, J. M., Beyerlein, R. A., and Vaughan, D. E. W., *J. Am. Chem. Soc.*, **109**, 4118–4119 (1987).

75. Newsam, J. M. and Vaughan, D. E. W. In *Zeolites: synthesis, structure, technology and application*, Stud. Surf. Sci. Cat. No. **24** (eds Drzaj, B., Hocevar, S., and Pejovnik, S.), pp. 239–248, Elsevier, Amsterdam (1985).

76. Terasaki, O., Thomas, J. M., and Millward, G. R., *Proc. Roy. Soc. (London)*, A**395**, 153–164 (1984).

77. Treacy, M. M. J., Newsam, J. M., Vaughan, D. E. W., Beyerlein, R. A., Rice, S. B., and DeGruyter, C. B. In *Microstructure and properties of catalysts*, MRS Symp. Proc. Vol. **111** (eds Treacy, M. M. J., White, J. M., and Thomas, J. M.), pp. 177–190, Materials Research Society, Pittsburgh PA (1988).

78. Terasaki, O., Thomas, J. M., Millward, G. R., and Watanabe, D., *Chem. Mater.*, **1**, 158–162 (1989).

79. von Ballmoos, R., *Collection of simulated XRD powder patterns for zeolites*, Butterworths, Surrey (1985).

80. Baerlocher, Ch., *Zeolites*, **6**, 325–333 (1986).

81. Rietveld, H. M., *J. Appl. Cryst.*, **2**, 65–71 (1969).

82. Smith, J. V., *Chem. Rev.*, **88**, 149–182 (1988).

83. Meier, W. M. and Villiger, H., *Zeit. Kristallogr.*, **129**, 411–423 (1969).

84. Newsam, J. M., *Science*, **231**, 1093–1099 (1986).

85. Newsam, J. M. and Liang, K. S., *Int. Rev. Phys. Chem.*, **8**, 289–338 (1989).

86. Toby, B. H., Eddy, M. M., Fyfe, C. A., Kokotailo, G. T., Strobl, H., and Cox, D. E., *J. Mater. Res.*, **3**, 563–569 (1988).

87. McKusker, L., *J. Appl. Cryst.*, **21**, 305–310 (1988).

88. Eisenberger, P., Newsam, J. M., Leonowicz, M. E., and Vaughan, D. E. W., *Nature (London)*, **309**, 45–47 (1984).

89. Jirak, Z., Vratislav, S., Zajicek, J., and Bosacek, V., *J. Catal.*, **49**, 112–114 (1977).

90. Ramdas, S., Thomas, J. M., Betteridge, P. W., Cheetham, A. K., and Davies, E. K., *Angewandte Chemie (Int. Edn.)*, **23**, 671–679 (1984).

91. Suffritti, G. and Gamba, A., *Int. Rev. Phys. Chem.*, **6**, 299–314 (1987).

92. Jackson, R. A. and Catlow, C. R. A., *Molec. Simulation*, **1**, 207–220 (1988).

93. Sauer, J. and Zahradnik, R., *Int. J. Quantum Chem.*, **26**, 793–822 (1984).

94. Kiselev, A. V., *J. Chem. Tech. Biotechnol.*, **29**, 673–685 (1979).

95. Baker, M. D., Ozin, G. A., and Godber, J., *Catal. Rev.*, **27**, 591–651 (1985).

96. Kevan, L. and Narayana, M. In *Intrazeolite chemistry*, ACS Symp. Ser. No. **218** (eds Stucky, G. D. and Dwyer, F. G.), pp. 283–299, American Chemical Society, Washington DC (1983).

97. Venuto, P. B. and Habib, E. T., *Fluid catalytic cracking*, Marcel Dekker, New York (1979).

98. Rabo, J. A., *Zeolite chemistry and catalysis*, ACS Monograph **171**, American Chemical Society, Washington DC (1976).

99. Weisz, P. B., *Pure Appl. Chem.*, **32**, 2091–2103 (1980).

100. Weisz, P. B., *Ind. Eng. Chem. Fundam.*, **25**, 53–58 (1986).
101. Csicsery, S. M., *Zeolites*, **4**, 202–213 (1984).
102. Vaughan, D. E. W., *Properties and applications of zeolites*, Chem. Soc. Spec. Publ. No. **33** (ed. Townsend, R. P.), pp. 294–328, The Chemical Society, London (1979).
103. Hutchings, G. J., *Chemistry in Britain*, 762–766 (1987).
104. Miesel, S. L., *Chemtech.*, 32–37 (1988).
105. Hölderich, W., Hesse, M., and Näumann, F., *Angewandte Chemie (Int. Edn.)*, **27**, 226–246 (1988).
106. Barrer, R. M., *Zeolites and clay minerals as sorbents and molecular sieves*, Academic Press, London (1978).
107. Keller, G. E., *Separations: new directions for an old field*, *A.I.Ch.E. Monograph Series No. 83*, pp. 1–52, AIChE, New York (1987).
108. Smith, J. V. and Dytrych, W. J., *Nature*, **309**, 607–608 (1984).
109. Kristmannsdottir, H. and Tomasson, J. In *Natural zeolites. Occurrence, properties, use* (eds Sand, L. B. and Mumpton, F. A.), pp. 277–284, Pergamon Press, Oxford (1978).
110. Haag, W. O. In *Proc. 6th. Int. Conf. on Zeolites* (eds Olson, D. and Bisio, A.), pp. 466–478, Butterworths, Surrey UK (1984).
111. Heinemann, H. In *Catalysis: science and technology. Vol. I* (eds Anderson, J. R. and Boudart, M.), pp. 1–41, Springer-Verlag, Berlin (1981).

8 Ferroics

C. N. R. Rao and K. J. Rao

8.1 INTRODUCTION

Ferromagnetism is a well-known phenomenon wherein solids exhibit spontaneous magnetization even in the absence of an external magnetic field. An important characteristic of ferromagnetic materials is the hysteresis loop found in the relationship between magnetization and magnetic field. The electrical analogue of ferromagnetic materials are ferroelectric materials[1] which show a hysteresis loop in the relationship between polarization and electric field, and exhibit spontaneous polarization in the absence of an external electric field. These materials therefore possess permanent dipole moments, the dipoles arising from the absence of a centre of symmetry.

A wide variety of compounds are known to exhibit ferroelectricity.[7] These include oxides of perovskite structure (e.g. $BaTiO_3$), hydrogen bonded solids (e.g. Rochelle salt, KH_2PO_4), tungsten bronze type structures, pyrochlores, simple salts (e.g. $(NH_4)_2SO_4$, $NaNO_2$, KNO_3), alums, organic compounds (e.g. thiourea, glycine sulphate) and binary compounds as simple as HCl, FeS, GeTe, and V_3Si.

If energy considerations favour an antiparallel arrangement of the permanent dipoles on adjacent planes instead of a parallel arrangement, then such a crystal is called an antiferroelectric.[1] A typical antiferroelectric material is $PbZrO_3$ which has the perovskite structure. Other examples are $NaNbO_3$, $BiFeO_3$, $PbCo_{0.5}W_{0.5}O_3$, $NH_4H_2PO_4$, and $Cu(HCOO)_2 \cdot 4H_2O$. Both ferroelectric and antiferroelectric materials show dielectric constant anomaly at a critical temperature, but the latter do not show polarization–electric field hysteresis loops. It should be noted that, like ferromagnetic materials, ferroelectric materials do not generally contain iron as one of the constituents. We shall discuss properties of ferroelectric and related materials in this chapter, but before doing so we shall examine a more general class of materials called ferroics[2,3] of which ferroelectric and ferromagnetic materials form a part.

A ferroic may be defined as a material possessing two or more orientation states or domains which can be switched from one to another through the application of one or more appropriate forces.[2] Thus, in a ferroelectric, the orientation state of spontaneous electric polarization can be altered by the application of an electric field; in a ferromagnet the orientation state of magnetization in domains can be switched by the application of a magnetic field; in a ferroelastic, the direction of spontaneous strain in a domain can be switched by the application of mechanical stress. Such transitions are described as ferroic transitions. In all the above cases, the boundaries of domains are moved by the application of force in order to accomplish changes in orientation.

$BaTiO_3$ is a typical ferroelectric, CrO_2 is a ferromagnet(ic) while $CaAl_2Si_2O_8$ is a good example of a ferroelastic. The extensive properties for which directionality has been changed in these examples, namely electric polarization, magnetic polarization, and elastic strain, are primary quantities in the sense that their magnitudes directly determine the free energy of the system. Ferroics governed by the switchability of these properties are therefore called primary ferroics. In secondary ferroics, these properties occur as induced quantities. The orientation states differ in derivative quantities which characterize the induced effects. Thus, the induced electric polarization is characterized by dielectric susceptibility, K_{ij}, induced magnetic polarization by magnetic susceptibility, χ_{ij}, and the induced strains by the elastic compliances, C_{ijkl}. The orientation states in secondary ferroics, therefore, differ in K_{ij}, χ_{ij}, and C_{ijkl}; these are tensor quantities and the rank of the tensor is equal to the number of subscripts. The induced effects such as polarization or magnetization can also result from cross-coupled effects such as stress induced polarization (piezoelectric), stress induced magnetization (piezomagnetic) or as a combined effect of two types of fields such as in elastoelectric or elastomagnetic effects. The directional change can then be visualized to occur in the corresponding derivative quantities. Following Newnham, we have listed different types of ferroic effects in Table 8.1 where we have also indicated the switching field and an example of each type of ferroic. The classification scheme of ferroics into primary, secondary, and so on is thermodynamic in origin as shown in the next section.

8.2 THERMODYNAMIC CLASSIFICATION OF FERROICS

The Gibbs free energy of a crystalline solid is written in the differential form as

$$dG = -S\,dT + P_i\,dE_i + M_i\,dH_i + \varepsilon_{ij}\,d\sigma_{ij} \qquad (8.1)$$

where the small variation in the free energy of a mole of material (dG) is the sum of the thermal ($S\,dT$) term and terms arising from various forces. In eqn (8.1), P_i and E_i stand for electric polarization and electric field, respectively, M_i and H_i are magnetic polarization (magnetization) and magnetic field, respectively, while ε_{ij} and σ_{ij} represent the elastic strain and the corresponding mechanical stress respectively; the subscripts i and j run from 1 to 3. When we examine ferroic phenomena of interest under

Table 8.1 *Primary and secondary ferroics*

Ferroic class	Ferroic property	Switching field	Example
Primary			
Ferroelectric	Spontaneous polarization	Electric field	$BaTiO_3$
Ferromagnetic	Spontaneous magnetization	Magnetic field	CrO_2
Ferroelastic	Spontaneous strain	Mechanical stress	$CaAl_2Si_2O_8$
Secondary			
Ferrobielectric	Dielectric susceptibility	Electric field	$SrTiO_3$
Ferrobimagnetic	Magnetic susceptibility	Magnetic field	NiO
Ferrobielastic	Elastic compliance	Mechanical stress	α-Quartz
Ferroelastoelectric	Piezoelectric coefficients	Electric field and mechanical stress	NH_4Cl
Ferromagnetoelastic	Piezomagnetic coefficients	Magnetic field and mechanical stress	$FeCO_3$
Ferromagnetoelectric	Magnetoelectric coefficients	Magnetic field and electric field	Cr_2O_3

isothermal conditions, the first term in eqn (8.1) may be taken as zero. The quantities P_i, M_i, and ε_{ij} include contributions from both spontaneous and induced effects and can therefore be written as

$$P_i = P_i^s + K_{ij}E_j + d_{ijk}\sigma_{jk} + \alpha_{ij}H_j \tag{8.2}$$

$$M_i = M_i^s + \chi_{ij}H_j + Q_{ijk}\sigma_{jk} + \alpha_{ij}E_j \tag{8.3}$$

$$\varepsilon_{ij} = \varepsilon_{ij}^s + C_{ijkl}\sigma_{kl} + d_{kij}E_k + Q_{kij}H_k. \tag{8.4}$$

In the above equations, terms with the superscript 's' represent spontaneous quantities, K_{ij}, χ_{ij}, and C_{ijkl} are the induced quantities referred to earlier, d_{ijk} the piezoelectric coefficients, Q_{ijk} the piezomagnetic coefficients, and α_{ij} the magnetoelectric coefficients. Equations (8.2), (8.3), and (8.4) may be substituted into eqn (8.1) and integrated in order to obtain the free energy of the system for a given orientation state. If $G(I)$ and $G(II)$ are the Gibbs free energies of two orientation states of interest, then, $\Delta G = G(II) - G(I)$ is given by

$$\Delta G = \Delta P_i^s E_i + \Delta M_i^s H_i + \Delta \varepsilon_{ij}^s \sigma_{ij}$$
$$+ \tfrac{1}{2}[\Delta K_{ij}E_iE_j + \Delta\chi_{ij}H_iH_j + \Delta C_{ijkl}\sigma_{ij}\sigma_{kl}]$$
$$+ 2[\Delta\alpha_{ij}E_iH_j + \Delta d_{ijk}E_i\sigma_{jk} + \Delta Q_{ijk}H_i\sigma_{jk}]. \tag{8.5}$$

Here Δ represents the difference in the relevant quantity between orientation states II and I. That is, $\Delta\varepsilon_{ij}^s = \varepsilon_{ij}^s(II) - \varepsilon_{ij}^s(I)$, etc., are the differences in the ij components of ε^s, etc., in the states II and I. When no external force is acting on the system, $\Delta G = 0$ and the orientation states are energetically degenerate.

Several possible ferroic phenomena are evident from eqn (8.5) depending upon the dominance of particular terms. In a material which has a large value of spontaneous polarization, other terms become unimportant and the free energy in an electric field is governed by the expression

$$\Delta G = \Delta P_i^s E_i = \Delta P^s E \tag{8.6}$$

where E represents the electrical field in a chosen direction (say Z). Similar expressions can be written for spontaneous magnetization and spontaneous strain when they are the dominant quantities and interact with the corresponding external fields. When $\Delta P_i^s = \Delta M_i^s = \Delta \varepsilon_{ij}^s = 0$, the quantities which determine the ΔG values arise from terms in the two sets of square brackets in eqn (8.5). The first set of quantities are ΔK_{ij}, $\Delta \chi_{ij}$ and ΔC_{ijkl}. When $\Delta K_{ij} \neq 0$ and $\Delta \alpha_{ij}$ and Δd_{ijk} make little or negligible contribution, the free energy is determined by the expression

$$\Delta G \simeq \tfrac{1}{2}\Delta K_{ij} E_i E_j \simeq \Delta K E^2. \tag{8.7}$$

Dominance of other terms give rise to similar expressions. Equation (8.6) and its analogues define the primary ferroics while eqn (8.7) and its analogues define secondary ferroics. Equation (8.7) represents a ferrobielectric.

A ferroelectric transition from one orientation state to another (observed through hysteresis loops) is electrically a first order transition. The order of the transition between domain states in ferroics is simply the sum of the exponents of the field terms in the free energy expression. Ferrobielastic, ferrobimagnetic, ferroelastoelectric, ferromagnetoelectric, and other such transitions are all second order. We should note here that the various coefficients are in themselves interdependent. For example, spontaneous polarization is associated with a field which in turn produces a strain through strong coupling to the lattice, thus activating electrostrictive or piezoelectric coefficients. Wherever a spontaneous polarization exists, therefore, spontaneous strain would also be present and *vice versa*. Whenever spontaneous polarization and spontaneous strain produce orientation states fully independent of each other, both ferroelastic and ferroelectric properties can be fully realized and the materials are referred to as fully ferroelectric, fully ferroelastic materials. Other interrelations of a similar nature can be readily visualized from the interdependence of various quantities implicit in eqn (8.5).

8.3 PROPER AND IMPROPER FERROICS

At high temperatures, ferroelectric materials transform to the paraelectric state (where dipoles are randomly oriented), ferromagnetic materials to the paramagnetic state and ferroelastic materials to the twinfree normal (para-elastic?) state. The transitions are conveniently characterized through order parameters.[4] These order parameters are characteristic properties parametrized in such a way that the resulting quantity is unity for the ferroic state at a temperature sufficiently below the transition temperature and is zero in the non-ferroic phase beyond the transition temperature. Polarization, magnetization, and strain are the proper order parameters for the ferroelectric, ferromagnetic, and ferroelastic transitions, respectively. Whenever transitions are governed by the expected variations of these order parameters, they are called proper ferroics. As we noted earlier, the coupled nature of ferroic phenomena is such that the order parameter which determines the transition can often be different from the 'proper' order parameter. Ferroics where the order parameter does not represent a 'proper' property are called improper ferroics. It is possible that the order parameter driving the transition in primary ferroics is none of the 'obviously' related quantities in the expression for free energy, and the ferroic can then be considered to be totally improper (or highly improper). A good example is terbium molybdate in which the order parameter is a condensed optic mode.[5] The optic mode causes a spontaneous strain which in turn causes a spontaneous polarization through piezoelectric coupling; terbium molybdate is therefore both an improper ferroelectric and an improper ferroelastic.

A hexagonal representation of proper and improper primary ferroics as proposed by Newnham and Cross[3] is given in Fig. 8.1. The proper order parameter appears on the diagonals of the hexagon while the sides of the hexagon represents improper ferroics. They indicate the cross-coupled origin of ferroic phenomena. An improper primary ferroic in this classification is distinguished from a true secondary ferroic by the appearance of the prefix 'ferro' only with the primary ferroic quantity and not for both of the coupled quantities. Thus, the term magnetoferroelectric (e.g. Cr_2BeO_4) implies that the material is an improper ferroic where as the term ferromagnetoelectric (e.g. Cr_2O_3) would mean that the material is a secondary ferroic.

8.4 FERROELECTRICS

Among the 32 crystal classes, 11 possess a centre of symmetry and are centrosymmetric and therefore do not possess polar properties. Of the 21 non-centrosymmetric classes, 20 of them exhibit electric polarity when subjected to a stress and are called piezoelectric; one of the non-centrosymmetric classes (cubic 432) has other symmetry elements which combine to exclude piezoelectric character. Piezoelectric crystals obey a linear relationship between polarization P and force F,

$$P_i = d_{ij}F_j$$

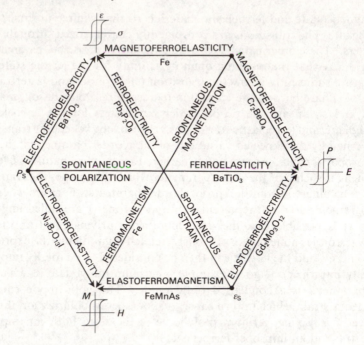

Fig. 8.1. Diagram showing several types of order parameters involved in proper and improper ferroics (after Newnham and Cross[3]).

where d_{ij} is the piezoelectric coefficient. An inverse piezoelectric effect leads to mechanical deformation or strain under the influence of an electric field. Ten of the 20 piezoelectric classes possess a unique polar axis. In non-conducting crystals, a change in polarization can be observed by a change in temperature, and they are referred to as pyroelectric crystals. If the polarity of a pyroelectric crystal can be reversed by the application of an electric field, we call such a crystal a ferroelectric. We thus see that a knowledge of the crystal class is sufficient to establish the piezoelectric or the pyroelectric nature of a solid, but reversible polarization is a necessary condition for ferroelectricity. While all ferroelectric materials are also piezoelectric, the converse is not true; for example, quartz is piezoelectric, but not ferroelectric.

One of the important characteristics of ferroelectrics is that the dielectric constant, D, obeys the Curie–Weiss law:

$$D = D_\infty + \frac{C}{T - T_c} \tag{8.8}$$

where D_∞ is the dielectric constant at optical frequencies, T_c the critical (Curie) temperature and C the Curie constant. Equation (8.8) is similar to

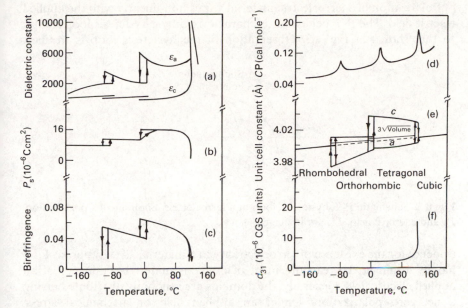

Fig. 8.2. Phase transitions in $BaTiO_3$ accompanied by changes in (a) dielectric constant, (b) spontaneous polarization, (c) birefringence coefficient, (d) heat capacity, (e) lattice dimensions, and (f) piezoelectric coefficient. Experimental points are not shown.

the equation relating magnetic susceptibility with temperature in ferromagnetic materials. In Fig. 8.2, the temperature variation of dielectric constant of a single crystal of $BaTiO_3$ is shown to illustrate the behaviour. Above 393 K, $BaTiO_3$ becomes paraelectric. Polycrystalline samples show less marked changes in D at the transition temperature.

Barium titanate which crystallizes in the perovskite structure has cubic symmetry above 393 K with Ba^{2+} in the body centre and TiO_6 octahedra in the corners. It undergoes a transformation to a tetragonal structure at 393 K, to an orthorhombic structure at 278 K and to a rhombohedral structure at 183 K (Fig. 8.2). Above 393 K, the material becomes paraelectric and the dipoles are randomized. Relative to the cubic phase, elongation occurs along one of the edges ([100] direction) in the tetragonal phase, along one of the face diagonals ([110] direction) in the orthorhombic phase, and along one of the body diagonals ([111] direction) in the rhombohedral phase. The Ti^{4+} ion moves in these three directions successively as the crystal is cooled from the cubic phase (which has Ti^{4+} in the centre of the octahedra). Besides the dielectric constant and polarization, other properties such as heat capacity also show anomalous changes at the three phase transitions of $BaTiO_3$ (see Fig. 8.2).

Polarization of a ferroelectric material varies non-linearly with the applied electric field. The P–E behaviour is characterized by a hysteresis loop similar to that shown in Fig. 8.3. Observation of the hysteresis loop is the best

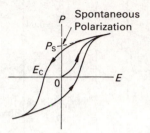

Fig. 8.3. Schematic P–E hysteresis loop of a ferroelectric. Spontaneous polarization, P_s, and coercive field, E_c, are indicated.

evidence for the existence of ferroelectricity in a material. The hysteresis loop has its origin in the rearrangement of domains under the influence of an applied electric field. Generally, the domains are randomly distributed giving a net zero polarization. Under an applied field or mechanical stress, favourably oriented domains grow at the expense of the less favourably oriented domains until a single domain configuration is obtained. The domain structure itself is related to the crystallography of the ferroelectric phase with respect to the paraelectric phase. Thus, in the tetragonal phase of $BaTiO_3$, adjacent domains may have their polar axes making angles of 90° or 180° (see Fig. 8.4).

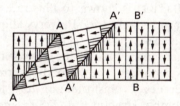

Fig. 8.4. Domain walls in tetragonal $BaTiO_3$. AA' is a 90° wall and BB' is a 180° wall.

Many of the ferroelectric materials exhibit softening of certain vibrational modes.[4] That is, at the ferroelectric–paraelectric transition temperature, the square of a normal mode frequency goes to zero:

$$\omega^2 = v(T - T_c). \tag{8.9}$$

Soft mode behaviour of ferroelectric materials has been investigated in detail by employing Raman spectroscopy and neutron scattering and the subject has been reviewed by Blinc and Zeks.[6]

Table 8.2 *Typical ferroelectric materials*

	T_c (K)	$P_s(T, K)$ (10^{-6} C/cm^2)	Other transitions (K)
BaTiO$_3$	393	26.0 (296)	183, 278, 1713
PbTiO$_3$	763	>50 (296)	173?
KNbO$_3$	73	12.0 (73)	627, 835, 913
LiNbO$_3$	1483	71 (296)	—
PbTa$_2$O$_6$	533	10.0 (298)	—
Cd$_2$Nb$_2$O$_7$	185	6.0 (88)	85
Bi$_4$Ti$_3$O$_{12}$	948	>30	—
Sm$_2$(MoO$_4$)$_3$	470	0.24 (323)	—
HCl	98	1.2 (83)	120
DCl	105	—	—
SbSI	293	25 (273)	—
FeS	410	0.7	200, 600
NaNO$_2$	436	8 (373)	168, 438
KH$_2$PO$_4$	123	4.75 (96)	—
KD$_2$PO$_4$	213	4.83 (180)	—
RbH$_2$PO$_4$	147	5.6 (90)	—
(NH$_4$)$_2$SO$_4$	224	0.62 (221)	—
(NH$_4$)HSO$_4$	270, 154	0.8 (155)	—
(ND$_4$)DSO$_4$	262, 158	—	—
RbHSO$_4$	258	0.65 (103)	—
NH$_4$Fe(SO$_4$)$_2 \cdot$12H$_2$O	88	0.40 (86)	—
K$_4$Fe(CN)$_6 \cdot$3H$_2$O	248.5	1.45 (223)	—
K$_4$Fe(CN)$_6 \cdot$3D$_2$O	255	1.50 (233)	—
NaKC$_4$H$_4$O$_6 \cdot$4H$_2$O (Rochelle Salt)	297, 255	0.25 (278)	—
NaKC$_4$D$_4$O$_6 \cdot$4D$_2$O	308, 251	0.35 (279)	—

In Table 8.2 we have listed typical ferroelectric materials. Hydrogen bonded ferroelectrics like KH$_2$PO$_4$ show deuterium isotope effects (with T_c increasing with deuterium substitution) showing the important role of hydrogen bonds.

Antiferroelectric materials show superstructure in the antipolar phase as well as a dielectric constant anomaly at T_c. Changes in other physical properties are also observed at T_c when the antiferroelectric phase transforms to the paraelectric phase. Antiferroelectric materials, however, do not show the P–E hysteresis loop. Since the energy difference between antiferroelectric and ferroelectric states is rather small, application of a large electric field, mechanical stress or compositional variation can induce antiferroelectric materials to become ferroelectric. All the known antiferroelectric materials have been tabulated in the literature.[7] Soft modes are also associated with antiferroelectric–paraelectric transitions.

In recent years, a large number of materials exhibiting other interesting properties besides ferroelectricity have been reported. Such materials have been tabulated and discussed in the literature.[7] Typical paired properties with examples are given below:

ferroelectric–ferroelastic, $Gd_2(MoO_4)_3$, $KNbO_3$,
ferroelectric–antiferromagnetic, $YMnO_3$, $HoMnO_3$,
ferroelectric–ferromagnetic, $Fe_3B_7O_{13}Cl$, $Bi_9Ti_3Fe_3O_{27}$,
antiferroelectric–antiferromagnetic, $BiFeO_3$, $Cu(HCOO)_2 \cdot 4H_2O$,
ferroelectric–semiconducting, FeS, reduced $SrTiO_3$, $YMnO_3$,
ferroelectric–superconducting, $SrTiO_3$, GeTe, V_3Si.

It is obvious that depending on the temperature and other conditions, one can have paired properties with materials in the paraelectric or paramagnetic phase as well. Paired properties of the kind shown above have important technological implications.

8.5 PRIMARY FERROICS

As mentioned earlier, the orientation states in ferroelectrics correspond to polarization directions in domains. We can classify ferroelectric materials based on the magnitude of the Curie constant, C. Large values of C ($\sim 10^5$ K) are evident in oxide ferroelectrics, particularly those containing ions Ti^{4+}, Nb^{5+}, and W^{6+} from Groups IV, V, and VI, respectively, of the periodic table and those containing lone pair ions Pb^{2+}, Bi^{3+}, and Sb^{3+} from Groups IV and V. Due to their electronic structure, these ions promote distortions giving rise to spontaneously polarized structures. The second category where C is around 10^3 K is comprised mainly of order–disorder type ferroelectrics where rotational ordering of dipolar ions (as in $NaNO_2$) or positional ordering of protons as in KH_2PO_4 gives rise to ferroelectricity. The third category with very low values of C (~ 1–10 K) is generally comprised of improper ferroelectrics such as $Gd_2(MoO_4)_3$. The dimensionality of atomic displacements also increases in the same order as the decrease in C values. While in titanates the displacements are one dimensional, giving rise to strong dipole moments, in hydrogen phosphates and $NaNO_2$ dipoles are ordered (two-dimensional displacements) in planes, decreasing the net effect of polarization. In improper ferroelectrics such as molybdates, displacements are three dimensional. It may be noted in passing that in ferroelectrics, the direction of spontaneous polarization is capable of being switched from one state to another by a coercive field which does not exceed the dielectric breakdown strength.

Ferroelectrics transform to a paraelectric state on heating ($T > T_c$). In general, the transitions are well understood on the basis of the Landau–Devonshire theory using polarization as an order parameter.[4] The ordered ferroelectric phase has a lower symmetry belonging to one of the subgroups of the high-symmetry disordered paraelectric phase. However, the exact structure to which the paraelectric phase transforms is determined only by energy considerations. Polar ferroelectric phases possess lattice modes of which a weak (low frequency) optic mode is associated with the ionic displacements responsible for the spontaneous polarization. This mode weakens further and vanishes as the displacements themselves vanish at T_c giving rise to the more symmetrical paraelectric phase. Ferroelectric–paraelectric transitions are therefore usefully interpreted in terms of soft modes.[4,6] It should, however, be recognized that soft modes characterize lattice instabilities, a feature that encompasses a whole range of structural transitions[4] and ferroelectricity is only incidental in such transitions.

Domain patterns in ferroics are affected by applied fields. In ferroelastics, domain walls (twin lamellae) can be moved by the application of stress. Spontaneous strains associated with ferroelastics are of the order of a few parts per thousand and the coercive stresses are not greater than 10^4–10^5 N/cm^2.

By the definition given above ferromagnetics encompass both ferromagnets (e.g. hematite) and ferrimagnets (magnetite) since they both possess spontaneous magnetization. Magnetic domains such as those in hematite can be observed by using the Faraday effect and observations of domain wall movements can be made under applied magnetic field. A further discussion of ferromagnetics is not relevant to the purpose of this chapter.

8.6 SECONDARY FERROICS

$SrTiO_3$ (which is an incipient ferroelectric) and $NaNbO_3$ (which is an antiferroelectric) may be considered to be ferrobielectric. The dielectric anisotropy in antiferroelectric domains can give rise to high values of induced polarizations which are orientationally different in different domains. Thus, they give rise to domain rearrangement under applied fields. Quartz is a classic example of ferrobielasticity. The twin structure in α-quartz is governed by rules of twin morphology. However, when an external stress is applied, the induced strains differ in different twins; as a consequence, the differential of the induced strains gives rise to twin wall movement in the stress field. Stress-induced movement of Dauphine twins (180° twins) in quartz was first observed by Aizu[8] in his pioneering work on secondary ferroics. Nickel oxide (NiO) which is a well-known antiferromagnetic material is also ferrobimagnetic; it is also known to be ferroelastic. Domain wall movements can be seen under applied magnetic field in polarized light.

Ammonium chloride is considered to be a ferroelastoelectric. In the low temperature (< 247 K) ordered, CsCl phase, NH_4Cl has domains in which values of d_{ijk} (elastoelectric or piezoelectric coefficients) are different and when both an electric field and a mechanical stress are applied simultaneously, orientational switching occurs. Ferromagnetoelasticity is exhibited by $FeCO_3$ (siderite) and CoF_2. It appears that potential candidates for ferrobielectric, ferrobimagnetic, and ferrobielastic behaviour may be generally expected from among those materials which are normally antiferroelectric, antiferromagnetic and antiferroelastic (those with internally compensated strains and have 180° twins), respectively. The antiferroic states suggest large susceptibilities which can be expected to give rise to pronounced induced effects.

8.7 IMPROPER FERROICS

Improper ferroics occur as a result of cross-coupled phenomena. Typical examples of improper ferroics are nickel iodine boracite ($Ni_3B_7O_{13}I$) which is an electroferromagnet[9] and lithium ammonium tartrate which is an elastoferroelectric[10]; Cr_2BeO_4 is a magnetoferroelectric.[11] Upon cooling, the cubic boracite undergoes a transformation in which the Ni^{2+} ions produce an antiferromagnetic ordering. This phase, which is an antiferromagnetic–piezoelectric, transforms on further cooling to an orthorhombic phase (at 64 K) which is polar and develops a spontaneous polarization. As a consequence of the polarization effect, the neutral spin alignment is destroyed and a weakly ferromagnetic state results. Since nickel iodine boracite is ferromagnetic because of its ferroelectricity, it is referred to as an electro-ferromagnet—an improper ferroelectric whose order parameter is electric polarization. At its magnetic transition, tetragonal (paramagnetic) Cr_2BeO_4 transforms to an orthorhombic antiferromagnetic phase. The antiferromagnetic coupling causes slight shifts and rearrangements in the positions of the Cr^{3+} ions in such a way that a polar axis develops, giving rise to a spon-taneous polarization. Cr_2BeO_4 is thus a ferroelectric where the driving force arises from magnetic interactions, and is therefore referred to as a magnetoferroelectric.

8.8 RELAXOR FERROELECTRICS

In passing, we shall introduce another class of materials known as relaxor ferroelectrics.[12] These materials show diffuse transitions extending over a large temperature range which are associated with large changes in dielectric constants. The magnitude of the peak dielectric constants decrease with increasing frequency and the peaks shift to higher temperatures suggesting features of a relaxational mechanism. It is generally understood from optical and electrical measurements that microdomains arising from compositional fluctuations are responsible for the phenomenon. The importance of these relaxor ferroelectrics stems from the fact that in many of them large permittivities are associated with very low electrostrictive coefficients, a feature desirable in many engineering applications such as in pressure gauges. It should be noted that electrostriction, as distinct from inverse piezoelectric effect, is a phenomenon in which the strain and the electrical field inducing the strain are related by $\varepsilon_{ij} = M_{ijk}E_k^2$ where M_{ijk} are electrostriction coefficients. Several relaxor ferroelectrics are known to possess very low thermal expansivities. Some good examples of relaxor ferroelectrics are $Pb(Mg_{1/3}Nb_{2/3})O_3$, $0.9Pb(Mg_{1/3}Nb_{2/3})O_3 \cdot 0.1PbTiO_3$ and $Pb(Zn_{1/3}Nb_{2/3})O_3$ investigated by Uchino *et al.*[13]

8.9 COMPOSITE FERROICS

In order to fully exploit the potential of ferroics, ingenious experiments are being performed with composites made from a ferroic and another material.[3, 14-16] For example, in a piezoelectric like PZT, the piezoelectric voltage coefficient g can be defined for a given direction (say $Z = 33$); thus $g_{33} = d_{33}/(\varepsilon_0 K_{33})$. There are situations (e.g. in hydrophones) where g_{33} is required to be high. Further, when a hydrostatic pressure (such as prevails in underwater conditions) is applied, the equation for g gets modified to $\bar{g} = \bar{d}/\varepsilon_0 \bar{K}$ and $\bar{d} = d_{33} + 2d_{31}$ ($\bar{K}$ has a similar expression); $\bar{d}$ is effectively zero under hydrostatic pressure since $d_{33} = -2d_{31}$. The composite ferroic strategy is to design a ceramic–polymer composite in such a way that d_{31} is eliminated and $\bar{K}$ is reduced. The reduction in $\bar{K}$ is obvious since introducing a polymer filler phase reduces the volume fraction of the high dielectric constant piezoelectric. The method by which reduction (or elimination) of d_{31} is achieved can be visualized from Fig. 8.5. The ceramic is exposed only in the $\langle 33 \rangle$ direction and being the less compressible phase (as compared to the polymer), takes up the entire pressure in the $\langle 33 \rangle$ direction upon itself. No hydrostatic pressure is directly experienced by the ceramic in the $\langle 11 \rangle$ and $\langle 22 \rangle$ directions since the polymer absorbs all the stress. Thus in the composite, $d_{31} \sim 0$, hence $\bar{d} = d_{33}$. On the other hand $\bar{K} = fK_{33}$ where f is the volume fraction of the ceramic. The $\bar{g}$ value would be $d_{33}/(\varepsilon_0 fK_{33})$, a value much higher than when a cubic piece of piezoelectric is used.

The composite structure in Fig. 8.5 is referred to as having 1–3 connectivity. That is, PZT is connected to itself in one direction while the polymer phase is interconnected in all three dimensions. Other types of connectivities in ceramic–polymer diphasic composites are now easy to visualize such as 2–3 or 3–3 composites. Various possible connectivities are represented in Fig. 8.6. In a 2–3 connected diphasic composite, we can hope to exploit two different properties of the ceramic in two different directions or the same property in an additive manner.

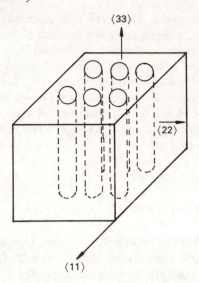

Fig. 8.5. Schematic of a (1 + 3) composite ferroic.

3–3 composites have been made by the so-called Replamine lost-wax technique. A natural template for 3–3 connectivities is provided by goniopera coral. Corals are first filled with casting wax and later the carbonate (coral) structure is dissolved away in acid. The negative blanks are now invested with PZT in place of the coral carbonate. It is then sintered, during which process the wax is lost. Finally, the sintered product is filled with the polymer.

8.10   GLASS CERAMICS

A composite in which the matrix or the binding phase is a glass is known as a glass ceramic. Large, ceramic, pore-free bodies can be made from glass ceramics in which the crystalline phase is a polar material such as a ferroelectric, a piezoelectric, or a pyroelectric.[17,18] Such polar glass ceramics containing ferroelectric phases like $BaTiO_3$, $NaNbO_3$, $PbTiO_3$, $Pb_5Ge_3O_{11}$, $LiTaO_3$, and $LiNbO_3$ have been studied extensively. Transparent ferroelectric glass ceramics in which the given sizes are controlled to $0.2\,\mu m$ possess excellent electro-optic properties. However, good ferroelectric materials are poor glass formers and the low dielectric constants of the glassy phases which bind the ferroelectric crystallites make it difficult to pole the products.

Polar glass ceramics containing non-ferroelectric phases have been reviewed by Hallial *et al.*[19] Reasonably good polar glass ceramics are formed from $Ba_2TiSi_2O_8$, $Ba_2TiGe_2O_8$, $Li_2Si_2O_5$, Li_2SiO_3, and $Li_2B_4O_7$ along with

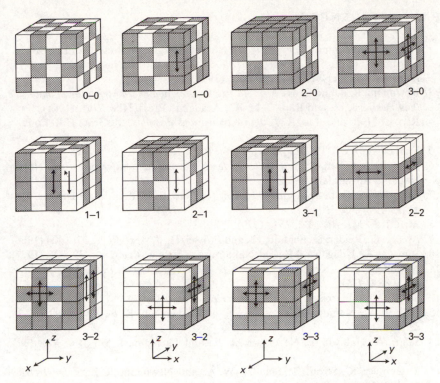

Fig. 8.6. Ten connectivity patterns for a diphasic solid. Each phase has 0-, 1-, 2-, or 3-dimensional connectivity to itself. In the 3–1 composite, for example, the shaded phase is three-dimensionally connected and the unshaded phase is one-dimensionally connected. Arrows are used to indicate the connected directions. Two views of the 3–3 and 3–2 patterns are given because the two interpenetrating networks are difficult to visualize on paper. The views are related by 90° counterclockwise rotation about Z. (After Newnham and Cross.[3])

suitable additions such as PbO, CaO, and ZnO, after giving carefully chosen heat treatment.

Many glass ceramics have been found suitable for pyroelectric applications such as vidicons. Although TGS possesses a higher figures of merit than $LiTaO_3$, its hygroscopic nature as well as low T_c makes TGS, $(NH_2CH_2COOH)_3 \cdot H_2SO_4$, less suitable for device applications as compared to $LiTaO_3$. Hallial *et al.*[20] have examined systems such as $Ba_2TiSi_2O_8$ and have obtained glass ceramics with high figures of merit for pyroelectro applications. Glass ceramics (fresonite compositions) containing $BaO–SiO_2–TiO_2$ possess piezoelectric properties with high values of d_{33} and g_{33}, best suited for hydrophone applications. Development of glass ceramics based on the same fresonites for SAW (surface acoustic wave) device applications have also been reported.[21]

8.11   REFERENCES

1. Jona, F. and Shirana, G., *Ferroelectric crystals*, Pergamon Press, Oxford (1962).
2. Newnham, R. E., *Am. Mineral.*, **57**, 906 (1974); also see in *Structure property relations*, Springer-Verlag, New York (1975).
3. Newnham, R. E. and Cross, L. E., in *Preparation and characterization of materials* (eds Honig, J. M. and Rao, C. N. R.), Academic Press, New York (1981).
4. Rao, C. N. R. and Rao, K. J., *Phase transitions in solids*, McGraw-Hill, London (1978).
5. Dorner, B., Axe, J. D., and Shirane, G., *Phys. Rev.*, **B6**, 1950 (1972).
6. Blinc, R. and Zeks, B., *Soft modes in ferroelectrics and antiferroelectrics*, North-Holland, Amsterdam (1974).
7. Subbarao, E. C., in *Solid state chemistry* (ed. Rao, C. N. R.), Marcel Dekker, New York (1974).
8. Aizu, K., *Phys. Rev.*, **B2**, 754 (1970).
9. Ascher, H., Reider, H., Schmid, H., and Stossel, H., *J. Appl. Phys.*, **37**, 1404 (1966).
10. Sawada, A., Udagawa, M., and Nakamura, T., *Phys. Rev. Lett.*, **39**, 829 (1977).
11. Newnham, R. E., Kramer, J. J., Schulze, W. A., and Cross, L. E., *J. Appl. Phys.*, **49**, 6088 (1978).
12. Setter, N. and Cross, L. E., *Ferroelectrics*, **37**, 551 (1981).
13. Uchino, K., Nomura, S., Cross, L. E., and Newnham, R. E., *Ferroelectrics*, **38**, 825 (1981).
14. Safari, A., Halliyal, A., Newnham, R. E., and Lachman, I. M., *Mat. Res. Bull.*, **17**, 301 (1982).
15. Rittenmyer, K., Strout, T., Schulze, W. A., and Newnham, R. E., *Ferroelectrics*, **41**, 189 (1982).
16. Lynn, S. Y., Newnham, R. E., Klicker, K. A., Rittenmyer, K., Safari, A., and Schulze, W. A., *Ferroelectrics*, **38**, 955 (1981).
17. Gardopee, G., Newnham, R. E., Hallial, A., and Bhalla, A. S., *Appl. Phys. Lett.*, **36**, 817 (1980).
18. Takashige, M., Mitsik, T., and Nakamura, T., *Jpn. J. Appl. Phys.*, **20**, L159 (1981).
19. Hallial, A., Bhalla, A. S., Newnham, R. E., and Cross, L. E., *Glasses and glass ceramics* (ed. Lewis, M. H.), Chapman and Hall, London and New York (1989).
20. Hallial, A. S., Bhalla, A., Cross, L. E., and Newnham, R. E., *J. Mat. Sci.*, **20**, 3745 (1985).
21. Ito, Y., Nagatsuma, K., Takeuchi, H., and Jyomura, S., *J. Appl. Phys.*, **52**, 4479 (1981).

Index

Note: Figures and Tables are indicated by *italic page numbers*